The Architecture of Matter

A complex of interrelated problems plagued the theory of matter during the seventeenth and eighteenth centuries: problems concerning matter's divisibility, composition, and internal architecture. Is any material body divisible to infinity? Must we posit atoms or elemental minima from which bodies are ultimately composed? Are the parts of material bodies themselves material concreta? Or are they merely potentialities or possible existents?

Questions such as these—and the press of subtler questions hidden in their amibiguities—deeply unsettled philosophers of the early modern period. They seemed to expose serious paradoxes in the new world view pioneered by Galileo, Descartes, and Newton. The new science's account of a fundamentally geometrical Creation, mathematicizable and intelligible to the human inquirer, seemed to be under threat. This was a great scandal, and the philosophers of the period accordingly made various attempts to disarm the paradoxes. All the great figures address the issue: most famously Leibniz and Kant, but also Galileo, Hobbes, Newton, Hume, and Reid, in addition to a crowd of lesser figures.

Thomas Holden offers a brilliant synthesis of these discussions and presents his own overarching interpretation of the controversy, locating the underlying problem in the tension between the early moderns' account of material parts on the one hand and the programme of the geometrization of nature on the other.

Thomas Holden is Assistant Professor of Philosophy at the University of California, Santa Barbara.

The Architecture of Matter

Galileo to Kant

Thomas Holden

CLARENDON PRESS · OXFORD

This book has been printed digitally and produced in a standard specification in order to ensure its continuing availability

OXFORD
UNIVERSITY PRESS

Great Clarendon Street, Oxford OX2 6DP

Oxford University Press is a department of the University of Oxford.
It furthers the University's objective of excellence in research, scholarship,
and education by publishing worldwide in

Oxford New York

Auckland Cape Town Dar es Salaam Hong Kong Karachi
Kuala Lumpur Madrid Melbourne Mexico City Nairobi
New Delhi Shanghai Taipei Toronto

With offices in

Argentina Austria Brazil Chile Czech Republic France Greece
Guatemala Hungary Italy Japan South Korea Poland Portugal
Singapore Switzerland Thailand Turkey Ukraine Vietnam

Oxford is a registered trade mark of Oxford University Press
in the UK and in certain other countries

Published in the United States
by Oxford University Press Inc., New York

© Thomas Holden 2004

The moral rights of the author have been asserted

Database right Oxford University Press (maker)

Reprinted 2008

All rights reserved. No part of this publication may be reproduced,
stored in a retrieval system, or transmitted, in any form or by any means,
without the prior permission in writing of Oxford University Press,
or as expressly permitted by law, or under terms agreed with the appropriate
reprographics rights organization. Enquiries concerning reproduction
outside the scope of the above should be sent to the Rights Department,
Oxford University Press, at the address above

You must not circulate this book in any other binding or cover
And you must impose this same condition on any acquirer

ISBN 978-0-19-920420-5

For Robin and Enid

PREFACE

This book presents a historical and critical study of a controversy that raged among all the leading figures of early modern natural philosophy. The issue at stake is the constitution and internal architecture of matter. On the one hand, an array of a priori arguments seems to show that matter must be infinitely divisible: it must be fundamentally continuous and altogether without simple first parts. On the other hand, an opposing battery of a priori arguments seems to show that matter cannot be divisible ad infinitum: it must be fundamentally discrete in its fine structure, resolving to a finite base of logically unsplittable first elements.

The conflict between these rival sets of arguments greatly discomfited philosophers of the seventeenth and eighteenth centuries. Since arguments on both sides of the question seemed compelling, the clash between them appeared to expose a paradox or antinomy in the new world-view pioneered by Galileo, Descartes, and Newton. The new science's account of a fundamentally geometrical Creation, mathematicizable and intelligible to the human inquirer, seemed to be under threat. This was of course a great scandal, and the philosophers of the period accordingly made various attempts to disarm the conflict and resolve the paradox. All the great figures address the issue: most famously Leibniz ('the labyrinth of the continuum') and Kant (the 'Second Antinomy'), but also Galileo, Descartes, Newton, and Hume, in addition to a crowd of lesser figures. This book ties together all of these discussions and offers an overarching interpretation of the controversy.

A pivotal interpretative thesis of this study is that the historical debate turns crucially on certain metaphysical doctrines concerning the status of material parts. The problems of infinite divisibility in fact brought out an underlying tension at the heart of early modern natural philosophy, between the geometrization of Nature on the one hand, and the dominant metaphysical account of material parts on the other. This interpretation can be contrasted with a rival view of the early modern controversy over material structure: the purely mathematical reading. According to this latter interpretation, the early modern debate simply concerns mathematical constructions of material body. It concerns the 'form' of body (its spatial, mathematically tractable structure), but not its 'matter' (its filling or content). On this view, purely formal, mathematical questions to do with geometry, topology, and the logic of infinity exhaust the dispute. And given this interpretation, the early moderns' travails over material structure can appear simply quaint. They seem to rest merely on the shortcomings of early modern mathematics. But this purely mathematical reading misconceives the debate, or so I shall argue. In fact the early modern controversy essentially depends on certain metaphysical theses concerning the content of matter—its 'stuffing', not just its geometrical structure. And once we appreciate that the controversy turns on these metaphysical theses (which concern the ontological status of parts and wholes, and foreshadow positions that have re-emerged in recent analytic metaphysics), we will see that the problems of material structure are much more interesting—and harder to resolve—than the straight mathematical reading allows.

In addition to defending this interpretation of the early modern debate, I mobilize the metaphysical reading in order to reassess the various period responses to the problems of matter's internal architecture. The paradox greatly stimulated early modern physical and metaphysical thought, generating a wide range of theoretical responses. For Galileo, the problems refuted the received Aristotelian account of infinity. For Hume, they mandated a radical atomism and a reassessment of the claims of geometrical knowledge. Dynamists like Boscovich and Kant argued that they showed the necessity of a dynamical account of matter. For Berkeley they amounted to a *reductio* on the extra-mental existence of matter and a demonstration of idealism; while for sceptics they embarrassed the programme of modern science and undermined reason's pretensions to investigate the world altogether. The metaphysical interpretation of the debate over material structure casts all of these responses in a new light.

Regarding methodology, my aims and hence my techniques are both historical and philosophical. I emphasize context and historical interpretation on the one hand, and philosophical assessment and criticism on the other. It is a truism that each of these approaches requires some measure of the other: if criticism without historical understanding is blind, historical interpretation without evaluation is certainly empty. But in terms of the precise balance between historical scholarship and philosophical assessment, it is worth distinguishing three (somewhat overlapping) phases in this study. First, the defence of my central interpretative thesis (the metaphysical reading of the early modern controversy) rests largely on close textual readings and historical contextualization of the period debates. While much of the recent commentary endorses the purely mathematical interpretation, I argue that this misses both the explicit metaphysical framework that can be found in the writings of various less well-known figures, and the implicit reference to this background in the more famous texts. (See especially Chapter 1, but also points of Chapters 2 to 4.) The second methodological phase of the book is more directly philosophical and critical in spirit. Here I examine the structure and assess the merits of the arguments that frame the early modern debate. This includes the arguments for and against the competing metaphysical systems of actual and potential parts (Chapter 2), the arguments for metaphysical atomism (Chapters 3 and 4), for infinite divisibility (Chapter 5), and for dynamism (Chapter 6). In each of these cases I document the arguments as they occur in the historical texts, and my criticisms are in the main immanent, developed within the terms of the Enlightenment debate rather than from the perspective of current philosophical agendas. Nevertheless, I do pass judgement on these various arguments, and on occasion I relate them to subsequent developments in philosophy. These first two phases set the terms of the book: first, my initial defence of the metaphysical interpretation of the early modern controversy, largely by historical scholarship; second, the examination of the way the debate (so construed) plays out, contextualizing but also assessing the historical arguments.

The third methodological phase is much smaller, occurring only in certain subsections of my discussion of the rival metaphysical systems of actual and potential parts (in Chapter 2). Here I pause to show the way in which the historical clash between these two competing systems prefigures a cluster of debates in current analytic metaphysics (concerning 'arbitrary undetached parts', 'scattered objects', 'stuff ontologies', and the like). It would be perverse not to document these connections; the early modern debate mirrors the current problem space remarkably closely, and yet neither historians of philosophy nor current metaphysicians have related the two in any sustained fashion. This seems a clear case where the history of philosophy can inform the current debate, and I have tried to bring this out. But these subsections are quite self-contained. No prior knowledge of the current controversies is required to follow them, and the more historically minded reader can simply skip them at no loss to the main interpretative and critical narrative. A more detailed chapter-by-chapter overview of the book can be found at the end of Chapter 1.

This book grew out of a dissertation written at the University at North Carolina at Chapel Hill under the invaluable guidance of Simon Blackburn and Don Garrett. I am deeply indebted to both of these philosophers for all their encouragement and assistance. I have been helped by many others along the way, but I would like to express especial thanks to Donald Baxter, John Cottingham, Tamar Szabó Gendler, Rolf George, Hans-Johann Glock, Patrick Miller, Daniel Nolan, David Fate Norton, David Raynor, Jay Rosenberg, Tad Schmaltz, Theodore Sider, Keith Simmons, and Galen Strawson. This book has also been greatly improved by suggestions made by two anonymous readers for Oxford University Press, and by the excellent editorial work of Peter Momtchiloff, Rebecca Bryant, and their staff at Oxford.

Some of this material appears in print elsewhere. Chapter 1 Section VI overlaps with my article 'Infinite Divisibility and Actual Parts in Hume's *Treatise*', *Hume Studies*, 28 (2002), 3–25. Chapter 2 makes free use of my 'Bayle and the Case for Actual Parts', *Journal of the History of Philosophy*, 42 (2004), 145–64. I am grateful to the editors of *Hume Studies* and the *Journal of the History of Philosophy*, and to Johns Hopkins University Press for permission to reprint.

T. H.

CONTENTS

List of Tables   xii

1. Mating Horses with Griffins: the Problems of Material Structure   1
2. Actual Parts and Potential Parts   79
3. The Actual Parts Doctrine and Shortcircuit Arguments   132
4. The Actual Parts Doctrine and the Argument from Composition   169
5. The Case for Infinite Divisibility   206
6. The Kant–Boscovich Force-Shell Atom Theory   236

Conclusion   273

Bibliography   280

Index   295

LIST OF TABLES

1. The classic paradoxes of infinite m-divisibility — 37
2. Arguments for actual parts and potential parts — 128
3. The argument from actual parts to a determinate number of parts — 136
4. The argument from the definiteness of parts to ultimate parts — 143
5. Reconstruction of Kant's *Physical Monadology* argument from composition — 177
6. A demonstration of K3 from the inadmissibility of relations without ultimate relata — 186

1

Mating Horses with Griffins: the Problems of Material Structure

I. Introduction

A complex of interrelated problems plagued the theory of matter during the seventeenth and eighteenth centuries: problems concerning matter's divisibility, composition, and internal architecture. Is any material body divisible ad infinitum? Must we posit atoms or elemental *minima* from which bodies are ultimately constructed? Are the parts of material bodies themselves material *concreta*? Or are they merely potentialities or possible existents? Questions such as these—and the press of subtler questions hidden in their ambiguities—deeply unsettled natural philosophers of the early modern period. They seemed to expose serious conflicts in the most basic metaphysical commitments of the new science. Fundamental doctrines that enshrine the new world-view appeared to be at odds with one another.

The simplest way to introduce this cluster of problems and the antinomies they threaten is by way of a dilemma. (This is also the way most early modern philosophers approached the issue.) We can begin with the fact that bodies occupy space and so are divisible into parts. Bodies—at least the medium-sized ones we are acquainted with in everyday life—are spatially extended: they occupy regions of space; they have volume. As such they are divisible, at least *logically* or *metaphysically*. Even if a body's spatially distinct subsections are *physically* indivisible (that is, they are so tightly bonded that they cannot be broken apart by natural forces), their rupture and separation is at least logically possible. For instance, God could destroy the one subsection while preserving the other, or could separate and distance them one from another. The mere fact that a body occupies space seems to guarantee that it is *metaphysically* divisible—that it has spatially distinct parts whose separation is logically possible. (From hereon I refer to this metaphysical divisibility as *m-divisibility* in order to distinguish it from physical divisibility or *p-divisibility*. This book addresses questions concerning m-divisibility—questions that are logical, topological, mereological, and metaphysical in nature—rather than the essentially empirical question of p-divisibility. A full account of this distinction between m-divisibility and p-divisibility, along with certain other varieties of divisibility is given in section III below.)

The fundamental dilemma that introduces our problems of material structure is as follows. Are bodies only finitely m-divisible, or are they m-divisible ad infinitum? The problem with the first horn is obvious: it seems hopelessly ungeometrical. Bodies are extended, and geometry, the science of extension, tells us that whatever is extended contains ever-smaller spatially distinct subparts ad infinitum. And however small they become, there seems no logical bar on these subparts being m-divided one from another. So long as we find parts of matter that are spatially distinct, we find parts that God could separate. So it seems we must say that bodies are m-divisible ad infinitum, and thus reject the first horn. (This argument from geometry will be the focus of Chapter 5.)

However, there are also problems with the second horn of the dilemma. At least according to most early modern natural philosophers, the doctrine that matter is m-divisible to infinity engenders a whole host of paradoxes. For instance, it seems to entail that all bodies have the same number of parts, or that they are all infinitely large. But these results are absurd; so, by *reductio*, body cannot be infinitely m-divisible. If the paradoxes are indeed telling, then, we must reject the second horn. (I set out these paradoxes in full, along with all the other classic paradoxes, in section VIII of the current chapter.)

A linchpin thesis of this study is that the paradoxes of infinite m-divisibility are metaphysical in nature, not purely formal or mathematical. They arise not when infinite divisibility is considered purely mathematically or abstractly, but only when it is placed alongside a substantive ontological thesis concerning matter and its parts. The paradoxes do not raise any general difficulty for the mathematician's purely formal constructions or models of infinitely divisible quantity *in abstracto*. They raise a problem only for the doctrine that the substantial in space—concrete corporeal matter, matter really existing—is m-divisible ad infinitum. My argument will be that the paradoxes in fact stem from an underlying conflict between infinite m-divisibility on the one hand and what I shall call *the actual parts doctrine* on the other. According to this metaphysical doctrine, basic to the new world-view of the seventeenth and eighteenth centuries, the parts of material bodies are each concrete existents in their own right. They each enjoy a distinct existence independently of the whole and prior to any act of division. As we shall see, the actual parts doctrine is a fundamental metaphysical axiom for most of the period's 'new philosophers', and a principle no more easily abandoned than the geometrical describability of the material realm that it conflicts with.

According to the interpretation I will prosecute in this chapter, then, the fundamental challenge facing early modern matter theory was the apparent conflict between the geometrization of nature on the one hand, and certain basic metaphysical commitments on the other. Geometry seemed to mandate matter's infinite divisibility; metaphysics (more particularly, the actual parts doctrine) to preclude it. This at least partially explains why the problems of material structure caused such disquiet in the period. Natural philosophers were beginning to grasp that the paradoxes were not merely local difficulties. Rather, they exposed a deep-rooted incoherence in the fundamental tenets of the new world-view.

This reading of the early modern controversy squares well with the interpretation of at least one central Enlightenment commentator. In both his early discussion of the problems of material structure, and again in the later critical period, Kant argues that the root problem bedevilling the theory of matter is the clash between the edicts of geometry on the one hand and the metaphysical doctrine of actual parts on the other.[1] This, I shall argue, is the right way to understand the problems of material structure. As Kant puts it in his playful metaphor, the central problem is one of effecting the nuptials of a betrothed yet antagonistic geometry and metaphysics, two surly partners who resist each other's advances.

---

[1] Kant's earlier view is set out in the *Physical Monadology* (1756), in *The Cambridge Edition of the Works of Immanuel Kant: Theoretical Philosophy 1755–1770*, ed. and tr. David Walford and Ralf Meerbone (Cambridge: CUP, 1992), 1: 473–87. (Here and hereafter I follow the traditional practice of citing the pagination from *Immanuel Kant gesammelte Schriften*, 29 vols. (Berlin, 1902–), which is given in the margins of the Cambridge edn.) Kant's critical period view is given in the discussion of the Second Antinomy in the 1781/7 *Critique of Pure Reason*, tr Werner S. Pluhar (Indianapolis: Hackett, 1996), B462–71 and B551–5 (here and hereafter I cite the pagination of the 1787 (B) edn.), and in the 1786 *Metaphysical Foundation of Natural Science*, in Immanuel Kant, *Philosophy of Material Nature*, tr. James W. Ellington (Indianapolis: Hackett, 1985).

4 / Problems of Material Structure

Metaphysics...which many say may be properly absent from physics is, in fact, its only support; it alone provides illumination. For bodies consist of parts; it is certainly of no little importance that it be clearly established of which parts, and in what way they are combined together...But how, in this business, can metaphysics be married to geometry, when it seems easier to mate griffins with horses than to unite transcendental philosophy with geometry?[2]

This sets the basic challenge facing early modern matter theory: to reconcile the claims of geometry and metaphysics, or at least to uphold the one at the expense of the other.

In this first chapter I have two basic (and interrelated) goals. First, I want to present a rational reconstruction of the problems of material structure as they faced natural philosophers of the seventeenth and eighteenth centuries. This reconstruction should identify and enumerate the paradoxes that plague early modern matter theory; it should bring out their structure and various hierarchical interrelationships; and it should show the way in which the paradoxes threaten the new world-view of a rational, mathematicizable Creation. It should also make good my claim that the fundamental problem is the underlying conflict between an infinite m-divisibility mandated by geometry on the one hand, and the actual parts metaphysic on the other. Lastly, in clarifying the deeper issues that underlie the problems of material structure, it should also isolate the pivotal questions that set the agenda for the rest of this study.

My second main goal in this chapter is to establish my interpretative thesis that the problems of material structure that preoccupied early modern natural philosophy are *metaphysical* in nature, not merely *mathematical* or *formal*. In particular, the paradoxes must be understood in the context of the actual parts metaphysic.

The chapter proceeds as follows. Section II is primarily historical rather than philosophical in content. It shows the degree to which the problems of material structure gripped thinkers of the seventeenth and eighteenth centuries as issues of great moment in natural philosophy. The problems are largely neglected today, though not necessarily with justice. They are thought of, if they are thought of at all, as somehow quaint or obscure. But in the early modern period they seemed to strike at the heart of the new geometrical world-view and were seen as a terrible threat to that system.

[2] Immanuel Kant, *Physical Monadology*, in *Works*, 1: 475.

I then turn from the history of the problems of material structure to their logic. Sections III and IV introduce the central concepts and terminology that I will be deploying in this chapter and throughout this work. Section III sets out a taxonomy of the varieties of division and divisibility. Section IV introduces the rival metaphysical doctrines of actual parts and potential parts.

Armed with these various concepts and doctrines, I then present the main argument of this chapter. This is a rational construction of the classic paradoxes of infinite divisibility that shows that the problems are inescapably metaphysical and not just mathematical. The argument develops as follows. In section V, I set out my interpretation and show how it contrasts with the standard reading of the paradoxes in the secondary literature. In section VI, I present a case study of one particular paradox, drawn from Hume's texts, in order to show in detail the merits of my reading and the shortcomings of the standard interpretation. Hume provides an exceptionally clear example of an early modern opponent of infinite divisibility who has been roundly—and unfairly—condemned and ridiculed by many commentators. But once we understand that his argument belongs in an actual parts context, we can rescue Hume from the objections stemming from their purely mathematical reading. In section VII I generalize this argument and show how the actual parts metaphysic underpins not just one particular paradox, but the full range of classic paradoxes of infinite divisibility. And, finally, in section VIII, I offer an organizing survey and systematic reconstruction of *all* the classic paradoxes. This will bring out their hierarchical structure and show us which paradoxes are most fundamental, and it will once again show how they all depend on the actual parts metaphysic. This completes the main argument of the chapter.

The last key part of this chapter is section IX, in which I present a survey of the various doctrinal camps that emerge in reaction to the problems of material structure. Here I marshal the natural philosophers of the Enlightenment into various factions and outline their diverse theoretical responses to the threatened antinomy. In section X I wrap up with an outline of the structure of this book and its subsequent chapters.

II. The Significance of the Controversy

> I believe that the smallest portion of matter may be practically divided *ad infinitum*: that equal qualities taken from equal qualities, are unequal

quality will remain: that two and two make seven: that the sun rules the night, the stars the day; and the moon is made of green cheese.

Tobias Smollet, *The History and Adventures of an Atom*, 1769

Nowadays the Enlightenment tumult over the problems of material structure can seem a somewhat quaint or eccentric business. When one hears of baroque scholars wrestling over the abstruse quandaries of infinite divisibility, the image conjured up can seem somewhat ridiculous. One is reminded, perhaps, of angels on pinheads, or of the rarefied debates of the Laputan scholars in *Gulliver's Travels*. And the more one learns of the extent and the ferocity of the dispute—of the 'racking of...brains' by the 'the most towering and subtle sort of speculators', of 'the very greate controversy in the schooles', of the 'hot and scarce ingenious Altercations' that infinite divisibility provokes 'among the gravest and leading Philosophers'—the more the whole drama can seem to slide towards burlesque.[3] Certainly Rabelais would have loved the mêlée of bewigged metaphysicians, all up in arms over the minutiae of the minute.

I certainly agree that there *is* something comical about the dispute, as indeed many of the parties to the Enlightenment debate also recognized. But this should not lead us to dismiss the controversy as peripheral, unimportant, or philosophically eccentric. In this brief section I want to suggest how important a familiarity with the dispute is to a proper understanding of early modern natural philosophy. There are two main points to appreciate here. First there is the sheer degree to which the issue preoccupied Enlightenment philosophy. As we shall see, this was not some sort of random craze. Rather it reflected the realization that the problems of material structure expose serious fractures in the basic assumptions of the new science, posing a severe threat to that system. So here we have the heat of the debate—a matter of historical interest—and the importance of the problems' challenge to the new science—a matter of philosophical interest. Second, setting aside the intrinsic interest of the problems of material structure and their threat to the new system, it is also important to appreciate the way in which the dispute stimulates theorizing in Enlightenment natural philosophy on a wide range of related issues.

---

[3] Kenelm Digby, *Two Treatises, in the one of which the Nature of Bodies is expounded; in the other, the Nature of Man's Soule; is looked into* (1644; facsimile edn.: Stuttgart: Friedrich Fromman Verlag, 1970), 10. Robert Boyle, *Selected Philosophical Papers*, ed. M. A. Stewart (Manchester: Manchester University Press, 1979), 217. Walter Charleton, *Physiologia Epicuro-Gassendo-Charletoniana* (London, 1654; facsimile edn.: New York: Johnson Reprint Co., 1966), 95.

The first main point to appreciate is that the paradoxes of material structure really were problems of the greatest moment for Enlightenment thinkers. For Leibniz, the problems constituted one of the 'two famous labyrinths' that forever challenge human reason (the other being the issue of free will). The problems stand at the very starting point of Leibniz's later work—the opening sections of the 1714 *Principes de la nature et de la Grace* and *Monadologie*, for instance—and give rise to the central argument for monads. Hume devotes the second part of book 1 of the 1739–40 *Treatise* to an extended treatment of the issue. This clearly testifies to his conviction of its fundamental importance: it is the very first substantive issue he addresses after the opening segment on empiricist methodology and the general system of impressions and ideas. And for Kant the issue is also of great significance. On his view, the problems of material structure give rise to one of the four inexorable antinomies that embarrass transcendental realism: he thinks that the problems undermine all pre-critical systems. In fact *all* the leading philosophers of the period address the problems: one could mention Galileo, Descartes, Arnauld, Spinoza, Malebranche, Bayle, Locke, Newton, Clarke, Berkeley, Euler, Boscovich, Reid—and many others beside. These issues are also accorded an important status in the classic physics textbooks of the period: for instance, in Descartes's 1644 *Principia Philosophiae*, in Jacques Rohault's 1671 *Traité de Physique* (which, translated and accompanied by Samuel Clarke's Newtonian footnotes, became the standard text at Oxford and Cambridge), and in Antoine Le Grand's 1672 *Institutio philosophiae*; likewise in the lectures of mathematicians like Isaac Barrow and John Keill. As I shall show, the great agitation over this '*opprobrium Philosophorum*'[4] is readily explained once one appreciates that it appears to expose an antinomy in the basic metaphysical presuppositions of the new science. Two fundamental principles of the new world-view— on the one hand, the geometrical nature of the material realm, and on the other, the actual parts metaphysic—seem to be in conflict with one another. This is at once a scandal and a terrible threat to the celebrated new system.

As an aside, it is also interesting to note the degree to which the dispute over the structure of matter spills beyond the borders of Republic of Letters and enters the wider educated culture. It becomes the subject

---

[4] Edward Stillingfleet, *Origines Sacrae, or a Rational Account of the Grounds of Christian Faith, as to the Truth and Divine Authority of the Scriptures, And the matters therein contained* (London, 1662), 380.

of numerous public lectures and disputations.⁵ It enters the fashionable talk of salons and royal courts, with both Cardinal Richelieu and Frederick the Great dabbling in the controversy.⁶ If Euler is to be believed, for a moment the debate even monopolizes the gossip of the guardroom.⁷ It certainly enters popular literature, providing satirical material for bestselling authors such as Cyrano de Bergerac, Jonathan Swift, Voltaire, Laurence Sterne, and Tobias Smollett.⁸ Eventually the controversy even precipitates an astonishing, highly politicized squabble in the Berlin Royal Academy with the 1746–7 prize competition on infinite divisibility and material structure. (As Laurence Bongie notes, 'it is rather difficult for us today to imagine the intense public agitation stirred up by the announcement of a formal European competition on this metaphysical issue'.⁹) The Academy contest quickly bitters with polemic and mutual recrimination— accusations of impiety and atheism are pressed by one faction and then the other—and ultimately explodes into something of a political crisis in German letters.¹⁰ (Oddly enough, there is something of a tradition of this sort of cut-throat episode in the history of the debate over infinite divisibility and material structure. For instance, there is the case of the wretched Pythagorean Hippasos of Metapontion, marooned and left to die by his fellow sectaries for disclosing the terrible secret of incommensurable magnitudes.¹¹ And then there is the formal condemnation of Wyclif at the Council of Constance in 1414. Among the articles enumerating the crimes

---

⁵ See the mocking commentary in Pierre Bayle, *Historical and Critical Dictionary*, ed. and tr. Richard H. Popkin (Indianapolis: Hackett, 1991), 362.

⁶ On Richelieu, ibid. 360 n. 60. Frederick II played a role pacifying the furore over the Berlin Academy dispute over the problem (see the references in Ch. 4 n. 10).

⁷ Leonhard Euler, *Letters on Different Subjects in Natural Philosophy* (1761), tr. Henry Hunter, ed. David Brewster (New York: J. and J. Harper, 1833), 39.

⁸ Cyrano de Bergerac, *Voyages to the Moon and the Sun*, tr. Richard Aldington (London: Routledge & Sons, 1923). On Jonathan Swift, see Leroy Loemker, introduction to G. W. Leibniz, *Philosophical Papers and Letters* (Chicago: University of Chicago Press, 1956), 1–62, 31. Voltaire, *Philosophical Dictionary*, tr. and ed. Theodore Besterman (Bungay, Suffolk: Penguin, 1971), 158, 295. Laurence Sterne, *Life and Opinions of Tristram Shandy, Gentleman* (Baltimore: Penguin Books, 1967), 160. Tobias Smollett, *The History and Adventures of an Atom* (Athens, Ga.: University of Georgia Press, 1989), 27.

⁹ Laurence L. Bongie, introduction to Etienne Bonnot de Condillac, *Les Monades* in *Studies on Voltaire and Eighteenth Century Philosophy*, 187 (London: Cheney & Sons, 1980), 7–107, 22.

¹⁰ On the history of this controversy, see the references given in Ch. 4 n. 10.

¹¹ Bertrand Russell, *Our Knowledge of the External World* (London: George Allen & Unwin, 1926), 168.

of this unruly theologian, we find—of all things—a denunciation of his unorthodox stance on infinite divisibility.)

The second main reason why an appreciation of the controversy over material structure is essential to a full understanding of Enlightenment natural philosophy is that the dispute played a crucial role in motivating a whole range of important theories in physics and metaphysics. Partly because of their menace, the problems of material structure greatly stimulated Enlightenment physical and metaphysical thought, generating a wide range of theoretical responses. For Galileo, the problems demanded the rejection of the orthodox, Aristotelian account of infinity and the resurrection of the actual infinite. For Hume they mandated a radical Epicureanism and a reassessment of the status of geometrical knowledge. Early field theorists like Boscovich and the younger Kant argued that they showed the necessity of a dynamical theory of matter and the admission of action at a distance—a shocking anathema to the mechanist. For Bayle and Berkeley, the problems amounted to a *reductio* on the extra-mental existence of matter and a demonstration of idealism. Kant similarly thought that they mandated transcendental idealism. Finally, radical sceptics found in the problems unanswerable paradoxes that undermined reason *tout court*. So, many important theories in early modern philosophy find their origins in—or at least draw additional support from—the problems of material structure. (See section IX of the current chapter for an organizing survey of these various responses.) The dispute over material structure can then hardly be dismissed as peripheral or inconsequential.

III. Four Varieties of Divisibility

In this section and the next I introduce terminology that will be employed throughout this study. The current section sets out a taxonomy of different varieties of division and divisibility.

In discussing the divisibility of extended entities, Enlightenment philosophers naturally invoke various distinctions between different types of division. Typically the central contrast is between, on the one hand, a division that is (in some sense) 'actual' or 'real' and, on the other, a division that is (in some sense) 'logical' or 'in the thing itself'. Thus 'a gross tearing or cutting one part from another' is contrasted with an 'intellectual division'; a 'severing or pulling asunder' with a mere 'diversity of consideration'; an 'actual separation of parts from one another' with 'the resolution of any magnitude into its parts, or their distinction and assignment'; and 'a division practicable by our strength and skill' with 'that which is possible in itself' or that which 'Divine Omnipotence is able to accomplish'.[12] However, while certain of these distinctions do clearly map onto one another, others turn out to chart subtly different taxonomies. The confusion of these distinct types of division can create much philosophical heavy weather if one is not careful: there is something of a history of philosophers led astray by this sort of terminological misunderstanding.[13] It is therefore crucial to present a clear taxonomy of the varieties of divisibility.

For the purposes of this study, I will be distinguishing between four types of divisibility. This taxonomy captures four philosophically interesting ways in which an extended entity could be said to be broken down into distinct subparts. (It may be that some of these varieties of division and divisibility end up collapsing into one another. But at this stage of the argument I do not wish to prejudge this question. There are certainly some early modern philosophers who think that all these distinctions bear up. In order to locate all these period philosophers' various positions in the logical space of the debate we will have to proceed, at least initially, as if these four types of divisibility can indeed all come apart.)

---

[12] Henry More, *The Immortality of the Soul*, ed. Alexander Jakob (Dordrecht: Kluwer, 1987), 27, 40. Thomas Hobbes, *De Corpore*, in *The Collected Works of Thomas Hobbes*, ed. William Molesworth, 11 vols. (London: 1839–40; facsimile edn.: London: Routledge Thoemmes Press, 1992), i. 96. John Keill, *An Introduction to Natural Philosophy* (London: Sennex, Innys, Manby, Osborn, & Longman, 1733), 20. Euler, *Letters*, 43.

[13] Consider two examples. (1) Henry Lee, in his *Anti-Scepticism* of 1702, criticizes Locke as follows: '[Locke's] thoughts are spun a little too fine, when he says, that *Space* can neither be *really* or *mentally* divided; for he owns, it may be consider'd partially (as, to be sure it may...)...Denying the *Divisibility*...of the Parts of Space, so much as in our manner of Conception, when we own that they are...*distinguishable*, is beyond my Comprehension.' But of course Locke simply meant that the parts of space, though distinguishable, could not be separated and distanced one from another, even in our thought—and Lee himself agrees with this just ten pages later. (Henry Lee, *Anti-Scepticism* (Hildesheim, New York: Georg Olms Verlag, 1973), 73, 83.) (2) There is much ado in the Leibniz–Clarke correspondence over whether Newtonian absolute space is divisible or not. Leibniz insists that it is; Clarke insists it is not. Despite the fact that it is clear to the reader—and surely to both parties—that Leibniz simply means that distinct regions of absolute space can be identified, whereas Clarke is arguing that these regions cannot exist independently of one another, the *ignoratio elenchi* runs on through three of their exchanges. (See Samuel Clarke and G. W. Leibniz, *The Leibniz–Clarke Correspondence*, ed. H. G. Alexander (Manchester: Manchester University Press, 1956), e.g. 21, 31–2, 48, 103.)

I present the taxonomy in detail in a moment. But first it may help if I introduce the four types of division by way of an illustration drawn in terms of the most immediately familiar of Enlightenment systems, the classical physics and metaphysics of the later Newton. Consider one of Newton's indestructible extended atoms: a miniature corpuscle, perfectly solid, operating in the void. This atom cannot be physically broken apart by any natural force: unlike the grosser bodies aggregated from atoms, it is *physically indivisible*. Nevertheless, since it is extended it has spatially distinct parts—for instance, it has a left and a right side. So it is *formally divisible*. Moreover, were someone to consider these halves, they could, in their thought, designate them as two entities—individuating them in mental representation, perhaps through diversity of consideration or selective attention. So it is *intellectually divisible*. Finally, notwithstanding their physical inseparability, there is no *logical* reason why those spatially distinct halves could not exist separated from one another. God, for instance, could rupture and distance the two halves, or perhaps annihilate the one part while preserving the other. So it is *metaphysically divisible*. Now consider Newtonian space. Like the atom, Newton's space is physically indivisible (it cannot be naturally broken apart), formally divisible (it has spatially distinct parts or subregions), and intellectually divisible (a mind could represent it as containing diverse parts). However, unlike the atom, Newtonian space is metaphysically indivisible. Not even God can separate and distance its parts: if one part is given, all are given in immutable, inseparable, and immoveable array. Here is the taxonomy in more detail.

1. Physical Divisibility

An extended entity is *physically divisible* (hereafter *p-divisible*) if and only if its spatially distinct parts can be broken apart by natural processes and separated from one another. The act of *p-division* thus involves the actual rupture and distancing of parts and can only occur within the confines of the laws of nature. For this reason, questions of p-divisibility are relatively unexciting philosophically. They clearly turn on various empirical matters—how tightly the parts of a given body are bonded, how strong and finely focused a force can be brought to bear on them to effect their rupture—rather than questions of a more metaphysical or logical nature.

2. Metaphysical Divisibility

An extended entity is *metaphysically divisible* (hereafter *m-divisible*) if and only if it is logically possible that its spatially distinct parts could exist separately from one another. Can a given part exist independently of the other parts, or is there some logical reason why they are necessarily co-dependent? The idea here is to abstract from the constraints of the laws of nature, and ask whether the parts of a given extended entity are logically separable from one another, regardless of whether or not they are physically divisible.

In the early modern literature there are two standard ways of approaching questions of an entity's m-divisibility. First, given the period's practice of taking conceivability as evidence of logical possibility, a standard device is to ask whether one part of an extended entity can be conceived as existing independently of the others. For instance, if a given part can be imagined separated from the others, or if that part can be imagined as continuing to exist while the others are annihilated, then the whole original can be said to be m-divisible. A second popular way of thinking about m-divisibility was by way of divine omnipotence. It was of course standard practice in the early modern period to appeal to God's power as a way of shifting debate away from matters of physical possibility within the laws of nature to focus instead on questions of logical possibility. And we find this same device employed in discussions of metaphysical divisibility: an entity is m-divisible if nothing prevents God from breaking it apart—this 'supernatural divisibility' (i.e. m-divisibility) being contrasted with 'natural divisibility' (i.e. p-divisibility).[14]

Correlative to this property of m-divisibility we can thus define the act of *m-division* as the actual rupturing and distancing of logically separable parts. Since the m-divisible whole that contains these logically separable parts may itself be p-indivisible, we may need to invoke God's supernatural power to overcome natural constraints and effect such an m-division.

I have already said that the later Newton holds corporeal matter to be m-divisible, even if not all of it is p-divisible (consider the extended corpuscles). Newtonian space, on the other hand, is m-indivisible: if one part

---

[14] Henry More, *Enchiridium Metaphysicum*, tr. Alexander Jacob (Hildesheim: Georg Olms Verlag, 1995), 72, 124–5. Samuel Clarke, *The Works of Samuel Clarke*, 4 vols. (London: John and Paul Knapton, 1738; facsimile edn.: New York: Garland, 1978), iii. 761, 762. Euler, *Letters*, 43.

of it is given, all the rest is necessarily given along with it in inseparable array.[15] The material plenum of Descartes and Spinoza presents a slightly more difficult case. Considered as a whole, the entire plenum is, like Newtonian space, m-indivisible. Since the plenum's defining essence is extension, the existence of one part necessarily implies the existence of the whole.[16] The similarity here between the entire plenum of Descartes and Spinoza and the infinite absolute space of Newton on this question should be clear: in each of these cases the framework of spatial reality, the entire theatre of material existence, is taken to be one m-indivisible whole.

However, notwithstanding the sense in which the whole plenum is m-indivisible, there is a clear sense in which *particular* bodies within the plenum are m-divisible. It is true that any given part of a body must be completely surrounded at all times by *some* other portions of matter. In this sense it cannot become radically distinct from the whole plenum. But Descartes and Spinoza nonetheless allow that the parts of any given body are separable one from another. Any given part can continue to exist without being adjacent to any other particular portion of matter. This is because the parts of the plenum that constitute bodies are moveable one from another (contrast the parts of Newtonian space, which are immobile). So, having conceded that the plenum considered as a whole is m-indivisible, and that no part of it is radically distinct from the rest of the plenum, I nonetheless wish to insist that particular bodies in the plenum should be classified as m-divisible. Any particular body can be broken apart, at least by God, and this is sufficient for it to be m-divisible.

We have seen that the entire theatre of material existence—be it Newtonian absolute space or the universal Cartesian *res extensa*—is

---

[15] Isaac Newton, *Principia Mathematica* (1687), tr. Andrew Motte (Amherst, NY: Prometheus Books, 1995), 15–16; *De Gravitatione*, in *Unpublished Scientific Papers of Isaac Newton*, tr. and ed. A. R. Hall and M. B. Hall (Cambridge: CUP, 1962), 136–7. See also Samuel Clarke's defence of this Newtonian view in *Leibniz–Clarke Correspondence*, 21, 31–2, 48, 103, and John Locke, *An Essay Concerning Human Understanding*, ed. P. H. Nidditch (Oxford: OUP, 1975), II. xiii. 13, II. xv. 10 (here and hereafter I follow the standard practice of referring to book, chapter and section number). For commentary see Edward Grant, *Much Ado About Nothing: Theories of Space and Vacuum from the Middle Ages to the Scientific Revolution* (Cambridge: CUP, 1981), 235, 251–2, esp nn. 412 and 413).

[16] Descartes, *Principles of Philosophy*, ss. I. 60, II. 4, in *The Philosophical Writings of Descartes*, tr. and ed. John Cottingham, Anthony Kenny, Dugald Murdoch, and Robert Stoothoff, 3 vols. (Cambridge: CUP, 1991), i. 215, 224. Spinoza, *Ethics*, in *The Collected Works of Spinoza*, ed. and tr. Edwin Curley (Princeton: Princeton University Press, 1985–), part 1, proposition 15 scholium.

m-indivisible. Examples of Enlightenment philosophers who allow *finitely* extended entities to be m-indivisible are harder to find, though there are a few. The extended, penetrable spirits of Henry More and Samuel Clarke are supposed to be 'indiscerpible' or m-indivisible.[17] Certain thinkers also allow there to be extended but m-indivisible *minima* of matter: More's 'physical monads', Cordemoy's atomic substances, the absolutely partless elements of neo-Epicureans like Charleton and the early Newton, the *minima* of Berkeley, Boscovich's and the pre-critical Kant's force-shell atoms. (It is important not to confuse these varieties of m-indivisible *metaphysical* atoms with the p-indivisible extended *physical* atoms that these thinkers also allow. In some systems—such as Boscovich's—the metaphysical atoms and physical atoms are one and the same. But in other systems—such as the Epicureans'—the physical atoms are constructed from coalitions of smaller metaphysical atoms.)

3. Formal Divisibility

An extended entity is *formally divisible* (hereafter *f-divisible*) if and only if it has parts that can be distinguished by their spatial properties, regardless of whether those parts can be separated from one another, either naturally or by God. *F-division* is then the (correct) discrimination of separately located parts: parts that differ in their respective spatial properties, such as their relations of adjacency and distance *vis-à-vis* other spatially located beings.

---

[17] My term 'm-divisibility' corresponds to Henry More's 'discerpibility'. When More introduces this term, his language can at times appear suggestive of p-divisibility rather than m-divisibility. (*Immortality of the Soul*, 27; *Enchiridium Metaphysicum*, tr. Alexander Jakobs (Hildesheim: Georg Olms Verlag), 72). But it is clear that he really intends this to stand for m-divisibility. Whenever reasons are advanced to show that an entity is indiscerpible, those reasons are logical or metaphysical in nature rather than empirical (*Immortality*, 6–7, 36–7, 40; *Enchiridium*, 72, 124–5); moreover, those reasons prohibit even God from breaking apart an indiscerpible entity (*Immortality*, 40; see also *The Philosophical Writings of Henry More*, ed. F. I. Mackinnon (New York and Oxford: OUP, 1925), 216–19.). More's term re-emerges in the texts of 18th cent. Newtonians such as Samuel Clarke and William Wollaston, where 'discerpibility' again stands for m-divisibility rather than p-divisibility. Clarke, *Works*, iii. 762. William Wollaston, *The Religion of Nature Delineated* (1724; facsimile edn.: Delmar, NY: Scholars' Facsimiles and Reprints, 1974), 74. The phrase is also used—again with the same meaning—in Richard Blome's 1694 tr. of Antoine Le Grand's 1672 *Institutio philosophiae* (*An Entire Body of Philosophy*, 2 vols. (London 1694), ii. 7).

It is a standard view—and certainly one common to Cartesians and Newtonians alike—that whatever is extended is f-divisible. All regions of space, and all entities that occupy regions of space, have spatially distinct parts and so are f-divisible. Thus even the m-indivisible extended spirits of Henry More and Samuel Clarke are f-divisible, as are More's extended but m-indivisible physical monads. A small minority of philosophers resist this view, however, claiming that there can be extended regions of space that are completely without spatially distinguishable subparts. This neo-Epicurean view is championed by Walter Charleton, the younger Newton, and Berkeley, each of whom resists the infinite (f-)divisibility of space and asserts instead the existence of extended but atomic spatial *minima*. These peculiar *minima* of space—and anything that occupies just one such *minimum*—lack spatially distinguishable parts and thus are f-indivisible.

4. Intellectual Divisibility

An extended entity is *intellectually divisible* (hereafter *i-divisible*) if and only if a mind could represent it in thought as containing diverse parts—regardless of whether those parts are separable in the thing itself (through either p-division or m-division), and regardless of whether those parts are genuinely spatially distinct (i.e. f-divisible). *I-division* is then defined as the discrimination of parts in mental representation. It need not involve a *separation* and *distancing* of parts in thought—all that is required is that parts be somehow discriminated or told apart through some mental operation such as diversity of consideration or selective attention. The parts can thus remain adjacent both in reality and 'in the mind's eye' while yet being i-divided.

It is a common view that something is i-divisible if and only if it is f-divisible (i.e. has spatially distinguishable parts). A very small minority of philosophers resists this however. Berkeley, for instance, claims that extended entities are only finitely i-divisible before one arrives at f-indivisible extended *minima*. (These are his famous *minima sensibilia*. Since the mind can perceive no smaller subparts within the *minima*, and *esse est percipi*, they lack spatially distinct subparts altogether.) But then he allows that a mind can represent any extension by an arbitrary number, from whence we can think of ever smaller parts of that extension simply by dividing that number any way we please. So there is at least one sense in which Berkeley thinks of any given extension as endlessly i-divisible, though he will immediately insist it is a mistake to think of this endlessly iterable mental operation as implying that there *actually exist* spatially distinct parts below the level of *minima sensibilia*.[18] (A similar position can, I think, be retrieved from the texts of neo-Epicureans like Charleton and the younger Newton, and from Hume.) Now, perhaps Berkeley is wrong about this and the distinction between f-divisibility and i-divisibility will ultimately collapse. But at this stage of the argument it will be useful to keep the two varieties of divisibility distinct in our taxonomy. The important thing to realize is that i-division merely requires a certain sort of mental operation, whereas f-divisibility requires the actual existence of spatially distinct parts.

IV. The Actual Parts and Potential Parts Doctrines

Here I outline the actual parts doctrine and the potential parts doctrine: two rival metaphysical theses concerning the nature of the parts of bodies. I discuss these two doctrines and the clash between them in much greater detail in Chapter 2 below. The present outline sketch is merely meant to introduce the competing theses in précis form so that I can invoke them in my reconstruction of the paradoxes of material structure.

The actual and potential parts doctrines offer conflicting accounts of the ontological status of the parts of bodies. Are such parts fully fledged concrete existents? Or are they merely *possibilities* or *potentialities* until actualized by a positive operation of division?

The Actual Parts Doctrine

According to this doctrine, the parts into which a body can be m-divided (that is, its logically separable parts, the parts God could break it into) are each a distinct existent. They each exist independently of the whole, and independently of all the other non-overlapping parts, and they each exist prior to any positive act of division. The parts are all *already* embedded in the architecture of the whole: division merely separates them, it does not create them. The whole body is thus an aggregate of independently existing parts, and its metaphysical structure is best described with count terms: it is a composite of *so many* distinct parts. Samuel Clarke sums up the doctrine nicely:

---

[18] George Berkeley, *Principles of Human Understanding*, ss. 123–7, in *The Works of George Berkeley*, ed. A. A. Luce and T. E. Jessop, 9 vols. (London: Thomas Nelson & Sons, 1948–9), ii. 97–100.

some of the *first* and *most obvious* Properties which we *certainly know of Matter* [include] its having partes extra partes, strictly and properly speaking, that is its consisting of such Parts as are *actually unconnected*, and are *truly distinct Beings*, and *can* (as we see by Experience) *exist separately*, and *have no dependance one upon another*.

Matter is thus, Clarke adds, essentially composite: 'a Being that consists of a Multitude of separate and distinct Parts'.[19] The doctrine is also clearly presented by Thomas Reid:

There seems to be nothing more evident than that all bodies must consist of parts, and that every part of a body is a body, and a distinct being, which may exist without the other parts... when [matter] is divided into parts, every part is a being or substance distinct from all the other parts, and was so even before the division.[20]

It is important to be clear that the actual parts doctrine states a *metaphysical* thesis about the ontological status of the parts of bodies (as does the rival potential parts doctrine). It must not be confused with the view that bodies have any particular internal *physical* organization or architecture of *physically* heterogeneous parts. For all that has been said here, bodies might be perfectly homogeneous in their physical structure, altogether lacking physically differentiated subparts. (Think, for instance, of a block of jelly that is, so to speak, homogeneous jelly 'all the way down'.) But, whether or not they have any heterogeneous physical structure, all bodies are, according to the current doctrine, nonetheless each a composite of so many ontologically independent parts at the metaphysical level.

As I shall show in Chapter 2, this account of the ontological status of the parts of matter is basic to the new world-view of Enlightenment natural philosophy. During the seventeenth century it is a contentious doctrine, resisted by the scholastic old guard, and championed by Galileo and the Young Turks of the new science. A central tenet of classical Newtonianism, it gains near universal acceptance in the eighteenth century with the sweeping triumph of that system. In many ways the doctrine represents new science's revival of Epicurean models of nature and its rejection of Aristotelian theories of substantial forms. I substantiate all this fully in Chapter 2.

---

[19] Clarke, *Works*, iii. 762, 790.
[20] Thomas Reid, *Essays on the Intellectual Powers of Man*, in *The Works of Thomas Reid*, ed. Sir William Hamilton, 2 vols. (1863; facsimile edn.: Bristol: Thoemmes Press, 1994), i. 323.

The Potential Parts Doctrine

According to the rival potential parts doctrine, the parts into which a body can be m-divided are not distinct existents prior to their being actualized by a positive operation of division. Division *creates* these parts as so many freshly minted beings—it does not simply separate pre-existing parts. Prior to the act of division that generates these parts, the whole is best described as containing *possible* or *potential* parts. Such 'potential parts' represent ways in which the whole could be broken down, but do not exist other than as aspects of the whole until an act of division actualizes them as independent entities. On this view, then, the whole is not a composite or aggregate structure, a construction from distinct actual parts. It has no particular inherent structure of distinct parts prior to division. Rather it is a metaphysically simple entity. Simplicio, the orthodox Aristotelian character in Galileo's *Dialogues Concerning Two New Sciences*, puts the doctrine as follows:

Parts cannot be said to be in a body which is *not yet divided* or at least *marked out*; if this is not done we say that they exist potentially.[21]

Sir Kenelm Digby (no less a party-line Aristotelian than Simplicio on this particular issue) also gives us a clear statement of the view. Any given body, he writes,

is but one *whole* that may indeed be cutt into so many severall partes: but those partes are not really there, till by division they are parcelled out: and then, the whole (out of which they are made) ceaseth to be any longer; and the partes succeede in lieu of it; and are, every one of them, a new *whole*.
[T]he partes which are considered in Quantity, are not diverse things: but are onely a vertue or power to be diverse thinges.[22]

Our final statement of the view is from Hobbes:

it is manifest, that nothing has parts till it be divided; and when a thing is divided, the parts are only so many as the division makes them.[23]

As with the rival actual parts doctrine, it is important to appreciate that this is a *metaphysical* thesis about the ontological status of the parts of bodies,

---

[21] Galileo Galilei, *Dialogues Concerning Two New Sciences*, tr. Henry Crew and Alfonso de Salvio (New York: Macmillan, 1914), 34. Emphasis added.

[22] Digby, *Two Treatises*, 10, 13.

[23] Hobbes, *De Corpore*, ch. 7.9, in *Works*, i. 97; see also *Leviathan*, in *Works*, iii. 677, where division again 'makes' parts.

not a thesis about the internal *physical* architecture of bodies. For all the potential parts doctrine asserts, bodies might be either completely homogeneous or highly complex in terms of their internal physical organization.

The potential parts doctrine is central to Aristotelian natural philosophy. Ratified by St Thomas, it becomes the orthodoxy of high medieval and Renaissance scholasticism. One or two of the new philosophers also endorse it (such as Hobbes, in the quotation above), though it is important to appreciate that on this issue they stand opposed to the overwhelming majority of the Enlightenment's new anti-scholastic philosophers. I substantiate all this in Chapter 2.

V. The Mathematical Reading of the Problems of Material Structure and the Actual Parts Reading

At the start of this chapter I outlined the basic dilemma that introduces the problems of material structure. On the one hand, finite m-divisibility seems ruled out by considerations drawn from geometry: whatever is extended, it seems, must be endlessly m-divisible. But then, on the other hand, infinite m-divisibility also seems ruled out, this time by metaphysical considerations: it leads to paradox when set alongside the actual parts doctrine. The fundamental problem facing Enlightenment matter theory, then, is to square the claims of geometry with the claims of the actual parts metaphysic, or at least rule in favour of the one at the expense of the other. In the next three sections of this chapter, I substantiate this interpretation and show the way in which the actual parts metaphysic is in tension with the edicts of geometry. But first I want to contrast my interpretation with another reading that dominates the current secondary literature. This should show why my reading is contentious, and set the challenge of showing why my interpretation is preferable to the received one.

According to the standard reading of the paradoxes in the secondary literature, the paradoxes of infinite divisibility are purely mathematical problems. They challenge *mathematical constructions* of the form or structure of infinitely divisible body, and purport to show that there is no workable mathematical model of divisibility ad infinitum. On this view, the paradoxes invoke premises that appeal exclusively to the structural or geometrical properties of extended, infinitely divisible body and purport to show that absurdities follow directly from these properties alone. In particular, at no point do the paradoxes appeal to any ontological or metaphysical theses about the 'stuffing' or 'content' of body. In Aristotelian terms, they simply concern the 'form' (structure) of body, not its 'matter' (filling); in Galilean or Cartesian terms, they simply concern its mathematically tractable spatial or 'primary' qualities. In fact, since the paradoxes invoke only the geometrical features of infinitely divisible body, they would equally apply to anything that shared those features, and not just to concrete matter. For instance, they would apply no less to the mathematician's formally defined infinitely divisible geometrical figure, or to any region of infinitely divisible space.

Given this standard interpretation, the paradoxes turn out to be quite easy to disarm. They seem to stem simply from the shortcomings of early modern mathematics. In particular, they rest on the period's halting and flawed mathematical understanding of the logic of limits and the infinite series. So the commentator inclined towards this sort of reading will then be quick to respond to the alleged paradoxes. These mathematical confusions may (perhaps) have been forgivable in the early modern period, but they are clear errors all the same, and the so-called paradoxes evaporate once we appreciate this. Or so the standard reading goes.[24]

But this purely mathematical interpretation of the paradoxes is quite mistaken, or so I shall argue. The paradoxes are *not* supposed to challenge infinite divisibility simply on mathematical grounds. They are not supposed to raise a purely formal problem for mathematical constructions of infinitely divisible body. For all that the paradoxes claim, this sort of mathematical model that considers the form or structure of infinitely divisible body *in abstracto*, divorced from all thought of the actual physical filling or stuffing of real concrete bodies, may indeed be perfectly unproblematic. Rather (on my interpretation) the paradoxes purport to challenge only the doctrine that entities with *actual parts* (parts that are concrete, independent existents) could be m-divisible ad infinitum. In short, the paradoxes arise only in the context of the actual parts metaphysic. They raise an objection only against the infinite divisibility of extended entities that have actual parts. And, as we have seen, for the new philosophers of the Enlightenment, this would certainly include actual physical bodies—concrete, corporeal objects, the substantial in space—since these are constructed from actual parts. This explains, as the standard reading cannot,

---

[24] For examples of this standard interpretation in the literature, see my discussions of the Hume literature in s. VI below, and of the Kant literature in Ch. 4, s. IV.

why natural philosophers of the period saw the infinite divisibility of *material body* as the key problem, and not the infinite divisibility of space or extension in general. This is because matter, for the new science, is constructed from concrete, independently existing actual parts in a way that space or extension in general need not be. (The exceptions that truly prove the rule are those philosophers, such as Pierre Bayle, who are working in a Cartesian framework that *identifies* matter with space and extension in general. Bayle *does* sometimes phrase the problem in terms of the divisibility of extension; but this is only because he identifies matter with extension and sees 'both' as constructed from actual parts.[25] Those philosophers, such as the Newtonians, who *do* draw a distinction between body on the one hand and space on the other phrase the problem in terms of the divisibility of body, not space.[26])

So Enlightenment natural philosophers understood that the paradoxes were restricted to those entities with actual parts. But this point is lost in much of the recent commentary. As I will show, the standard interpretation both gets the intellectual history wrong, misinterpreting the actual arguments that preoccupied the Enlightenment, and seriously underestimates the philosophical importance of the controversy. If I am right that the arguments against infinite divisibility crucially depend on the actual parts metaphysic, then it is hardly surprising that, once they are divorced from this metaphysical setting and reconstructed without their actual parts premises, they lose whatever force they might originally have had. But once we place them in their proper context and rearm them their actual parts

---

[25] Bayle, *Dictionary*, 135–9 (the 'Leucippus' article on the vacuum), 359–58 (the 'Zeno of Elea' article on the problems of continua). Hume is another principled exception, since he holds that *all* continuous quantities (including matter, space, and time) are constructed from actual parts. See s. VI for details.

[26] The clearest example of this comes in Kant's 'Comment on the Second Antinomy', where he elaborates on one of the paradoxes of infinitely divisible matter. The paradox only arises, he says, in the case of a substantive whole—this being the composite proper, i.e., the contingent unity of the manifold, that, *given separately* (at least in thought), is put into a reciprocal linkage and thereby makes a unit'. So the paradox only applies to entities with actual parts (such as material bodies) and not to space: 'Space should, properly, be called not a *compositum*, but a *totum*, because its parts are possible only in the whole, not the whole through the parts.... Space is not a composite made up from substances.' Immanuel Kant, *Critique of Pure Reason*, tr. Werner S. Pluhar (Indianapolis: Hackett, 1996), B466, B468. For an earlier, pre-Newtonian example, see Stillingfleet, *Origines Sacrae*, 380, where the paradoxes of infinite divisibility again threaten 'extended matter' while explicitly leaving 'empty space' untouched.

premises, the paradoxes regain much of their menace. And once we appreciate that genuinely metaphysical questions are at stake—matters of ontology properly philosophical—we see that the problems cannot be simply set aside as founded on an outmoded mathematics. Perhaps post-Enlightenment developments in mathematics can indeed play a part in resolving the various problems. Nevertheless, the problems, in raising substantive questions in metaphysics proper, demand answers that are themselves inescapably metaphysical, and not simply mathematical or formal.

I have said (contra the standard reading) that the paradoxes of infinite m-divisibility are not purely mathematical but rather essentially metaphysical. They turn on the conflict between infinite m-divisibility and the doctrine of actual parts. The best way to make this case, I think, is to present a rational reconstruction of the full range of classic paradoxes pressed against infinite m-divisibility in the Enlightenment literature, showing that these paradoxes crucially depend on metaphysical actual parts premises. Such a reconstruction would show that the actual parts doctrine underpins each and every one of the classic paradoxes of infinite m-divisibility. Without this substantive metaphysical doctrine, no purely mathematical argument can generate the paradoxes.

I set out this full reconstruction of all the classic paradoxes in sections VII and VIII below. But I begin, in section VI, with a case study that focuses in detail on one particular textual statement of one particular paradox. I think it will be helpful to give a detailed example of how my reading is supposed to work and where the standard reading goes wrong. My hope is that this sort of case study will bring out my argument more clearly and will show the appeal of my interpretation. Once I have shown this detailed case of my reading in action, I then generalize my argument and (in my comprehensive reconstruction) show systematically how it applies to *all* the classic paradoxes.

VI. A Case Study of One of the Classic Paradoxes: Hume's Argument Against Infinite Divisibility

In this section I examine Hume's main argument against infinite m-divisibility in book 1, part 2, of the *Treatise*. Using this argument as a particular example, I want to embarrass the standard mathematical reading and to bring out the case for my metaphysical interpretation of the paradoxes of infinite m-divisibility. (In the next two sections I generalize this interpretative argument and show how *all* the classic paradoxes warrant a metaphysical reading that invokes actual parts. The current section focuses in some detail on Hume, and since it is really a specific application of a more general type of interpretative argument, the hurried reader could skip it and advance to the next section without too much loss to the overall narrative.) I focus on Hume's argument in particular, since here there is a wealth of commentary, and an exceptionally clear case of a major Enlightenment philosopher accused—quite unjustly—of the most grotesque mathematical blundering.

Although there are several arguments against infinite divisibility in the notoriously thorny book 1, part 2, of the *Treatise on Human Nature*, commentators have—naturally enough—focused on the one argument that heads his discussion and that Hume clearly sees as the centrepiece of his case.[27] In sections 1 and 2, Hume frames this lead argument in terms of the divisibility of our *ideas* and *impressions* of extended entities (i.e. our mental representations of extended entities). But in section 4 he makes it clear that similar reasoning is supposed to apply to extended entities in the extra-mental physical world.[28] And, as has been well pointed out in

---

[27] John Laird, *Hume's Philosophy of Human Nature* (D. W. Livingstone and J. T. King (eds.), Archon Books, 1967), 66–7. Anthony Flew, 'Infinite Divisibility in Hume's *Treatise*', in D. W. Livingstone and J. T. King (eds.), *Hume: A Re-evaluation*, (New York: Fordham University Press, 1976), 257–69. Robert Fogelin, 'Hume and Berkeley on the Proofs of Infinite Divisibility', *Philosophical Review*, 97 (1988), 47–69. Marina Frasca-Spada, 'Some Features of Hume's Conception of Space', *Studies in History and Philosophy of Science*, 21 (1990), 371–411. Donald L. M. Baxter, 'Hume on Infinite Divisibility', in Stanley Tweyman (ed.), *Hume: Critical Assessments*, 6 vols. (London: Routledge, 1995), iii. 16–22. For commentary that lays a greater emphasis on the role of Hume's other arguments against infinite divisibility, see C. D. Broad, 'Hume's Doctrine of Space', Dawes Hicks Lecture on Philosophy, *Proceedings of the British Academy*, 47 (1961), 171. Dale Jacquette, 'Hume on Infinite Divisibility and Extensionless Individuals', *Journal of the History of Philosophy*, 34 (1996), 61–78.

[28] Whatever Hume's considered opinion about the existence of an external reality in book 1, part 2, s. 6, here in s. 2 he is prepared to apply his reasoning to external world continua, admitting their existence at least for the sake of the argument. Compare the title of s. 1 ('Of the infinite divisibility of our ideas of space and time') with the title of s. 2 ('Of the infinite divisibility of space and time'), where the lead argument occurs. And in s. 4: 'no idea of extension or duration consists of an infinite number of parts or inferior ideas, but of a finite number, and these simple and indivisible; 'Tis therefore possible for space and time to exist conformable to this idea: And if it be possible, 'tis certain they actually do exist conformable to it; since their infinite divisibility is utterly impossible and contradictory.' David Hume,

the literature,[29] the lead argument is indeed general: it purports to show that *no* finite thing can be infinitely divisible, and that *every* finite thing must resolve to a finite number of first elements. So we can follow the commentators in bracketing the fact that Hume introduces his argument in terms of mental representations of extended things. It would apply no less to extended entities in the physical world.

Hume's lead argument is brisk. It runs as follows:

(H1) '[W]hatever is capable of being divided *in infinitum*, must consist of an infinite number of parts'; 'Every thing capable of being infinitely divided contains an infinite number of parts' (*Treatise*, 1. 2. 1. 2, 1. 2. 2. 2; SBN 26, 29)

(H2) '[T]he idea of an infinite number of parts is individually the same idea with that of infinite extension;... no finite extension is capable of containing an infinite number of parts' (*Treatise*, 1. 2. 2. 2; SBN 30)

Therefore: (H3) '[N]o finite extension is infinitely divisible' (*Treatise*, 1. 2. 2. 2; SBN 30)

In short: (i) whatever is infinitely divisible has an infinite number of parts; (ii) whatever has an infinite number of parts is infinitely large; so (iii) nothing finitely extended is infinitely divisible. Now, I think that this argument is essentially incomplete or abbreviated. It should be read in the context of the actual parts metaphysic and, were it stated in full, actual parts premises that are currently implicit would become explicit. But for the moment let us put my interpretation to one side and entertain the standard reading: argument (H1)–(H3) is not a metaphysical enthymeme with suppressed actual parts premises, but rather a *complete* argument against the *mathematical* possibility of infinitely divisible quantity.

A Treatise of Human Nature, ed. David Fate Norton and Mary J. Norton (Oxford and New York: OUP, 2000), 1. 2. 4. 1. Reference is to book, part, section, and paragraph number. Hereafter additional references abbreviated 'SBN' give the corresponding page numbers in David Hume, *A Treatise of Human Nature*, ed. L. A. Selby-Bigge, 2nd edn. with text revised and notes by P. H. Nidditch (Oxford: OUP, 1978). Here the SBN reference is 39.) As Frasca-Spada puts it, 'Hume's discussion implies some sort of presupposition about the existence of external reality. Such a presupposition is not developed, in fact it is not even enunciated, but simply underlies Hume's discussion line after line' (Marina Frasca-Spada, 'Reality and the Coloured Points in Hume's *Treatise*: Part 2: Reality', *British Journal for the History of Philosophy*, 6 (1998), 25–45, 37).

[29] Flew, 'Infinite Divisibility', 266. See also Frasca-Spada in n. 28.

So interpreted, argument (H1)–(H3) seems to exhibit a variety of flaws. And it is here that commentators, taking these premises as purely formal, mathematical principles about the geometrical properties of the extended, have been quick to find fault with Hume's elementary mathematical abilities. First, premise (H1)—the principle that whatever is infinitely divisible has an infinite number of parts—is woefully misguided. Anthony Flew writes that this 'ruinous premise' is 'mistaken twice over' and 'without qualification false'; Robert Fogelin that it amounts to 'a conceptual confusion'; Marina Frasca-Spada, that there are 'many excellent reasons for saying that [(H1)] is entirely mistaken'.[30] Anthony Flew states the 'absolutely crucial' objection to (H1) clearly: 'to say that something may be divided *in infinitum* is not to say that it can be divided into an infinite number of parts. It is rather to say that it can be divided, and sub-divided, and sub-divided as often as anyone wishes: infinitely, without limit. That this is so is part of what is meant by saying: "Infinity is not a number!"'[31] Infinite divisibility, Flew insists, just involves the ability to endlessly divide and sub-divide: it decidedly does not involve the ability to divide something into a greater-than-finite number of parts. In short, it requires only a potential infinity of parts (an ever-increasable but always actually finite number of parts), not an actual infinity (a completely given greater-than-finite number of parts).[32]

Now, in addition to this first objection to (H1), Flew adds a second criticism that he thinks 'less important': 'to say that something is divisible into so many parts is not to say that it consists of—that it is, so to speak, already divided into—that number of parts. A cake may be divisible into many different numbers of equal slices without its thereby consisting in, through already having been divided into, any particular number of such slices.'[33] This is interesting, since Flew is here brushing up against the

---

[30] Flew, 'Infinite Divisibility', 259, 260. Robert Fogelin, *Hume's Scepticism in the Treatise of Human Nature* (London and Boston: Routledge & Kegan Paul, 1987), 27; see also his 'Hume and Berkeley', 51. Frasca-Spada, 'Some Features', 397.

[31] Flew, 'Infinite Divisibility', 260.

[32] On actual and potential infinities, see David Bostock, 'Aristotle, Zeno, and the Potential Infinite', *Proceedings of the Aristotelian Society* (1972–3), 37–51. Jonathan Lear, 'Aristotelian Infinity', *Proceedings of the Aristotelian Society* (1979–80), 188–210. Fred D. Miller, Jr., 'Aristotle against the Atomists', in Norman Kretzman (ed.), *Infinity and Continuity in Ancient and Medieval Thought* (Ithaca, NY: Cornell University Press, 1982), 87–111. Maurice Clavelin, *The Natural Philosophy of Galileo* (Cambridge, Mass.: MIT Press 1974), 39–45.

[33] Flew, 'Infinite Divisibility', 259–60.

substantive actual parts–potential parts controversy. So he is at least somewhat aware that there are metaphysical issues in the background of (H1)–(H3). Notice, however, that his objection is underdeveloped: he seems to think that Hume's way of conceptualizing division and parthood is just obviously wrong, and that the cake example shows this.[34] Moreover, he continues to insist that the 'crucial' criticism of (H1) is not this second 'less important' point, but rather the former, mathematical objection shared by Fogelin and Frasca-Spada. Flew is still thinking of Hume's argument as fundamentally a mathematical one, just as they are.

Now to premise (H2): the principle that whatever consists of an infinite number of parts must be infinitely large. Commentators have also singled this premise out as mathematically mistaken. Certainly something that consists of an infinite number of *same-sized* parts will be infinitely large. But if those parts are of proportionately diminishing size, then the whole need not be infinite in size. Robert Fogelin presses this objection to (H2) as follows: 'It is true that if we take a finite extension (however small) and repeat it *ad infinitum*, we will get an infinite extension. That, however, is quite beside the point, because the argument for infinite divisibility depends on the possibility of constructing ever smaller finite extensions, as in the sequence [1/2, 1/4, 1/8, etc.] whose sum approaches, not does not exceed, 1.'[35] Now, Hume does mention this objection in a footnote to his argument, but promptly dismisses the distinction between 'proportional' parts and 'aliquot' (same-sized) parts as 'entirely frivolous'.[36] But most commentators have been inclined to think that this distinction is not half so frivolous as Hume claims, and that the current objection to (H2) is a good one.[37]

We have, then, two distinct objections to Hume's argument. Contra (H1), whatever is infinitely divisible need *not* consist of an infinite number of parts; and, contra (H2), something that consists of an infinite number of parts need *not* be infinite in size. In addition to these two particular objections—each of which identifies a specific mathematical mistake—we might

---

[34] See my complaint about this type of example in Ch. 2, s. VI, under argument (P1).
[35] Fogelin, 'Hume and Berkeley', 51.
[36] Hume, *Treatise* 1. 2. 2. 2 n.; SBN 30.
[37] See, in addition to Fogelin, Laird, *Hume's Philosophy*, 65; Flew, 'Infinite Divisibility', 264. J. M. M. H. Thijssen, 'David Hume and John Keill and the Structure of Continua', *Journal of the History of Ideas*, 54 (1992), 271–86, 280. H. Mark Pressman, 'Hume on Geometry and Infinite Divisibility in the *Treatise*', *Hume Studies*, 23 (1997), 227–44, 241.

also add the more general complaint that, since there certainly *is* a mathematically coherent conception of infinite divisibility, there must be a mistake *somewhere* in Hume's complaint against it. James Franklin puts this last general objection as follows: 'The infinite divisibility of space and time is possible. (This is because there exists a consistent model which incorporates infinite divisibility, namely the set of infinite decimals.) It follows that all supposed proofs of the impossibility of infinite divisibility, whether mathematical or philosophical are invalid.'[38]

My reaction to all this should, by now, be no surprise to the reader. I think that this interpretation of Hume's argument, which takes Hume to be raising a mathematical challenge to infinitely divisible quantity *in abstracto*, is entirely mistaken. Hume's argument is not supposed to raise a formal problem for mathematical models of infinite divisibility, divorced from all thought of the stuffing or filling of actual physical continua. Rather (on my interpretation) Hume's argument must be set in the context of the actual parts account—an account that was generally accepted as an analysis of matter in Hume's day, and that was accepted by Hume as an analysis of *all* continuous quantities (as we shall see below). On this reading, the argument raises an objection to the infinite divisibility of *physical quantities with actual parts*.

I will substantiate Hume's commitment to the actual parts metaphysic and provide further evidence for this reading in a moment. But first let us look at the way this interpretation disarms the standard mathematical objections.

First, the objection to (H1) immediately disappears, since infinitely m-divisible continua with actual parts must indeed have an actual infinity (a greater-than-finite number) of parts pre-given. According to the actual parts doctrine, the parts into which the whole can be m-divided each *already* exists prior to division. So if a continuant can be m-divided without end, it cannot merely have an indefinite, ever-growing potential infinity of parts. Rather it must already have a greater-than-finite number of them laid up in it from the very start.[39] (Here the obvious contrast is with the

---

[38] James Franklin, 'Achievements and Fallacies in Hume's Account of Infinite Divisibility', *Hume Studies*, 20 (1994), 85–101 87. Franklin's claim here echoes a similar analysis in Bertrand Russell, *Our Knowledge of the External World* (London: George Allen & Unwin, 1926), 137, 159.

[39] Donald L. M. Baxter makes this same point in his paper 'Hume on the Simplicity of Moments' (unpublished draft). Thanks to Baxter for showing me this draft.

28 / Problems of Material Structure

potential parts metaphysic, where parts are created as division proceeds. On the potential parts account, infinite divisibility *does* only entail an indefinite potential infinity of parts.)

Second, the objection to (H2) is also much too quick. If a continuant such as a body has actual parts, then it is a composite or compound entity: a structured aggregate of so many distinct parts. And, at least according to the standard orthodoxy of the early modern period, such a composite structure will depend for its existence on the ontologically prior existence of its parts. If these parts are themselves also composite, then they in turn will depend for their existence on the ontologically prior existence of their subparts, and so on. But (according to this traditional argument) given that the whole original composite structure exists, the regress of ontological dependence here set off cannot go on forever without a ground floor. There must eventually be an atomic base of noncomposite first parts from which the whole original ultimately derives its existence. So continua with actual parts must have *ultimate* parts—atomic elements that, assuming their uniformity, will be *aliquot* (same-sized) *simples*.[40] Perhaps one can successively divide a body, from the top-down, into *proportional* parts (1/2, 1/4, 1/8...), but it must ultimately be built, from the bottom-up, from a foundation of atomic *aliquot* parts all the same. Given the actual parts metaphysic, then, if one also accepts the ontological regress argument to simple first parts and the homogeneity of those atomic simples, the proportional parts objection to (H2) is indeed as 'entirely frivolous' as Hume insists. And, once again, we can note the contrast with the potential parts metaphysic, where the proportional parts objection to (H2) would make perfect sense. (Hume presents this 'very strong and beautiful' ontological regress argument from composites to simples at *Treatise*, 1. 2. 2. 3 (SBN 30), where it is clearly intended to support (H2). It is, of course, a famous pattern of reasoning—most familiar, perhaps, from the opening sections of Leibniz's *Principles of Nature and Grace* and *Monadology*, or from Kant's Second Antinomy in the *Critique of Pure Reason*. I discuss it in detail in Chapter 4.)

Third, the general mathematical objection of Franklin also misfires once Hume's argument is placed in its proper actual parts context. If there

---

[40] Donald L. M. Baxter has stressed the crucial role of this argument for ultimate parts in Hume's overall case against infinite divisibility, and has also noted that it presupposes the principle that 'anything divisible is composed of parts': 'that *divisible* entails *not unitary*' (i.e. a system of actual parts). Baxter, 'Hume on Infinite Divisibility', 19.

is a coherent mathematical model of infinite divisibility, this merely shows that there can be no purely *formal* or *mathematical* complaint against it. It certainly does not show that there could be no *metaphysical* complaint against that model being translated into an actually existing physical structure.

In fact, the only objection that remains a genuine challenge to the argument once it is set in the actual parts context is Flew's 'less important' criticism of (H1)—actually, an objection of the utmost importance. The objection was that physical continua need not contain as many parts as they can be divided into. This constitutes an outright rejection of the actual parts doctrine, and of course this *would* constitute a serious challenge to Hume. Given my metaphysical reading, Hume's argument depends crucially on the actual parts doctrine and will collapse if that doctrine is false. However, this does not compromise *my* argument that Hume's case against infinite divisibility is a metaphysical one that turns crucially on the actual parts doctrine; in fact, of course, it upholds it. (Notice also that Flew offers no real argument against the actual parts view. He simply takes the cake example to establish the potential parts account, but of course this simply begs the question against the actual parts advocate, who will maintain that the slicing of the cake does not involve the literal creation of new parts but rather the separation of pre-existing parts. See my assessment in Chapter 2, section VI, under argument (P1).)

It is clear then that the standard objections misfire if we place Hume's argument against the actual parts metaphysic and understand it as an attack on the suggestion that material bodies (or, indeed, any continuous quantity with distinct, actual parts) could be divisible ad infinitum. This gives us one powerful reason to think that my metaphysical reading of Hume's case against infinite divisibility is correct, and that the orthodox mathematical reading in the secondary literature is mistaken. But perhaps it will be insisted, on behalf of the mathematical reading, that Hume's text does *not* explicitly set out the supposed actual parts context of his reasoning. And this may seem to favour the straight mathematical interpretation of the argument.

If I am right that Hume's argument is intended to apply only to physical quantities with distinct, actual parts, it certainly must be confessed that Hume failed to make this explicit. However, there are a number of supporting considerations that bolster my interpretation and cause further trouble for the purely mathematical reading.

First, we should stress just how appalling Hume's argument would have to be, given the mathematical interpretation. On this reading, Hume stands charged with failing to appreciate certain absolutely elementary facts about fractions: for instance, that a proportional division like 1/2, 1/4, 1/8, 1/16, etc., can be iterated as many times as one likes but its sum at any point will always be less than (though ever closer to) a finite limit. (Notice here that, whatever one thinks of the period's travails over fluxions, infinitesimals, and the problematic notion of a *completed* infinite series, this basic fact about *potentially* infinite series of proportional divisions was widely understood long before Hume's time. The elementary point here is in Eudoxus, Aristotle, and Euclid, and it was pressed by scholastics over and over during the centuries before the calculus, for instance.)

Second, to treat Hume's argument as a purely mathematical one divorced from the actual parts metaphysic is to fail to appreciate its connection to the neo-Epicurean tradition from which it is drawn. As I show in my full reconstruction of all the classic paradoxes below, the self-same argument is sponsored by Henry More, Joseph Glanvill, Isaac Newton, George Berkeley, and is reported by Isaac Barrow and John Keill (see paradox (3A) in section VIII below). Moreover—and most importantly for our purposes—it is also presented by Kenelm Digby, Walter Charleton, and Pierre Bayle, each of whom gives the abbreviated enthymeme captured by (H1)–(H3), *and then sets it in the actual parts context*, expressly identifying its additional metaphysical premises.[41]

Charleton's text is exceptionally gratifying here, since he first sets out the abbreviated version captured by (H1)–(H3),[42] and then turns to address the traditional Aristotelian and Stoic objections—'*Sophisms* framed upon design to evade it: among which we find only *Two*, whose plausibility and popular approbation seem to præscribe them to our præsent notice.' And these two objections turn out to be precisely the same mathematical complaints pressed against (H1) and (H2) by Flew, Fogelin, *et al.* in the recent literature: first, contra (H1), 'that the division of finite body into infinite

---

[41] Digby, *Two Treatises*, 10–12. Charleton, *Physiologia*, 91–3. Bayle, *Dictionary*, 362, 356. Charleton and Bayle endorse the actual parts doctrine and hence this argument against infinite divisibility. For Digby, the conclusion that bodies cannot be infinitely divisible constitutes a *reductio* on the actual parts doctrine.

[42] 'If in a Finite Body, the number of Parts, into which it may be divided, be not Finite also; then must the Parts comprehended therein be really Infinite: and, upon Consequence, the whole Composition resulting from their Commixture, be really Infinite; which is repugnant to the supposition.' Charleton, *Physiologia*, 91.

parts doth not make it actually infinite, because the parts are not actually, but only potentially infinite'; and, second, contra (H2), that 'by admitting an infinity of parts in a Finite Continuum a Continuum doth not become infinite; because that results properly not from *Proportional*, but *Aliquotal* parts'.[43] Charleton's response is that each of these objections ('subterfuges') is easily blocked by the invocation of the actual parts doctrine. Once the argument is clearly placed in its actual parts context, it is rescued from 'the Sophistry of the most specious Recesses invented to assist the Contrary opinion' and stands as 'perfectly Apodictical and...inoppugnably victorious'.[44] Now, setting aside Charleton's lavish prose, the moral here is clear. Hume's argument must be placed against a tradition in which that argument is invoked over and over. And in that tradition it is clear that the argument is understood to rest on the actual parts doctrine.

Third, the stock mathematical objections so popular in the recent literature—and reported by Charleton—would certainly have been familiar to Hume. Aside from the fact that the argument was traditional and the objections to it generally well-known, we know that Hume was definitely acquainted with Barrow's *Mathematical Lectures* and Bayle's famous article on Zeno of Elea in the *Dictionnaire historique et critique*. It is also extremely likely that he was familiar with Keill's highly influential *Introduction to Natural Philosophy*.[45] Now, in Barrow's text and again in Keill's, we find each of the two standard mathematical objections clearly set out. And in Bayle's we find at least the objection to (H1) addressed.[46] So Hume would certainly have been aware of the stock mathematical objections deployed by Flew, Fogelin, *et al*. This, once again, bolsters the suggestion that Hume could not have been guilty of such obvious errors, and that more is going on in his argument than the standard mathematical reading allows.

Fourth, as was noted above, Hume *does* explicitly address the stock objection to (H2) in one of his rare footnotes. He suggests that, at least for the

[43] Ibid. 92, 93.
[44] Ibid. 94, 91.
[45] On Hume's familiarity with Barrow's *Lectures* and Bayle's 'Zeno' article, see Norman Kemp Smith, *The Philosophy of David Hume* (London: Macmillan, 1941), 325–6, 343. See also Richard H. Popkin, 'So, Hume Did Read Berkeley', *Journal of Philosophy*, 61 (1964), 773–8: 775. On Keill's fame and influence and at least a 'very tenuous' link to Hume, see Thijssen, 'Hume and Keill', 272–3.
[46] Isaac Barrow, *The Usefulness of Mathematical Learning Explained and Demonstrated*, tr. John Kirkby (1734; facsimile edn.: London: Frank Cass, 1970), 157. Bayle, *Dictionary*, 356. Keill, *Introduction*, 37, 40–1.

purposes of this argument, the distinction between proportional and aliquot parts is 'entirely frivolous'. As we have seen, my metaphysical reading makes good sense of this dismissive attitude. On the other hand, the mathematical reading can make no sense of this. The commentators have had to claim that Hume's offhanded insouciance here is simply unjustified. But they can offer no plausible explanation of *why* Hume thought the distinction frivolous.

Fifth, and finally, we should note that Hume's arguments throughout part 2 address either the divisibility of ideas and impressions of extended things, or the divisibility of actually extended things out there in the real world. And Hume clearly subscribes to an actual parts metaphysic concerning both these types of thing. In the mental realm, ideas and impressions of extended things are constructed from distinct parts ('that compound impression, which represents extension, consists of several lesser impressions'[47]). Likewise, when Hume turns from the divisibility of the idea of space to the divisibility of space itself, an actual parts account is again in play ('A real extension, such as a physical point is supposed to be, can never exist without parts, which are different from each other'[48]). All this follows from Hume's more general principle that all entities that are separable or distinguishable are distinct existents: 'whatever objects are separable are also distinguishable, and . . . whatever objects are distinguishable are different'.[49] And this is simply one direction of the famous biconditional known as Hume's Separability Principle. This fundamental principle of Hume's thus incorporates a commitment to actual parts. The moral here is that Hume's argument (H1)–(H3), whether directed towards ideas and impressions of extended things, or towards extended things out there in a mind-independent reality, in either case functions in a context where Hume is committed to a metaphysic in which continua have distinct, actual parts. It must thus be understood in that metaphysical context. Contra the standard interpretation, Hume is decidedly *not* directing an argument against the mathematician's purely formal, abstract notion of infinitely divisible quantity.[50]

---

[47] Hume, *Treatise*, 1. 2. 3. 15; SBN 38. Similarly, 'taking [substance] for *something, that can exist by itself*, 'tis evident every perception is a substance, and every distinct part of a perception a distinct substance'. Hume, *Treatise*, 1. 4. 5. 24; SBN 244.

[48] *Treatise*, 1. 2. 4. 3; SBN 40.

[49] *Treatise*, 1. 1. 7. 3; SBN 18. See also 1. 4. 5. 5; SBN 233.

[50] For more on this interpretation of Hume's argument, including an examination of its relation to the Humean doctrine of *minima sensibilia*, see my 'Infinite Divisibility and Actual Parts in Hume's *Treatise*', *Hume Studies*, 28 (2002), 3–25.

VII. How the Actual Parts Doctrine Underpins the Classic Paradoxes

In the previous section I presented a case study of one particular argument against infinite m-divisibility found in Hume. I argued that the particular paradox pressed in this argument must be understood in the context of the actual parts metaphysic. Now I want to generalize this interpretation and show that not just Hume's argument but *all* the classic paradoxes of infinite divisibility are underpinned by the actual parts doctrine. To this end, I now present a systematic and comprehensive reconstruction of the range of classic paradoxes that both enumerates and orders them hierarchically, and shows the way in which they each depend on the actual parts metaphysic.

The reconstruction involves two main stages. First, I show that the actual parts doctrine was traditionally thought to entail two direct corollaries: (*a*) that bodies have a determinate number of parts, and (*b*) that bodies are constructed from metaphysical atoms: ultimate, m-indivisible first parts. (Without the actual parts doctrine, no purely mathematical argument can hope to establish either of these things.[51]) There is clearly at least a prima-facie conflict between infinite m-divisibility and each of these corollaries of the actual parts account. In the second stage of this reconstruction, I then show in detail how the actual parts doctrine, via these two corollaries, generates the classic array of paradoxes of infinite m-divisibility. Again, no purely mathematical argument can engender the paradoxes.

Two Widely Accepted Corollaries of the Actual Parts Doctrine

1. *Bodies have a determinate number of parts*

According to the actual parts doctrine, *all* the parts into which a body can be m-divided are already present in the body prior to division. The entire collection of parts into which it can be m-divided is given in advance as a complete totality. As opposed to the potential parts doctrine (according to which 'the parts are not really there, till by division they are parcelled

---

[51] I discuss two objections to this claim in Ch. 3, ss. V and VI.

out'[52]), there is nothing indefinite or open-ended about the number of parts: each and every one is there from the start; no additional parts are subsequently generated. It follows that the number of parts is determinate. Kant puts the point clearly: 'as soon as something is assumed as *quantum discretum*, then the multitude of units within it is determinate and hence, by the same token, is always equal to some number'.[53] The actual parts doctrine, in asserting the distinct existence of each part of matter, takes them as *quanta discreta*, and hence as determinate in number.

Of particular relevance to our present study is how this bears on the case of *infinitely* m-divisible matter. Here, again, one will have to say that all the parts into which a body can be m-divided are given as a determinate collection prior to division. So, if bodies are m-divisible *ad infinitum*, they must then have an *actual infinity* of parts (a completely given greater-than-finite number of parts). They cannot merely be said to have an Aristotelian *potential infinity* of parts (a number of parts that is always finite at any stage, but that may be increased without limit). This indeterminate potential infinity would only make sense if parts were generated as division proceeds. But if the parts are already one and all laid up in the whole prior to division, then they must be an actual infinity of pre-existent parts if the whole is endlessly divisible. Toward the end of our period, Kant offers the following neat summary of this reasoning, attributing it to the 'dogmatic metaphysician'—that is (in this context) someone who admits actual parts. 'If... matter is infinitely divisible, then (concludes the dogmatic metaphysician) it consists of an infinite multitude of parts; for a whole must in advance contain within itself all the parts in their entirety into which it can be divided.'[54]

This argument from actual parts to a determinate number of parts is also endorsed by Galileo, Charleton, More, the younger Newton, Bayle, Berkeley, and Hume. I document it and discuss it in detail in Chapter 3.

2. *Bodies are constructed from metaphysical atoms*

The second corollary that was thought to follow from the actual parts doctrine is that bodies must resolve to ultimate m-indivisible first parts:

---

[52] Digby, *Two Treatises*, 10.   [53] Kant, *Critique Pure Reason*, B555.

[54] According to the critical period Kant, the actual parts doctrine is unavoidable for all transcendental realists. For more detail, see s. IX, 'faction (1)' in the current chapter, and Ch. 2, s. III.

metaphysical atoms. There are two traditional arguments from actual parts to metaphysical atoms in the Enlightenment literature.

The argument from the definiteness of parts. The first argument from actual parts to metaphysical atoms takes its cue from the previous corollary. Given the actual parts doctrine, the entire collection of parts in any body is pre-given in determinate totality. No part awaits subsequent generation; one and all are fully present in the body from the start. But if we already have the complete set of parts fully given, then (according to the current argument) we must have the elements of that set that are the smallest members. Since *all* the parts are given, the parts that are the smallest must be given. The idea here is that one cannot take refuge in the claim that there is no smallest part, but rather an indefinitely expandable, open-ended series of successively smaller parts. This would be to appeal to an open-ended potential infinite. But what we have here is a completely given determinate totality of parts. So even if division can proceed endlessly, and no successive, top–down process of division and subdivision will ever *reach* smallest parts, such parts must nonetheless already be there, laid up in the whole. And of course, these smallest parts must be *m-indivisible atoms* (if they were m-divisible, they would not be the smallest). We find this first sort of argument in Galileo, Digby, and Spinoza; it is also at least implicit in Charleton, More, Bayle, Newton, Berkeley, and Hume. I document and discuss it in Chapter 3.

The argument from composition. The second of our two arguments from actual parts to metaphysical atoms is, perhaps, more familiar. This argument starts with the fact that (given the actual parts doctrine) bodies are essentially composite or compound entities: structured aggregates of their parts. It then adds the claim that composites are ontologically derivative entities, depending for their existence on the ontologically prior existence of their parts. And if these parts are themselves also composite, then they in turn depend for their existence on the ontologically prior existence of *their* parts. But (according to the current argument) if the whole original is to exist at all, this ontological regress cannot go on forever with no ground floor. It must terminate in noncomposite elements whose existence is not derivative. And, of course, these noncomposite, simple first elements must be *m-indivisible* (were they *m-divisible*, they would, given the actual parts doctrine, be composite). So we have an ultimate ground floor constitution of m-indivisible metaphysical atoms. This sort of argument is most well known from the works of Leibniz and Wolff. But a wide range of diverse

Enlightenment thinkers also sponsor it, including Bruno, Charleton, Newton, the younger Kant, and even supposedly positivistic, anti-metaphysical thinkers like Hume and Condillac. I document and discuss it in Chapter 4.

We have, then, in at least a preliminary outline form, the case for our two corollaries of the actual parts doctrine. Given the actual parts account, it was thought that bodies must (*a*) have a determinate number of parts and (*b*) have an ultimate constitution of metaphysical atoms. (Notice that the potential parts account mandates neither of these two doctrines.) It should already be obvious that each of these corollaries poses at least a prima-facie challenge to the notion that bodies are infinitely m-divisible. How could bodies be divisible without end, if they have a specific, fixed number of parts, or if they have ultimate, last parts? In the next section, I flesh out this challenge by showing how these two corollaries generate the classic array of paradoxes of infinite divisibility. This section will then complete my rational reconstruction of the case against infinite divisibility, showing that it is the actual parts doctrine—via these two traditional corollaries—that underpins the classic paradoxes.

VIII. The Classic Paradoxes of Infinite Divisibility

The array of paradoxes that is supposedly derivable from the doctrine of infinite divisibility can be somewhat overwhelming. Any study of the case against the infinite divisibility of matter will rapidly uncover a cacophony of such alleged paradoxes in the Enlightenment literature. Different philosophers press different antinomies. Some of their arguments overlap; some do not. My central aim in this section is to impose an organizing structure on this initially somewhat bewildering array of paradoxes. I want to marshal the distinct types of paradox and enumerate the tokens of each type. In addition to distinguishing the various types of paradox in this fashion, I want to bring out the hierarchical structure of the relationships between them. Certain paradoxes turn out to piggyback on others, and so—by way of this sort of rational reconstruction—we can identify the most fundamental ones. Naturally this will help us get clearer on the basic issues at stake.

The classic paradoxes of infinite divisibility—the ones that occur over and over in the Enlightenment literature, and also the ones that prove the most difficult to disarm—all derive from the attempt to square the infinite m-divisibility of matter with the actual parts doctrine concerning its parts. I should say at the outset that the actual parts doctrine is not always explicitly identified as a premise in the generation of the paradoxes of infinite divisibility. Often it *is*—but equally often it functions as a suppressed premise. This is more common in the eighteenth century than the seventeenth, since—as the new science takes hold and scholastic natural philosophy falls away—the actual parts account shifts from a philosophically contentious status to become increasingly taken for granted.

The basic hierarchical structure of the classic paradoxes is charted in the table. Notice that the paradoxes all derive from the attempt to grant infinite m-divisibility alongside the two corollaries of the actual parts doctrine: either (1) alongside metaphysical atomism, (2) alongside a determinate number of parts, or (3) alongside *both* metaphysical atomism and a determinate number of parts at once. All these classic paradoxes are then engendered only within the framework of an actual parts account of the ontological status of the parts of matter. They do not arise within the potential parts framework, since that framework mandates neither metaphysical atomism nor a determinate number of parts.

This maps the traditional paradoxes that are found over and over as standard in the Enlightenment literature. Now, I do not claim that this list captures every last objection pressed against infinite m-divisibility in the

Table 1 The classic paradoxes of infinite m-divisibility

---
1. The paradox of infinite m-divisibility alongside metaphysical atomism
2. Paradoxes from infinite m-divisibility alongside a determinate number of parts
   (The basic problem: this entails an actual infinity of parts.)
   2A. An actual infinite is incoherent in concept
   2B. Problems of equal but unequal infinities
   2C. The impiety of admitting an actual infinite
3. Paradoxes from infinite m-divisibility alongside *both* metaphysical atomism *and* a determinate number of parts
   (The basic problem: this entails an actual infinity of m-indivisible atoms in each body.)
   3A. Zeno's metrical paradox applied to bodies: Bodies should either be infinitely large or of no size whatsoever
   3B. Bodies should all be the same size
   3C. No ultimate part is contiguous to any other

Enlightenment period. One will occasionally find an alleged paradox that does not—like the ones in this architectonic—turn crucially on the corollaries of the actual parts doctrine. (For the sake of completeness, I list such further paradoxes that I am aware of in an appendix to this chapter.) But these further paradoxes are much rarer in the literature and can fairly be described as idiosyncratic. Moreover, they are much more readily disarmed than the ones charted here. The current list certainly captures all the paradoxes that could fairly be called classic or traditional in the period; it also captures the most recalcitrant problems.

1. The Paradox of Infinite M-Divisibility alongside Metaphysical Atomism

As we have seen, it was traditionally thought that the actual parts doctrine entails that bodies are constructed from ultimate m-indivisible first parts or metaphysical atoms. But the existence of m-indivisible atoms seemed (to most early moderns) to rule out infinite m-divisibility. When the process of division reaches ultimate m-indivisible parts it must then terminate, so the process cannot continue ad infinitum.

The basic antinomy of infinite m-divisibility alongside a metaphysical atomism mandated by the actual parts doctrine frames the fundamental conflict in Kant's discussions of material structure in both the pre-critical *Physical Monadology* (1756) and the famous Second Antinomy of the *Critique of Pure Reason* (1781/87). It is also the basic paradox for Nicolas de Malezieu, the Sun King's mathematician royal: according to his 1705 *Elemens de Geometrie*, one must admit each of these two theses, but at the same time confess them incompatible. The antinomy is also reported in John Keill's Oxford lectures of 1700 (published in English in 1745), and in Thomas Reid's 1785 *Essay on the Intellectual Powers of Man*. More generally, a basic premise shared by the overwhelming majority of our period's natural philosophers is that there is a fundamental, irreconcilable conflict between infinite m-divisibility on the one hand, and metaphysical atomism on the other. For instance, Walter Charleton, Henry More, Joseph Glanvill, Simon Foucher, Pierre Bayle, David Hume, and Leonhard Euler (among others) each explicitly take the one thesis to rule out the other, and the dichotomy between them thus to frame an exclusive dilemma. As Euler puts it, 'in the system of divisibility in infinitum, the term ultimate particle is absolutely unintelligible... whoever maintains that body is divisible in infinitum, or without end, absolutely denies the existence of ultimate particles.... The term can only mean such particles as are no longer divisible—an idea totally inconsistent with the system of divisibility in infinitum.' For most thinkers of our period, then, metaphysical atomism and infinite m-divisibility are mutually exclusive doctrines. If atomism is indeed mandated by the actual parts doctrine, we then have an objection to infinite m-divisibility.[55]

One way of attempting to resist this and to reconcile metaphysical atomism with infinite m-divisibility is the suggestion that bodies might be built up from a greater-than-finite number of metaphysical atoms. If bodies have only a finite number of m-indivisible ultimate elements, then clearly m-division cannot proceed ad infinitum. But if bodies have a greater-than-finite number of such parts, then division could go on endlessly without ever exhausting them. (This strategy is Galileo's response to the current paradox. In some ways it also prefigures Adolf Grünbaum's twentieth-century account of continua as aggregates built up from non-denumerably many point parts.[56]) However, as we shall see (under paradoxes (2) and (3) below), this notion of a completely given greater-than-finite number of parts faces problems of its own.

2. Paradoxes of Infinite M-Divisibility alongside a Determinate Number of Parts

We just saw that one might be tempted to posit an actual infinity of ultimate parts in order to disarm paradox (1): it may seem one way of reconciling metaphysical atomism and infinite m-divisibility. And in fact we already have another more direct argument from actual parts and infinite m-divisibility to an actual infinity of parts. One of our two corollaries of the actual parts doctrine was that bodies must have a determinate

---

[55] Kant, *Physical Monadology*, in *Works*, 1: 473–87, especially proposition IV (1: 479); *Critique of Pure Reason* B462–71. For the relevant extract from Nicolas de Malezieu's *Elemens de Geometrie*, see Kemp Smith, *Philosophy of Hume*, 341–2. Keill, *Introduction*, 40–1. Reid, *Works*, i. 323. Charleton, *Physiologia*, 90–1. Henry More, *An Antidote against Atheism*, in *Philosophical Writings of Henry More*, 11. Joseph Glanvill, *The Vanity of Dogmatizing* (1661; facsimile edn.: Hildesheim: Georg Olms Verlag, 1970), 53. Simon Foucher, extract quoted in G. W. Leibniz, *Philosophical Essays*, tr. and ed. Roger Ariew and Daniel Garber (Indianapolis: Hackett, 1989), 146. Bayle, *Dictionary*, 361–2. Hume, *Treatise*, 1. 2. 3. 10; SBN 33. Euler, *Letters*, 47–8, 49; the quotation is from 47–8.

[56] Galileo, *Two New Sciences*, 34. Grünbaum, 'A Consistent Conception of the Extended Linear Continuum as an Aggregate of Unextended Elements', *Philosophy of Science*, 19 (1952), 288–306.

number of parts. And, as we saw, this corollary, when married to infinite m-divisibility, implies that bodies must have an *actually infinite* number of parts: a completely given totality of parts that is greater-than-finite in number (see section VII).

Most Enlightenment philosophers found this notion of an actually infinite collection thoroughly paradoxical. They had resisted actual infinities in arithmetic and geometry, and in thinking about the overall size of the universe (here they typically claimed that potential infinities or at least epistemically indefinite series would suffice[57]). But now it seemed that two of the period's central doctrines in natural philosophy—the actual parts account of matter and the infinite m-divisibility of body—jointly entailed the existence of actually infinite collections, and thus jointly entailed all the paradoxes that arise with such collections.

There were three basic types of paradox threatened here, which I will categorize as (2A), the conceptual incoherence of an actual infinite, (2B) problems arising from the apparent equality and inequality of actual infinities, and (2C) the impiety of admitting an actual infinite. These three paradoxes purport to embarrass *any* notion of an actual infinite collection. They are general objections to the concept of—or at least the physical instantiation of—any such collection. (I deal with more specific objections to actually infinite collections serving as the elemental base of material bodies under (3) below.)

Paradox 2A. *An actual infinite is incoherent in concept*

The most abrupt objection to the actual infinite was that it is conceptually incoherent. The concept of infinity is the concept of something essentially open-ended and incompletable, something essentially unlimited. The very notion of an actual infinite—a completely given infinite—is therefore an antilogy. Leonhard Euler, for instance, writes that 'When we say that matter is infinitely divisible we are not simply affirming that any division to whatever point it may be continued would never arrive at particles which could not be further subdivided, but we are formally denying that there are any such particles. If they actually existed, they would be

---

[57] For a useful survey, see J. D. North, 'Finite and Otherwise: Aristotle and Some Seventeenth Century Views', in William Shea (ed.), *Nature Mathematized* (Dordrecht: D. Reidel, 1983), 113–48. As North notes, Newton is a great exception here, since he maintains that space is actually infinitely large. Another major exception (not mentioned by North) is Galileo: see my account of Galileo's admission of actual infinities in s. IX ('faction (2)') below.

determinate, which is contrary to infinity.'⁵⁸ Kant also states the argument exceptionally clearly: 'One cannot grant that matter, or even space, consists of infinitely many given parts (because there is a contradiction in thinking of an infinite number as complete, inasmuch as the concept of an infinite number already implies that it can never be wholly complete).'⁵⁹

In short, the only coherent notion of an infinite is that of the Aristotelian, ever-increasable *potential* infinite: essentially an open-ended, indeterminate finite. The completely given, larger-than-finite collection of parts required by the actual parts doctrine alongside infinite m-divisibility is a self-contradictory nonsense. This line of argument dates back to Aristotle's classic treatment and was the traditional orthodoxy of the seventeenth-century scholastic. Among the new philosophers, it is endorsed by Hobbes and Locke, as well as Kant.⁶⁰

Paradox 2B. *Problems of equal but unequal infinities*

The previous argument immediately ruled out all talk of an actually infinite collection as so much nonsense: the concept is transparently self-contradictory. More common than this extremely brisk objection was the strategy of provisionally admitting such a collection for purposes of *reductio*. An actual infinite collection is assumed to exist; various absurd results are then generated on this assumption; and the original assumption is thrown out by indirect proof.

The standard *reductio* arguments against actually infinite collections proceed on the basis that such collections must all be equal in size to one another, but can at the same time be unequal. Since this is absurd, there can be no actually infinite collections. The equality of actually infinite collections is often taken as an axiom, although it is sometimes defended

---

⁵⁸ Leonhard Euler, quoted in Polonoff, *Force, Cosmos, Monads*, 84.

⁵⁹ Kant, *Metaphysical Foundations of Natural Science*, in Kant, *Philosophy of Material Nature*, tr. and ed. J. W. Ellington (Indianapolis: Hackett, 1985) 53. See also the *Critique of Pure Reason*, B555, where Kant writes that, in thinking of 'a multitude of parts that is in itself determinate but also infinite, thereby one contradicts oneself. For this infinite involution is regarded as a series never to be completed (i.e., as infinite), and yet also—when gathered together—as nonetheless completed.'

⁶⁰ Aristotle, *Physics*, books 3 and 6. Hobbes, *De Corpore*, ch. 7.11, in *Works*, i. 98–9. Locke, *Essay*, II 17.7–9. In our era, this line of reasoning is developed by the intuitionist school. Following Kronecker and Brouwer, intuitionists argue that the only intelligible mathematical concepts are those that can be constructed by the intuitive activity of counting. Potential infinities are thus intelligible; but actual infinities are not.

with the following arguments. Since nothing can be larger than an actual infinite, one actual infinite must be the same size as any other. Or: since the members of an actually infinite collection can be paired off against the members of any other actually infinite collection, and the one will never exhaust the other, it follows that all such collections must be the same size. However, on the other hand, it seems that one actually infinite collection can be unequal to another. For instance, one actually infinite collection can be a proper part of another. Given an infinite chessboard, with an actually infinite number of squares, the sub-collection of white squares will be no less an actual infinite than the total collection of all squares. But it also seems that it must be smaller, since it is a proper part of the collection of all squares. So actual infinities can be unequal. Another way one can show that actual infinite collections are unequal is by the possibility of adding to or subtracting from them. It seems that if we add to such a collection, then the collection after our addition cannot be the same size as it was before. So infinite collections can be unequal in size. But (so the argument goes) they must also be equal in size. Thus they cannot exist.[61]

I think it is fair to say that most Enlightenment thinkers who pressed this sort of argument thought of it as constituting a *reductio* on the very concept of an actual infinite. The *reductio* thus backs up or substantiates the suggestion of (2A): the actual infinite is a *contradictio in adjecto*. But it is interesting to note another possible interpretation of these arguments: perhaps they embarrass, not the abstract mathematical concept of an actual infinite, but rather the suggestion that such a collection could be physically instantiated in the material world. Read this way, the *reductio* arguments would then constitute an indirect proof against the provisional assumption that an actually infinite collection could exist in the real world; they need not show the abstract notion of an actual infinite to be conceptually or mathematically self-contradictory. Certain commentators have defended this sort of real-world version of the *reductio* arguments in the recent literature on actual infinities.[62] But we need not get into the

---

[61] Since Cantor's ground-breaking work, it has become commonplace to dismiss this sort of objection to actual infinities as rooted in prejudices or habits of mind instilled by our familiarity with the arithmetical properties of finite numbers—properties that do not apply to infinite numbers. See, for instance, Bertrand Russell, *Our Knowledge of the External World* (London: Allen & Unwin, 1925), 187–8. For a different reading of these arguments, see the references in the next note.

[62] For instance, William Lane Craig argues that the classic paradoxes do refute the claim that an actual infinite collection could exist in material reality, but not the idea that it is a

issue of whether or not this is the most plausible way to read the arguments here. It is enough for our purposes to note that, if the *reductio* arguments are valid, then—whether they turn on the abstract concept or on the physical instantiation of an actual infinite—either way they show that there could be no completely given greater-than-finite collection in material reality.

This sort of *reductio* proof against actually infinite collections is found in our period in the works of Robert Boyle, George Berkeley, Edmund Law, and William Drummond. It is also reported (though not endorsed) by Galileo, Isaac Barrow, Newton, Jacques Rohault, Antoine Le Grand, John Keill, and Samuel Clarke.[63] Finally, it also stands behind the more specific complaint that, if bodies have an actual infinity of parts, they must all be the same size ((3B) below).

Paradox 2C. *The impiety of admitting an actual infinite*

A final Enlightenment concern about actually infinite collections was that it might be *impious* to allow their existence. Apparently the worry was that to admit such a collection would be to attribute a positively infinite property to something other than God, when such an actually

'fruitful and consistent concept in the mathematical realm'. 'Finitude of the Past and God's Existence', in William Lane Craig and Quentin Smith's *Theism, Atheism and Big Bang Cosmology* (Oxford: OUP, 1993), 9–16, 12. See also Pamela S. Huby, 'Kant or Cantor? That the Universe, if Real, must be Finite in Both Space and Time', *Philosophy*, 46 (1971), 121–32, 126, 129. For a rival view, see José A. Bernadete, *Infinity: An Essay in Metaphysics* (Oxford: OUP, 1964), 109–13.

[63] Robert Boyle, *Selected Philosophical Works of Robert Boyle*, ed. M. A. Stewart (Manchester: Manchester University Press, 1979), 213, 235. Berkeley, *Notebook B* s. 322, in *Works*, i. 39. Edmund Law, *An Enquiry into the Ideas of Space, Time, Immensity and Eternity* (Cambridge 1734; facsimile edn.: New York: Garland, 1976), 98–9. William Drummond, *Academical Questions* (1805; facsimile edn.: Delmar, NY; Scholars' Facsimiles and Reprints, 1984), 68. Galileo, *Two New Sciences*, 33. Barrow, *Usefulness*, 159–60. Isaac Newton, Letter to Bentley, 17 Jan. 1693, in *Sir Isaac Newton's Letters and Papers on Natural Philosophy and Related Documents*, ed. A. R. Hall and M. B. Hall (Cambridge: CUP 1978), 15–16, 21. Jacques Rohault, *System of Natural Philosophy*, tr. John Clarke and Samuel Clarke (New York: Johnson Reprint Co., 1969), 33. Antoine Le Grand, *An Entire Body of Philosophy*, tr. Richard Blome (London, 1694), i. 97. Keill, *Introduction*, 38–9. Samuel Clarke, *A Demonstration of the Being and Attributes of God and Other Writings*, ed. Ezio Vialiati (Cambridge: CUP, 1998), 9. To summarize the replies given to paradox (2B) in these texts: Keill and Wallis (whose views are reported in Newton's letter to Bentley) argue that actual infinities can be unproblematically unequal; they do not have to all be equal. Galileo, Barrow, Newton, Rohault, Le Grand, and Clarke all argue that the concepts of equality and inequality do not apply at all to infinities. However, it must be said that, with the exception of Galileo, these thinkers all seem to be confusedly thinking of *potential* rather than *actual* infinities. For instance, they typically appeal to the 'indefinite' nature of infinities when arguing that they are neither equal nor unequal.

infinite property—rather than a merely potentially infinite one—could belong to God alone. Now, I can find no example of a specific Enlightenment text that explicitly endorses this strange argument.[64] But both Pierre Bayle and Leonhard Euler go out of their way to report this sort of complaint, if only in order to respond to it.[65] Euler in fact devotes the entirety of one of his letters on natural philosophy to rebutting the objection. Opponents of infinite divisibility, he tells us, 'pretend that by divisibility in infinitum we are obliged to ascribe to bodies an infinite quality, whereas it is certain that God alone is infinite'. The insinuation is that it is Wolffian monadists that press this charge of apostasy: 'they accused us of atheism, and now they also charge us with polytheism, alleging that we ascribe to all bodies infinite perfections.... [W]e are accused of paying homage to all bodies, as so many divinities.'[66] So this objection had at least some currency during the Enlightenment. It certainly chimes well with the widespread scholastic concern that the infinitely *large* space of the new science might constitute a heretical rival to God.[67]

3. Paradoxes from Infinite M-Divisibility alongside both Metaphysical Atomism and a Determinate Number of Parts

The third class of paradox takes in those that depend upon *both* our corollaries of the actual parts doctrine: metaphysical atomism *and* a determinate number of parts taken together. First, infinite m-divisibility alongside

---

[64] This worry is sometimes obliquely suggested by Descartes. For instance: 'Our reason for using the term "indefinite" rather than "infinite" in these cases [concerning the extension of the world, the division of matter, the number of stars "and so on"] is, in the first place, so as to reserve the term "infinite" for God alone.' (*Principles of Philosophy*, part I, s. 27, in *The Philosophical Writings of Descartes*, i. 202.) But even here, the explicit reasons Descartes gives for reserving the term for God concern our own epistemological limitations: we positively recognize God is greater-than-finite; in these other cases we simply fail to discover any limits. To my knowledge Descartes never explicitly states that there is anything impious in the very idea of an actual infinite outside of God.

[65] Pierre Bayle, *Response aux Questions D'Un Provincial*, in *Œuvres Diverses*, 5 vols. (1731; facsimile edn.: Hildesheim: Georg Olms, 1968), iii. 546. Euler, *Letters*, 61–3.

[66] Euler, *Letters*, 61.

[67] For a useful account of these scholastic concerns, see Grant, *Much Ado*, chs. 7 and 8. A form of this complaint also re-emerged to haunt Cantor's theory of the actual infinite in the 1880s–1890s. Cantor was forced to reply to the concerns of Dominican theologians that his theory, in admitting actually infinite sets, amounted to a form of pantheism officially condemned by papal decree in 1861. Although himself a Lutheran, Cantor took the worry

Problems of Material Structure / 45

a determinate number of parts mandates an actual infinity of parts, as we have seen. If we then add the thesis that bodies are ultimately constituted from an atomic base of m-indivisible first parts, we then have the result that bodies are constituted from an actual infinity of metaphysical atoms. One may well think that this result is already inevitable, given *either one* of our two corollaries alongside infinite m-divisibility. How could an infinitely m-divisible body be built out of metaphysical atoms unless there were an actual infinity of them? And how could an infinitely m-divisible body have a determinate number of given parts unless it was an actual infinity of metaphysical atoms? These more abrupt arguments did indeed appeal to some philosophers. But for our purposes we need not address them. As we have already seen, it was a standard view in the early modern period that the actual parts doctrine entailed *both* metaphysical atomism *and* a determinate number of parts. And these two corollaries, taken together with infinite m-divisibility, certainly jointly entail an actual infinity of metaphysical atoms. In any event, under the current rubric of paradox category (3) I discuss those paradoxes that depend on an actual infinity of metaphysical atoms in a body.

This result that bodies are constructed from actually infinite collections of m-indivisible first parts has seemed paradoxical to many. The classic charges are: (3A) that this would make bodies either infinitely large or of no size whatsoever, depending on whether their atomic parts are extended or unextended; (3B) that this would make all bodies of equal size; and (3C) that there could be no contiguity between the ultimate parts.

Paradox 3A. *Zeno's metrical paradox applied to bodies: bodies must be either infinitely large or of no size whatsoever*

Perhaps the most popular objection of all to infinite m-divisibility was a version of the 'metrical paradox' commonly attributed to Zeno of Elea. This paradox starts with the premise that any continuous, infinitely divisible entity can be partitioned (at least in thought) into an actually infinite collection of same-sized parts. Now, this premise has often been resisted. For instance, it does not seem true of such continua as have *potential parts*—since here the number of parts is indexed to the successive process

seriously and argued that, while his theory did admit the actual infinite outside of God, it reserved for God a different, higher notion of the absolute infinite of all transfinite numbers. See Joseph Dauben, *Georg Cantor: His Mathematics and Philosophy of the Infinite* (Cambridge, Mass. Harvard University Press, 1979), 141–8.

of division, and so is only potentially infinite. But at least if the corollaries of determinate parts and ultimate parts do indeed follow from the actual parts doctrine, then it does seem to be true of continua that have distinct *actual* parts. Such continua would indeed each be built up from an actual infinity of elemental parts. And if we then add the plausible assumption that those elemental parts or metaphysical atoms are all alike in size, then we do have the starting point of Zeno's paradox: an actually infinite collection of same-sized parts. Zeno then lays down the following dilemma: are these elemental parts extended or are they unextended? If they are extended, then we have an actually infinite number of extended parts and it seems the whole body constructed from them must be infinitely large. On the other hand, if each atomic part is unextended, then it seems that the whole body must itself be unextended—for how could extensionless parts build an extended whole? Both results are absurd, and thus we have a *reductio* on the notion that body—or indeed any continuous quantity that has actual parts—is infinitely m-divisible.[68]

Running down all the way from Zeno through Epicurus and Sextus Empiricus to the Enlightenment, this argument was the traditional stock objection to the infinite m-divisibility of matter.[69] In the early modern period (and indeed in Epicurus and Sextus) it typically appears in the following heavily abbreviated form: (i) if a body is infinitely divisible, it must have an infinite number of parts; (ii) whatever has an infinite number of parts is infinitely large; therefore (iii) no body can be infinitely divisible. This abbreviated version is presented in our period in the writings of Giordano Bruno, Henry More, Joseph Glanvill, the younger Newton, and Berkeley; it is also reported (though not endorsed) by Isaac Barrow and John Keill.[70] It

---

[68] On the structure of this paradox, see Adolf Grünbaum, *Modern Science and Zeno's Paradoxes* (Middletown: Wesleyan University Press, 1967), 115–40. Brian Skyrms, 'Zeno's Paradox of Measure', in R. S. Cohen and L. Lauden (eds.), *Physics, Philosophy, and Psychoanalysis* (Dordrecht and Boston: D. Reidel, 1983), 223–54.

[69] Epicurus, *Letter to Herodotus*, s. 57, in *The Epicurus Reader*, tr. and ed. Brad Inwood and L. Gerson (Indianapolis: Hackett, 1994), 11. Sextus Empiricus, *Outlines of Scepticism*, tr. Julia Annas and Jonathan Barnes (Cambridge: CUP, 1994), 154 (book 3, s. 44).

[70] On Bruno, see P. H. Michel, *The Cosmology of Giordano Bruno*, tr. R. E. W. Maddison (Ithaca, NY: Cornell University Press, 1973), 132–3. Henry More, *Antidote against Atheism*, in *Philosophical Writings*, 11. Joseph Glanvill, *Vanity of Dogmatizing*, 53. Isaac Newton, 'Certain Philosophical Questions', tr. and ed. J. E. McGuire and Martin Tamny, in McGuire and Tamny, *Certain Philosophical Questions: Newton's Trinity Notebook* (Cambridge: CUP, 1983), 330–489, 341. Bayle, *Dictionary*, 362. Berkeley, *Principles*, s. 47, and in *Notebook B*, s. 352, in *Works*, i. 42, ii. 283. Barrow, *Usefulness*, 157. Keill, *Introduction*, 35–7.

is—of course—also Hume's 'lead argument' against infinite divisibility (H1)–(H3) that I examined in the case study of section VI above.

Stated in this abbreviated form, the reasoning is clearly problematic, as many recent commentators have pointed out. However, once we place the argument in the appropriate metaphysical setting and fill in the missing premises—(*a*) the argument assumes actual parts, and hence an actual infinity of m-indivisible ultimate parts; (*b*) those ultimate parts are all the same size; and (*c*) they could not be extensionless—the argument is much more compelling than the recent literature has allowed. (We saw this in detail in the case study on Hume.) But why was it typically stated as such an over-abrupt enthymeme? I think the answer is that—at least for the new philosophers of the Enlightenment—the suppressed steps were either thought too self-evident to belabour openly, or perhaps were simply made unconsciously. Certainly these implicit assumptions (*a*), (*b*), and (*c*) were each thought manifestly true by the overwhelming number of Enlightenment philosophers, and are nearly always explicitly endorsed elsewhere in the writings of the argument's advocates.[71] But in case it is still thought that this reconstruction is excessively speculative or charitable, we can turn to the fuller statements of the argument found in the seventeenth-century texts of Kenelm Digby, Walter Charleton, and Pierre Bayle. These three philosophers each present the argument in a much more developed form than the abbreviated version set out above. In particular, each makes it clear that the argument depends essentially on the actual parts doctrine. So these thinkers each identify the crucial role of premise (*a*). In addition,

---

[71] I show that premise (*a*) was a fundamental tenet of Enlightenment natural philosophy in Ch. 2, s. II, where I discuss the actual parts doctrine in full. Premise (*b*) was also regarded as self-evident: all those who admitted m-indivisible *metaphysical* atoms made the natural assumption that they would all be the same size. Similarly with premise (*c*): it just seemed an obvious arithmetical truth that unextended elements could not sum to form an extended whole: 'though you putt never so greate a number of them together, still they will drown themselves all in one indivisible point' (Digby, *Two Treatises*, 11–12). '[T]hat a *Line* will consist of *Points* Mathematically so-called, that is, *purely Indivisible*...is the grandest absurdity that can be admitted in Philosophy, and the most contradictious thing imaginable' (More, *Immortality*, 37). '[P]ersons of the slightest depth can comprehend with complete certainty, if they give the matter a little attention, that several nonentities of extension joined together will never make up an extension.' (Bayle, *Dictionary*, 359–50, see also 394). See also Barrow, *Usefulness*, 149. Antoine Arnauld and Pierre Nicole, *Logic, or The Art of Thinking*, tr. and ed. Jill Vance Buroker (Cambridge: CUP, 1996) 232. Leibniz, *Philosophical Essays*, 139, 162, 206, 229. Roger Boscovich, *A Theory of Natural Philosophy*, tr. J. M. Child (Cambridge, Mass.: MIT Press, 1966), s. 3, 7, 138, 372. Reid, *Works*, i. 308. Kant, *Critique of Pure Reason*, B467.

all three explicitly bring out premise (*c*), and Charleton expressly adds (*b*) as well.⁷² So we certainly find the full-blooded argument for (3A) in the works of these three philosophers. Moreover (it seems to me), once we appreciate that the actual parts metaphysic sets the implicit and understood context of the more abbreviated statements of the argument, we can see that the full argument should indeed be read between the lines of even these briefer versions. (I defended this line of interpretation further in my case study on Hume in section VI above.)

Paradox 3B. *Bodies should all be the same size*

The admission of an actual infinity of metaphysical atoms in every body was also thought to generate a second paradox: all bodies should be the same size. Once it is granted that bodies are constructed from elemental m-indivisible atoms, it seems that differences in the sizes of bodies would have to be accounted for by different bodies being built out of different numbers of atoms (at least if we assume that the atoms are all alike). But if we are then told that each and every body is alike constructed from an actual infinity of m-indivisible atoms, then we seem to lose this way of accounting for the different sizes of different bodies. It seems as if all bodies should be the same size.⁷³ (Notice that this paradox ties closely with paradox (2B) above and the view that all actual infinities are equal. The two paradoxes are often presented side by side, with (3B) given as an example of the general problem (2B).)

The current argument for (3B) is often confused or run together with the suggestion that the sheer fact of infinite m-divisibility *alone* (bracketing

---

⁷² Kenelm Digby, *Two Treatises*, 10–12. Charleton, *Physiologia*, 91–3, and—on premise (*c*)—107–8. Bayle, *Dictionary*, 362; on premise (*a*) see 356, on premise (*c*) see 359–60. And see Nicholas of Autrecourt, *The Universal Treatise* (Milwaukee, Wis.: Marquette University Press, 1971), 77–8, for a clear 14th-cent. statement of this argument and its actual parts context.

⁷³ In the recent literature, Dean W. Zimmerman reports (without endorsing) a paradox stemming from an actual infinity of metaphysical atoms that is fundamentally this same problem *in reverse*. '[I]f an extended body were nothing more than a continuous manifold of simple parts, one for each point in the region occupied by the body, then it should be possible for the same set of parts to be rearranged so as to form a body of *any* size you like' (Dean W. Zimmerman, 'Indivisible Parts and Extended Objects: Some Philosophical Episodes from Topology's Prehistory', *The Monist*, 79 (1996), 148–80, 155). Rather than concluding that all bodies are the same size, this variant has it that a body of a given size can be reconstructed to be any size one chose. But the essential paradox is the same: all bodies are supposedly built up from the same number of atomic parts, and this seems to conflict with the fact that they are different sizes to one another.

the actual parts doctrine and the admission of an actual infinity of metaphysical atoms) already shows that bodies must all be the same size, since the same number of parts can be found in any two bodies. However many parts a wall can be divided into, a brick in that wall can be divided into the same number. So the wall and the brick should both be the same size. But of course this latter argument is clearly invalid. Antoine Le Grand gives the obvious response: 'supposing two *unequal Bodies*, should be divided into an equal number of *Parts*, yet would it not follow thence that those two *Bodies* were Equal, because the *Parts* of one *Body*, would proportionably be greater than the Parts of the other *Body*'.[74]

The former argument from infinite m-divisibility to the conclusion that all bodies are the same size—which crucially depends on an actual infinity of atoms—is pressed by Giordano Bruno and Walter Charleton, and is reported (though not endorsed) by Leonhard Euler. Robert Boyle presses the latter, clearly invalid argument. In addition, Joseph Glanvill and William Drummond both state that infinite m-divisibility entails that bodies are all the same size, but they do not say *why* this follows and hence fail to differentiate explicitly between the two arguments for this conclusion. Jacques Rohault, Antoine Le Grand, Pierre Bayle, and John Keill each reply to the clearly invalid argument—though it is not clear that they appreciate there are two distinct arguments here and thus that their response is no answer to Bruno and Charleton.[75]

Paradox 3C. *No ultimate part is contiguous with any other*

Our final paradox stemming from an actual infinity of metaphysical atoms is the result that no atom or ultimate part of a body is contiguous with any other. We can see this if we first identify one atom, specifying it by its location: say an atom exactly six inches along a twelve-inch ruler. Now, we then locate a second atom by location: say an atom at the seven inch mark; or at the six and a half inch mark; or at the six and a quarter inch mark; or... Now, no matter where we specify this second atom is—no matter how close we locate it to the first—there will always be an infinity of atoms between it and the first one. Try as we might, we cannot locate

---

[74] Le Grand, *Entire Body*, i. 97.

[75] On Bruno, see Michel, *Cosmology of Giordano Bruno*, 132–3. Charleton *Physiologia*, 92. Euler, *Letters*, 46. Glanvill, *Vanity of Dogmatizing*, 55. Drummond, *Academical Questions*, 68. Rohault, *System*, 33. Le Grand, *Entire Body*, i. 97. Pierre Bayle *Système*, in *Œuvres Diverses*, iv. 304. Keill, *Introduction*, 38.

an atom that is not separated from the first by an infinite number of intervening atoms. Thus (the argument goes) no atomic part of an infinitely divisible body is contiguous to any other. But this is paradoxical, for how could these ultimate parts concatenate to build continuous bodies if they do not touch one another?[76] Bayle presents the argument in the following way. (In order to make sense of this extract, one has to realize, first, that Bayle is identifying matter with space or extension, in keeping with the then-dominant Cartesian physics, and, second, that he is (surely) talking of *ultimate* parts.)

[I]f there is no body that does not contain an infinity of parts, it is evident that each particular part of space is separated from all the others by an infinity of parts, and that the immediate contact of two parts is impossible... Since, then the existence of extension necessarily requires the immediate contact of its parts, and since this immediate contact is impossible in an extension that is divisible to infinity, it is evident that the existence of this extension is impossible.[77]

This completes my reconstruction of the classic paradoxes. My hope is that I have brought out their structure and imposed an organizing hierarchy the tumult of paradoxes that crowds the Enlightenment literature. More than this, I hope that my rational reconstruction has also brought out the fact that the classic paradoxes crucially depend on either the admission of metaphysical atomism, or on the admission of a determinate number of parts, or on both. Now, if one accepts the actual parts doctrine, then one must admit metaphysical atomism and a determinate number of parts. (Or such at least was the standard view in the early modern period.) So in the context of the actual parts metaphysic, the classic paradoxes do indeed loom large. If one conceives of body (or indeed, of any continuous quantity) as having parts that are each concrete, distinct existents, then the paradoxes threaten. But they do not threaten the rival potential parts metaphysic (since here the number of parts is indeterminate, and metaphysical atomism is not mandated); nor do they threaten the mathematician's purely formal notion of infinitely divisible quantity *in abstracto*.

---

[76] This sort of argument against an actual infinity of atomic parts in an extended entity is developed further by Brentano. See Zimmerman, 'Indivisible Parts', 157.

[77] Bayle, *Dictionary*, 363.

IX. Reactions to the Problems of Material Structure: Distinguishing the Factions

> The composition of bodies, whether it be of *Divisibles* or *Indivisibles*, is a question which must be rank'd with the *Insolvibles*: For though it hath been attempted by the most illustrious *Wits* of all *Philosophick* Ages; yet they have done little else, but shewn their own *divisions* to be almost as *infinite*, as some suppose those of their Subject.
>
> <div align="right">Joseph Glanvill, <i>The Vanity of Dogmatizing</i>, 1661</div>

In this section I present a survey of the various doctrinal camps that emerge in reaction to the problems of material structure. The search for a resolution greatly stimulated Enlightenment natural philosophy, generating a host of responses. What I want to do here is impose an organizing structure on this range of responses, and to marshal the period's philosophers into distinct factions engaged over certain pivotal questions.

Begin with the central conflict that underlies the paradoxes of material structure. On the one hand, we have the doctrine—apparently mandated by geometry—that matter is m-divisible ad infinitum. On the other, we have the doctrine that matter is constructed from actual parts—a thesis which (we have seen) would seem to preclude infinite m-divisibility. Starting from this apparent conflict, we can then demarcate four different types of reaction. A first group upholds the doctrine of infinite m-divisibility and accordingly rejects the actual parts doctrine: bodies only have potential parts. A second faction upholds the actual parts doctrine and accordingly rejects infinite m-divisibility: bodies are constructed of finite arrays of m-indivisible first parts. The third group endorses *both* the doctrine of infinite m-divisibility and the actual parts doctrine, attempting to show that the apparent conflict between them can be disarmed. Finally, a fourth faction likewise endorses *both* infinite m-divisibility and actual parts, but then maintains that each doctrine indeed rules out the other: we then have a fully fledged antinomy and a genuine crisis of reason.

Faction (1). Affirm Infinite M-Divisibility; Reject the Actual Parts Doctrine

Our first group of philosophers maintains that body is infinitely m-divisible and rejects the actual parts doctrine. These philosophers then endorse the *potential parts* account of matter and accordingly avoid the various classic paradoxes of infinite m-divisibility. Given the potential parts view, bodies do not have a determinate number of parts (rather, they each have as many parts as one chooses to actualize through division), and bodies need not have an atomic base of first parts. The classic paradoxes, therefore, do not arise.

Aristotle stands as the patron saint of this first faction. According to his classic account of the structure of continua in the *Physics*, all continuous quantities are infinitely divisible, but their parts are each posterior to the whole and each require some positive operation to actualize. Their parts are therefore merely *potentially* infinite in number.[78] Those philosophers who fall into this first faction, endorsing the potential parts account of material structure, are—in essence—simply applying Aristotle's general account of continua to the particular case of material objects. The main bastions of this first faction in the seventeenth century were Oxford and the Sorbonne, and more generally the schools of the orthodox Thomist establishment. Faced with the subversive new science's advocacy of actual parts, scholastics mounted a vigorous and sustained defence of the potential parts account of matter, both in written works, and in numerous public disputations.[79]

In addition to the schoolmen, there are certain new philosophers who stand apart from the mainstream of the new science on this issue and endorse the potential parts response to the problems of material structure. Kenelm Digby's 1644 *Two Treatises*, despite its corpuscularian tendencies and rejection of substantial forms, firmly endorses the traditional Aristotelian doctrine of potential parts, as does Hobbes's 1655 classic *De Corpore*.[80] The mathematician John Keill at least toys with the potential parts doctrine as one possible answer to the paradoxes of infinite m-divisibility in his published and widely read Oxford lectures of 1700.[81] And Kant, in his later, critical period, also adopts the potential parts response—adding an interesting idealist twist. He argues that the *transcendental realist* cannot avoid

---

[78] Aristotle, *Physics*, books 3, 6 and 8, tr. R. Hardie, in *The Complete Works of Aristotle*, ed. Jonathan Barnes (Princeton: Princeton University Press, 1984), 315–46. Aristotle, *On Generation and Corruption*, book 1, tr. H. H. Joachim, ibid. 512–54. For commentary, see the references listed in note 32 above.

[79] See, for instance, the irreverent accounts in Bayle, *Dictionary*, 362.

[80] Digby, *Two Treatises*, 10–14. Hobbes, *De Corpore*, 7. 9, in *Works*, i. 97.

[81] Keill, *Introduction*, 41.

admitting actual parts and hence a determinate number of ultimate, simple parts, and thus that they run into contradiction when this result is set against the geometrical demonstration of infinite m-divisibility. This constitutes a *reductio* on their system. (It is the 'Second Antinomy', one of the four conflicts that plague transcendental realism in the 1781/7 *Critique of Pure Reason*). Only the *transcendental idealist*—who thinks of bodies as progressions of appearances structured by the forms of intuition—can adopt the potential parts view (Kant says) and thus avoid the antinomy.[82] (For more on these figures and their various versions of the potential parts account, see Chapter 2, section III.)

Faction (2). Affirm the Actual Parts Doctrine; Reject Infinite M-Divisibility

Our second faction rejects the infinite m-divisibility of matter and upholds the actual parts doctrine. These philosophers argue that the classic paradoxes of infinite m-divisibility set out in section VII amount to a compelling *reductio* on infinite m-divisibility. Accepting that bodies have actual parts and that this entails that bodies cannot be m-divisible ad infinitum, they therefore maintain that bodies are only finitely m-divisible. M-division must eventually terminate in a finite array of m-indivisible atomic parts.

There are two main subgroups within this faction, which we can distinguish by considering the way in which they each reply to the following challenge. At least on the face of it, if bodies are built out of a finite array of m-indivisible atoms, then those atoms must be extended. But then it is hard to see why these supposed atoms are not m-indivisible further. If they are extended, then don't they have spatially distinct parts? (That is, aren't they f-divisible?) And if they have spatially distinct parts, then don't they have logically separable parts? (That is, if they are f-divisible, then aren't they *ipso facto* m-divisible?) Our two subgroups each attempt to block this objection in a different way.

The first subgroup attacks the first part of the objection. They argue that space itself has minimal, least parts and therefore an atom of matter

---

[82] Kant argues that transcendental realists must admit actual parts in the following places: *Critique of Pure Reason*, B440, B462–70 (the Second Antinomy), *Metaphysical Foundations of Natural Science*, 53, 55. But given transcendental idealism, there are only potential parts: see the *Critique of Pure Reason*, B552 and B333, and *Metaphysical Foundations of Natural Science*, 54. I discuss Kant's position in detail in Ch. 2, s. II.

that occupies just one such minimal space lacks spatially distinct subparts. Space itself resolves to f-indivisible granules, and so an atom of matter filling one such granule is itself f-indivisible. And since it lacks spatially distinct parts, it cannot then be m-divided. This first camp is then radically Epicurean: all extended physical continua—including space itself—resolve to granular first parts. Faced with the geometer's demonstration of the infinite divisibility of extension, these neo-Epicureans will deny that the classical axioms of geometry apply to physical space. At best they are true of the geometer's imaginary realm of pure, mathematical space, where one can stipulate whatever system of geometrical postulates one will. But real, physical space does not conform to the axioms and theorems of classical geometry. This first group includes Walter Charleton, whose 1654 tract *Physiologia Epicuro-Gassendo-Charletoniana* presents just such a radical granular system: 'not only Place and Time', but all 'Natural Quantity' resolve to 'insectile' *minima*; geometrical demonstrations, moreover, prove nothing about extended physical quantities.[83] Charleton is followed by the younger Newton, whose early Trinity notebooks (1664) echo the *Physiologia* (often word-for-word) in arguing that physical space and matter are both constructed from f-indivisible least parts.[84] George Berkeley also stands in this first subgroup, although his system of course adds idealism to Epicureanism. Roughly speaking, all extended continua in his 'physical' world are ideas that are constructed from arrays of contiguous f- and m-indivisible *minima sensibilia*.[85] Finally we can add David Hume, who maintains that ideas and impressions of extended continua, and external world extended continua are built up from f- and m-indivisible first parts. An important difference that sets Hume apart from Charleton, Newton, and Berkeley is that, while these latter philosophers suppose extended continua to be constructed from *extended* f- and m-indivisible granules, Hume maintains that extended continua are made up of finite arrays of *unextended* f- and m-indivisible first parts.[86] (For more on the approach of these so-called 'ungeometrical philosophers', see Chapter 5, section III.)

---

[83] Charleton, *Physiologia*, 94, 95–6; see also 85, 264.

[84] McGuire and Tamny, *Certain Philosophical Questions*, 27, 62, 74–5.

[85] See Harry M. Bracken, 'On Some Points in Bayle, Berkeley and Hume', *History of Philosophy Quarterly*, 4 (1987), 435–46, 437–8. Douglas Jesseph, *Berkeley's Philosophy of Mathematics* (Chicago: University of Chicago Press, 1993), 57.

[86] See, for instance, the exegesis given in Harry M. Bracken, 'On Some Points', 438. Frasca-Spada, 'Some Features', 402. Baxter, 'Hume on Infinite Divisibility', 18. For an early criticism of

The second subgroup takes a different approach. They also assert that bodies are made up of finite arrays of m-indivisible atoms, but conceive of these atoms as each an unextended core *punctum* that projects an extended shell of force, resisting the approach of neighbouring *puncta*. (It seems possible to describe these atoms as either extended or as unextended, depending on whether one thinks of the shell of force as included in the atom proper or not.) Since the core is altogether unextended, it cannot itself be either f- or m-divided. The shell of force, on the other hand, is extended and straightforwardly f-divisible. But, crucially, since it is nothing more than a force projected through space from the core, it cannot itself be broken apart into logically separable parts. So the shell is f-divisible but m-indivisible. Notice that this theory—unlike the Epicurean system of Charleton, Newton, Berkeley, and Hume—can allow that classical geometry applies to physical space and that space is accordingly f-divisible ad infinitum. It simply maintains that this infinite f-divisibility of space, and of all extended objects in space, does not establish the m-divisibility of the force shells.

This ingenious account of m-indivisible material atoms—essentially an embryonic field metaphysic—was first proposed by the younger, pre-critical Kant in his 1756 *Physical Monadology*, and by Roger Boscovich in his 1758/63 *Theory of Natural Philosophy*. So far as I can tell, these two philosophers each developed the system independently. Kant expressly formulates the theory as a solution to the conflict between actual parts and the claims of geometry: it effects a reconciliation by showing that, given force-shell atoms, geometry entails only the infinite f-divisibility of matter, not its infinite m-divisibility.[87] Boscovich, by contrast, arrived at the theory by way of a consideration of elastic impacts: he postulates force-shell atoms to make good the shortcomings of the Newtonian model of perfectly rigid, inelastic physical (p-indivisible) atoms. But he too states that an additional virtue of the theory is that it does away with the paradoxes of infinite m-divisibility.[88] (I discuss the pre-critical Kant's and Boscovich's atomic force-shell theory in detail in Chapter 6.)

Hume on this point, see Thomas Melvill's 'Observations on Light and Colours', in *Essays and Observations, Physical and Literary. Read before a Society in Edinburgh, and Published by them* (Edinburgh: G. Hamilton & J. Balfour, 1756), ii. 12–91, 71–2. The relevant section is reprinted in *Early Responses to Hume's Metaphysical and Epistemological Writings*, ed. James Fieser, 2 vols. (Bristol: Thoemmes Press, 2000), i. 135–6.

[87] Kant, *Physical Monadology*, in *Works*, 1: 473–87

[88] Boscovich, *Theory of Natural Philosophy*, s. 90 (quoted in Ch. 6, n. 22); see also ss. 138, 142, 372, and the Synopsis 12, 13.

Faction (3). Affirm *both* the Actual Parts Doctrine *and* Infinite M-Divisibility: the Two can be Reconciled

We come now to our third faction: those philosophers who endorse *both* the actual parts doctrine *and* infinite m-divisibility, maintaining that the two theses can somehow be squared with one another. Here we can usefully discriminate between three subgroups.

The first subgroup takes in those philosophers who seem altogether unaware of any supposed tension between the actual parts doctrine and infinite m-divisibility. As I have argued by way of my rational reconstruction of the classic paradoxes, there is at least an apparent conflict between these two theses. This is because the actual parts doctrine seems to entail both a determinate number of parts and atomic first parts, and each of these corollaries is in at least prima-facie conflict with infinite m-divisibility. Now, it is not that philosophers in this first subgroup acknowledge the apparent conflict and then try to disarm it, or even that they pooh-pooh it without argument. Alas—and despite my best attempts to find a more charitable explanation—they seem to be simply oblivious to the issue. The classic case here is the philosopher who fully endorses the actual parts account of matter in one place, but then, when turning in another place to address the question of m-divisibility, claims that infinite m-divisibility entails only an unparadoxical potential infinity of parts. But, of course, what such a philosopher misses is that one cannot *both* endorse the actual parts doctrine and then take refuge in an indeterminate potential infinity of parts. Now, the philosophers of this first group *do* all discuss objections to infinite m-divisibility, and often do so at length. But they will then give the brisk mathematical responses *à la* Flew, Fogelin *et al.* (say, appealing to indeterminate potential infinities of parts, or invoking proportional parts), while failing to see that the actual parts doctrine they themselves accept forestalls such moves.

But to name names. Isaac Barrow, Newton's famous predecessor in the Lucasian Chair at Trinity, discusses infinite divisibility at length in his *Mathematical Lectures* (originally given in 1664–6, then published in Latin in 1683 and in English in 1734). He expressly sponsors an actual parts analysis of all extended quantities (and of matter in particular) but then immediately claims that infinite divisibility entails only a potential infinity or an 'indefinite' number of parts. Barrow does not try to reconcile these two theses: he simply fails to acknowledge any conflict between them.⁸⁹ Ralph Cudworth's optimistically titled *True Intellectual System* (composed in 1671, published in 1678) stumbles in exactly the same place: bodies have actual parts and are m-divisible ad infinitum, but then all infinities are treated as potential.⁹⁰ Again, no tension is acknowledged. Jacques Rohault also commits the mistake. His celebrated 1671 Cartesian *Traité de Physique* first affirms that matter is constructed from actual parts, but then that its parts are indeterminate in number.⁹¹ The younger Pierre Bayle runs into the same trouble as Barrow, Cudworth, and Rohault. His is a particularly interesting case, since it shows the possibility of rehabilitating this sort of offender. The early *Système Abrégé de Philosophie* (1675–7) presents the period's most vigorous affirmation and defence of the actual parts doctrine; but, alas, it then defends infinite m-divisibility by invoking straightforwardly mathematical responses (such as the proportional parts gambit), failing to notice that the actual parts doctrine blocks such a move. By his later *Dictionnaire historique et critique* (1697), however, Bayle has learnt the error of his ways: the actual parts doctrine is still demonstrable, but now it *does*

---

⁸⁹ Barrow endorses the actual parts account of all extended quantities (see *Usefulness*, 148, 151, 162) and expressly applies this to the case of matter: 'all imaginable Geometrical Figures are really inherent in every Particle of Matter...if the Hand of an Angel (at least the Power of God) should think fit to polish any Particle of Matter without Vacuity, a Spherical Surface would appear to the Eyes of a Figure exactly round; not as created anew, but as unveiled and laid open from the Disguises and Covers of its circumjacent Matter' (76–7). He then explicitly endorses the orthodox Aristotelian account of infinity and hence only an 'indefinite' number of parts in a body. This leads him to speak, for instance, of 'the infinite, or rather indefinite (*i.e.* more than any Determinate Number of) Parts of a Body' (150, see also 142).

⁹⁰ Cudworth endorses actual parts: '[the nature of corporeal substance] is nothing but aliud extra aliud, "one thing without another," and therefore perfect alterity, disunity, and divisibility. So that no extensum whatsoever, of any sensible bigness, is truly and really one substance, but a multitude or heap of substances, as many as there are parts, into which it is divisible'; 'no body or extended thing is one but many substances; every conceivable or smallest part thereof being a real substance by itself' (*The True Intellectual System of the Universe*, 3 vols. (London: Thomas Tegg, 1845); iii. 393, 395; see also iii. 392). However, but for the pious exception of God, Cudworth rejects actual infinities and follows Aristotle in admitting only a 'potential infinity, or indefinite increasableness' in natural things (ibid. ii. 527; see also ii. 528, 540).

⁹¹ Rohault endorses the actual parts doctrine: 'When we consider a determinate Portion of Matter without Prejudice, and compare it with other Portions of Matter with which it is encompassed, we easily conceive that its particular Existence is wholly independent of those that are near it, and that it does not cease to be what it is, by being joined or united to other Portions of Matter' (*System of Natural Philosophy*, 31). But then, we are told, it is absurd to suppose that matter 'contain a certain determinate Number of...Parts' and that 'Matter is indefinitely divisible' (ibid., 33, 32).

rule out infinite m-divisibility, since the two would jointly entail a paradoxical actual infinity of *aliquot* parts.[92] Finally we can add Leonhard Euler and Thomas Reid to this first subgroup. Where Barrow, Cudworth, Rohault, and the younger Bayle fail to react to the argument that the actual parts doctrine entails *a determinate number of parts*, these two eighteenth-century philosophers fail to react to the arguments that the doctrine entails *ultimate parts* (i.e. the argument from the definiteness of parts and the argument from composition, outlined in section VII above). In Euler's *Letters on Natural Philosophy* (1761) he clearly accepts that there is a conflict between infinite m-divisibility and ultimate parts, but then, despite endorsing the actual parts doctrine, makes no attempt to respond to the arguments from actual parts to ultimate parts.[93] Likewise Reid in the *Essay on the Intellectual Powers of Man* (1785): he admits that infinite m-divisibility clashes with ultimate parts, and that bodies have actual parts; but then he does not see any reason to admit ultimate parts.[94] Now, it is not that Euler and Reid properly acknowledge the arguments from actual parts to ultimate parts and then offer a response; rather they simply seem to be unaware of—or at least unmoved by—these arguments. (Obviously, if these arguments fail, then Euler and Reid's position may ultimately be perfectly respectable: perhaps one *can* wed the actual parts doctrine with the theory of atomless gunk. Still, one would have liked them to have addressed these

---

[92] Bayle's *Système* defends actual parts in *Œuvres Diverses*, iv. 298–9; his *Historical and Critical Dictionary*, on 356 and 360. (I discuss these arguments at length in Ch. 2, s. IV.) The *Système* responds to the classic paradoxes (3A) and (3B) (from my rational reconstruction in s. VIII above) with the proportional parts gambit on 301–2. But the *Dictionary* then recognizes that the actual parts doctrine and infinite m-divisibility jointly entail an actual infinity of parts (356), and thus that paradoxes (3A) and (3B), along with (3C) rule out infinite m-divisibility (see 362–4).

[93] Euler accepts actual parts: '[Monadists] denominate [bodies] compound beings, which no one can deny, as extension necessarily supposes divisibility, and consequently a combination of parts which constitute bodies.' But he then sees no reason to accept the monadists' (i.e. Leibniz's and Wolff's) principle that 'a being can be compounded only so far as it is made up of simple beings'. (*Letters*, 59).

[94] Reid clearly avows the actual parts doctrine: see *Works*, i. 323 (quoted in s. IV above), 308. He also admits that infinite m-divisibility clashes with ultimate parts: 'There is, indeed, a principle long received as an axiom in metaphysics, which I cannot reconcile to the divisibility of matter; it is that, every being is one, *omne ens est unum*. By which, I suppose, is meant, that everything that exists must either be one indivisible being, or composed of a determinate number of indivisible beings' (*Works*, i. 323). But he then simply rejects the 'axiom' as Thomist twaddle. He does not acknowledge that there are popular arguments from the actual parts doctrine to 'a determinate number of indivisible beings' that need to be answered.

arguments from actual parts to metaphysical atomism, which seemed so compelling to their dialectical foes, including such prominent figures as Galileo, Leibniz, Hume, and Kant.)

The second subgroup of the current faction, by contrast, *does* recognize the traditional arguments from the actual parts doctrine to a determinate number of ultimate parts. Coupled with infinite m-divisibility, this then entails an actual infinity of ultimate parts in each body. Now, for the first (Aristotelian) faction and the second (Epicurean) faction this result was inadmissible. It showed that the actual parts doctrine and infinite m-divisibility could not both be correct, since they would jointly entail the full range of classic paradoxes. The current subgroup of philosophers attempts to brazen this out. They admit both actual parts and infinite m-divisibility, and thus that *bodies are indeed constructed from actual infinities of ultimate parts*: each body is a structured aggregate or compound built up from a greater-than-finite number of first parts. Contra the (Aristotelian) first faction, bodies do have actual parts, and indeed have an atomic base of actual first parts. And contra the (Epicurean) second faction, bodies are m-divisible ad infinitum and are each constructed, not from a finite collection of first parts, but from an actually infinite collection.

Galileo Galilei is at once the father figure of this approach and its most sophisticated advocate until Bernard Bolzano's *Grössenlehre* (1837).[95] Galileo's *Dialogues Concerning Two New Sciences* (1638) sets out the strategy, with his spokesman Salviati expounding the approach to the bumbling scholastic patsy Simplicio. Salviati's essential claim is that, since all continuous quantities have actual parts,[96] their infinite divisibility entails that they are constructed from actual infinities of m-indivisible atoms. Extended continua are

built up from an infinite number of indivisible quantities because a division and a subdivision which can be carried on indefinitely presupposes that the parts are infinite in number, otherwise the subdivision would reach an end; and if the parts are infinite in number, we must conclude that they are not finite in size, because an infinite number of finite quantities would give an infinite magnitude. And

---

[95] See Bernard Bolzano, *Paradoxes of the Infinite*, tr. D. A. Steele (London Routledge & Kegan Paul, 1950). For useful commentary, see Zimmerman, 'Indivisible Parts', 155–7.

[96] In fact Galileo attempts to sidestep the language of the whole actual parts–potential parts debate and claims that his argument does not turn on actual parts. But his argument does implicitly assume actual parts against the scholastic potential parts alternative, as I argue in detail in Ch. 3, s. VI.

thus we have a continuous quantity built up from an infinite number of indivisibles.[97]

As Salviati says, the m-indivisible first parts 'are not finite in size' (*'non quante'*), on pain of the whole being infinitely large, *à la* Zeno's metrical paradox. Elsewhere he describes these 'infinitely small elements' or 'points' as 'immeasurably small'.[98] (There is something of a controversy over whether Galileo means his m-indivisible atoms to be altogether *extensionless*—of zero size—or whether he truly intends them to be *infinitesimal*— neither determinately, finitely extended, nor altogether unextended.[99] But I cannot address this interpretative issue here.)

The obvious challenge to this sort of approach comes from the battery of paradoxes I have already set out above. Isn't the notion of an actually infinite collection inescapably paradoxical (see paradoxes (2A), (2B), and (2C))? And what about the specific paradoxes that arise when we conceive of a body as built up from an actual infinity of first parts (see paradoxes (3A), (3B), and (3C))?

Galileo does not explicitly respond to *all* of these problems. But he does answer the main challenges, and does so with astonishing rigour and audacity. He sees the issues raised by the classic paradoxes clearly and is quite open about the various bullets he must bite if he is to maintain his account. In many ways, his account anticipates the new logic of infinity that is developed in the nineteenth century by Cantor, Dedekind, and Frege. On the question of the general paradoxes of actual infinities, he boldly asserts that the mathematical properties of finite numbers that are traditionally carried over to actually infinite numbers to show their paradoxical nature simply do not apply to actual infinities at all. Actual infinities conform to a whole new mathematics, whose laws differ from that of finite number. We should be warned 'against the serious error of those who attempt to discuss the infinite by assigning to it the same properties which we employ for the finite, the natures of the two having nothing in common'. First, 'the attributes "larger", "smaller", and "equal" have no place either in comparing infinite quantities or in comparing infinite with

---

[97] Galileo, *Two New Sciences*, 34.

[98] Ibid. 36, 31, 34.

[99] For instance, Pietro Rebondi argues that they are altogether extensionless in his *Galileo: Heretic*, tr. Raymond Rosenthal (Princeton: Princeton University Press, 1987), 19. On the other hand, Zev Bechler argues that they have a positive though non-finite size in his *Newton's Physics and the Conceptual Structure of the Scientific Revolution* (Dordrecht: Kluwer, 1991), 131.

finite quantities'. Second, one actual infinite can contain another as a proper part—as, for instance, the set of natural numbers contains the set of square numbers: 'the totality of all numbers is infinite,... the number of squares is infinite,... neither is the number of squares less than the totality of all numbers'. Third, actual infinities are not constructed through any successive process of addition or division: they do not represent a target to be reached through a consecutive series of operations. Rather, they must be completely given 'at a single stroke' or not at all.[100] Anticipating Frege, actually infinite collections must be given all at once; they are decidedly *not* given through enumeration or other successive procedures. Galileo thus makes several bold steps toward the new logic of infinities that Cantor, Dedekind, and Frege will ultimately advocate in the nineteenth century.

Having faced down these classic paradoxes of actual infinite collections in general, Galileo then mobilizes his new machinery to reply to the paradoxes of bodies constructed from actual infinities of first parts. First, versus Zeno's metrical paradox ((3A) above): 'we reply that a divisible magnitude cannot be constructed out of two or ten or a hundred or a thousand indivisibles, but requires an infinite number of them'.[101] Contra Enlightenment orthodoxy, an infinite number of *parti non quante* can (Galileo says) sum to form an extended whole. And, second, versus the suggestion that all extended continua will be the same size if they are each constructed from an actual infinity of atoms (paradox (3B) above): 'When Simplicio introduces several lines of different lengths and asks me how it is possible that the longer ones do not contain more points than the shorter, I answer him that one line does not contain more or less or just as many points as another, but that each line contains an infinite number.'[102] Thus, just as his general theory of actual infinities anticipates certain of Cantor's, Dedekind's, and Frege's crucial moves, so too his account of extended continua built up from actually infinite numbers of *parti non quante* anticipates Adolf Grünbaum's twentieth-century deployment of the Cantorian theory to build extended continua from points.[103] Now, of course Galileo's theory lacks the technical sophistication and in particular the set theoretic basis of Cantor's and Grünbaum's accounts. But, nonetheless, the essential inspiration of their strategies is already there in Galileo. *A la* Cantor, the

---

[100] Galileo, *Two New Sciences*, 38, 33, 32, 48.   [101] Ibid. 31.   [102] Ibid. 33.
[103] Adolf Grünbaum, 'A Consistent Conception of the Extended Linear Continuum as an Aggregate of Unextended Elements', *Philosophy of Science*, 19 (1952), 288–306.

traditional paradoxes of actually infinite collections do not show the *impossibility* of such collections; rather they simply show that such collections must be subject to different mathematical laws than finite collections. And, *à la* Grünbaum, extended continua are constructed from actually infinite collections of atomic *parti non quanti*.

There are other period philosophers in this second subgroup of faction (3), though none that improve on the clarity and rigour of Galileo. In his early *Antidote Against Atheism* (1653) the Cambridge Platonist Henry More argued that the problems of material structure were inescapably paradoxical, but in his later *Immortality of the Soul* (1659) and *Enchiridium Metaphysicum* (1671), he seems to adopt a variant of Galileo's solution. Material bodies are constructed from coalitions of m-indivisible ('indiscerpible') first parts: '*minima corporalia*' or 'physical monads'.[104] Now, in neither of these later texts does More face up to the question of *how many* such physical monads it takes to build a body. (It is hard to avoid the impression that More is simply suppressing this difficult question—how different from Galileo!) But he does assert that his physical monads are 'infinitely little' and 'of such infinite thinness that they, like infinite magnitude, lack all figure'.[105] This at least would suggest that he might have Galileo's solution in mind: an actually infinite collection of infinitesimal first parts in each body. And, lastly, the mathematician John Keill also entertains Galileo's solution as a logically respectable response to the problems of material structure in his Oxford lectures of 1700 (though it is not the response he ultimately honours as the 'true answer').[106]

The third and final subgroup of faction (3) takes in the radical twin-world solution of Leibniz. There has been an enormous amount of scholarship devoted to Leibniz's handling of the problems of material structure, exploring the way his position shifts across time and contesting various of the central interpretative issues back and forth.[107] I cannot hope to add to

---

[104] More, *Antidote Against Atheism*, in *Philosophical Writings*, 11–12; *Immortality*, 6–8, 40–1; *Enchiridium Metaphysicum*, tr. Alexander Jacob (Hildesheim: Georg Olms Verlag, 1995), i. 71–2.

[105] More, *Immortality*, 41; see also 8. *Enchiridium Metaphysicum*, i. 72. More claims that his *minima* are 'infinitely small' and *therefore* m-indivisible, but he also asserts that they are f-divisible. It is hard to make sense of this.

[106] Keill, *Introduction*, 40.

[107] For a useful introduction to the issue, see Samuel Levey, 'Leibniz on Mathematics and the Actually Infinite Division of Matter', *Philosophical Review*, 107 (1999), 49–96. (In a subsequent paper, Levey argues that Leibniz is committed to an actually infinite number of parts in any body, and is committed to a constructivist mathematics that makes all infinities merely potential, but fails to address the tension between these two theses. I am not persuaded that

this scholarly literature here. My hope is merely to sketch in the briefest of outlines the essence of the Leibnizian strategy, at least as it appears in the later texts, and to locate it as a (radical) variant of the current faction's approach.

First, Leibniz is quite clear that extended continua such as bodies are m-divisible ad infinitum.[108] And he is certainly drawn towards the Aristotelian potential parts solution of faction (1). In continuous quantities 'the parts are indeterminate and one can obtain parts in an infinite number of ways'.[109] Parts are *assignable* to continua, not pre-given in it. Continua thus have only *potential parts*, and hence their parts are only *potentially infinite* in number.[110] But then, at the same time, Leibniz shies away from treating the Aristotelian potential parts doctrine as the full solution to the problems of material structure. This is because he is strongly drawn by actual parts intuitions when it comes to the analysis of 'actual things'.

[I]n actual things, there is only discrete quantity, namely a multitude of monads or simple substances... The continuum, of course, contains indeterminate parts. But in actual things nothing is indefinite, indeed every division has already been made in them. Actual things are composed as number is of unities, but ideal things are composed of fractions: there actually are parts in a real whole, but not in an ideal whole.'[111]

So 'actual things' have actual, not potential, parts. And 'actual things' would seem to include material bodies: 'Everyone agrees that matter has parts, and consequently that it is a multitude of many substances, as would be a herd of sheep'; 'A body is not a substance but an aggregate of substances, since it is always further divisible, and any given part always has another part, to infinity'.[112] So we have, at the same time, the claim that extended continua warrant a potential parts analysis, and the claim that actual things—including material bodies—warrant an actual parts

---

Leibniz did miss the problem, but if Levey is right on this point, then Leibniz really belongs in faction (3) subgroup (1) in my current taxonomy: he thinks he can endorse the actual parts doctrine along with infinite divisibility, and yet admit only potential infinities. See Samuel Levey, 'Leibniz's Constructivism and Infinitely Folded Matter', in Rocco J. Gennaro and Charles Huenemann (eds.), *New Essays on the Rationalists*, (Oxford: OUP, 1999), 134–62.)

[108] Leibniz, *Philosophical Essays*, 95, 179, 307. [109] Ibid. 251.
[110] Ibid. 145–6, 182, 185. [111] Ibid 185
[112] Leibniz, *Die philosophischer Schriften von Gottfried Wilhem Leibniz*, ed. C. I. Gerhardt (Hildesheim: Georg Olms Verlag, 1978), v.i. 552–3, quote and tr. from Donald Rutherford, *Leibniz and the Rational Order of Nature* (Cambridge: CUP, 1995), 220; Leibniz, *Philosophical Essays*, 103.

analysis.¹¹³ It is clearly difficult to see how these two theses can be squared with one another. Aren't material bodies also extended continua? But Leibniz's solution seems to be that extended continua in some sense belong to a realm that is phenomenal, whereas actual things—or at least the ultimate parts of actual things (the monads)—belong to a realm that is intelligible but not sensible. Thus the monads are not parts of extended continua in the same way that Galileo's indivisible atoms were parts of his extended continua. They do not sum with one another mathematically to construct the whole: indeed, they do not even belong to the same realm. Now, there is a huge secondary literature devoted to attempting to decipher Leibniz's precise meaning here. But for our purposes it is sufficient to appreciate that Leibniz wants to endorse *both* an actual parts account of matter (and its corollary of ultimate parts) *and* infinite m-divisibility, and thinks that he can reconcile the two by way of a twin-world metaphysic. As Kant puts it (focusing on the make-up of space, rather than body—but the point is the same): 'it was not Leibniz's intention, as far as I comprehend, to explain space by the order of the simple entities side by side, but rather to juxtapose this order as corresponding to space while yet belonging to a merely intelligible (for us unknown) world'.¹¹⁴

This completes my survey of the three subgroups of faction (3). Each of these subgroups affirms both the actual parts doctrine and infinite m-divisibility, and maintains—each in their own way—that the two are compatible. Before leaving this faction, I must note that the philosophers in this group do typically admit that there are difficulties in seeing how infinite m-divisibility is possible, notwithstanding their endorsement of it. But they will then claim that these difficulties—which are often attributed to our inability to fully understand infinity—do not rise to the level of positive contradictions. Isaac Barrow serves as a good example of this very common sort of attitude. He admits that 'Labyrinths, Difficulties and Inconveniences' seem to follow from infinite m-divisibility.¹¹⁵ But then:

I deny not but it is difficult to be understood, how every single Part can be divided so as all not to be actually reduced by the Division to Indivisibles, or to Nothing

---

¹¹³ One complication here is that, although Leibniz thinks that bodies are composites of actual parts—and indeed actual parts *in infinitum* all the way down—he does not strictly endorse the actual parts doctrine as I have presented it, since he does not think there are *arbitrary* actual parts: actual parts answering to every possible m-division. Here I overlook this complication. For more on this issue, see Ch. 2 n. 24.

¹¹⁴ Kant, *Metaphysical Foundations*, 56.   ¹¹⁵ Barrow, *Usefulness*, 141.

or what is next to Nothing: Nor yet do I think that, by Reason of the Imperfection of the Mind of Man and the Smallness of our Capacities, that therefore the Truth is to be deserted, when proved by so many evident Tokens, and supported by so many strong Arguments.... Though we are not able to comprehend how this indefinite Division can be performed, yet we ought not therefore to doubt, but it may be performed; because we clearly perceive it to follow of necessity from the Nature of Matter, a Thing most manifestly known to us, and we do also perceive it to be of the kind of Things, which cannot be comprehended by our Minds, as being finite.[116]

Similarly, John Keill writes of the problems of infinite m-divisibility as follows:

I acknowledge that there is something in the Nature of Infinites, that seems not to be adequately comprehended by the human Intellect; and therefore it is no wonder, if some things follow from it, which the Mind of Man, involved in thick Darkness, is not able to conceive, and especially in our present Question, there are many things that may seem as Paradoxes, and incredible to such Philosophers, who are less conversant in these Matters. However, nothing thence follows, that implies either a Contradiction or is repugnant to any Axiom or Demonstration.[117]

Very similar admissions appear in Galileo, Descartes, Arriaga, Stillingfleet, the younger Pierre Bayle, Antoine Arnauld and Pierre Nicole, Nicolas de Malezieu, Nicolas Malebranche, Henry Lee, Antoine Le Grand, Samuel Clarke, and Isaac Watts.[118] They each allow that 'difficulties' appear to follow from infinite m-divisibility, but then affirm that this should not disturb us overmuch since, first, our grasp of infinity is weak, and, second, there are compelling demonstrations of the doctrine of m-divisibility ad

---

[116] Ibid. 162; see also 152.
[117] Keill, *Introduction*, 33.
[118] Galileo, *Two New Sciences*, 26, 30. René Descartes, *Principles of Philosophy*, part 1, s. 26, in *Philosophical Writings of Descartes*, i. 201–2. On Arriaga, see the quotation given in Bayle's *Dictionary*, 361. Bayle, *Système Abrégé*, in *Œuvres Diverses*, iv. 292. Arnauld and Nicole, *Logic*, 231–3. On Nicolas de Malezieu, see the quotations given in Frasca-Spada, 'Some Features', 385–6. Nicolas Malebranche, *The Search after Truth*, tr. Thomas M. Lennon and Paul J. Olscamp (Columbus, Ohio: Ohio State University Press, 1980), 204–5. Lee, *Anti-Scepticism*, 88. Le Grand, *Entire Body*, ii. 8. Clarke, *Demonstration*, 8–9; see also his *Works*, iii. 799, 449–50. Isaac Watts, *An Inquiry Concerning Space*, in *The Works of the Reverend and Learned Isaac Watts*, 6 vols. (London: John Barfield, 1810), v. 520. Compare Robert Boyle, in *The Works of the Honourable Robert Boyle*, 6 vols. (London: J. and J. Rivington, 1772), i. 185—but see also my discussion of Boyle under faction (4) below. Compare also John Norris, *An Account of Reason and Faith: In Relation to the Mysteries of Christianity* (London, 1697), 327.

infinitum on the other side of the question. Some of them then go on, like Arnauld and Nicole, to draw a lesson in pious humility:

> This is why it is good to tire the mind on these subtleties, in order to master its presumption and to take away its audacity ever to oppose our feeble insight to the truths presented by the Church, under the pretext that we cannot understand them. For since all the vigour of the human mind is forced to succumb to the smallest atom of matter, and to admit that it clearly sees that it must be infinitely divisible without being able to see how this can be, is it not obviously to sin against reason to refuse to believe the marvelous effects of God's omnipotence, which is in itself incomprehensible, for the reason that the mind cannot understand them?[119]

Now, anyone who thinks, contra the current faction, that there *is* a genuine conflict between the actual parts doctrine and infinite m-divisibility, will of course find all this talk of non-lethal 'difficulties' a studied evasion. The suspicion is that genuine contradictions are being downplayed or explained away with pious reference to the puniness of our comprehension. The later Pierre Bayle (a reformed ex-member of faction (3)), George Berkeley, Arthur Collier, and David Hume each press this complaint. Collier, for instance, argues in his *Clavis Universalis* (1713) that the so-called 'difficulties' are full-scale *reductios* against the doctrine of infinite m-divisibility. He then replies explicitly to Arnauld and Nicole, who have 'drest up formal nothing into the shape of an objection... and that with an unusual air of gravity'.

> [F]or can anything be more evident than that finite and infinite are exclusive of each other; and that an idea which implies both is an impossibility in fact?... Should we doubt in this manner, if the subject spoken of were a circular square, or triangular parallelogram?... where then is the difficulty, supposed by the aforementioned author [he means the 'author' of *The Art of Thinking*, that is, Arnauld and Nicole], in the question about the divisibility of matter, &c. wherewith it is so good to fatigue our presumptuous minds?[120]

And Hume is no less disgusted:

> I doubt not it will readily be allow'd by the most obstinate defender of the doctrine of infinite divisibility, that these arguments [he has just presented his versions of paradox (1) and (3A)] are difficulties, and that 'tis impossible to give

---

[119] Arnauld and Nicole, *Logic*, 233. Malebranche, Boyle, and Watts also toe this line: see the references in n. 118.

[120] Arthur Collier, *Clavis Universalis* (1713; facsimile edn.: New York: Georg Olms Verlag, 1974), 54, 56. See also Bayle, *Dictionary*, 361–2; Berkeley, *A Treatise Concerning the Principles of Human Knowledge*, s. 129, in *Works*, ii. 100.

any answer to them that will be perfectly clear and satisfactory. But here we may observe, that nothing can be more absurd, than this custom of calling a *difficulty* what pretends to be a *demonstration*, and endeavouring by that means to elude its force and evidence.... A demonstration, if just, admits of no opposite difficulty; and if not just, 'tis a mere sophism, and consequently can never be a difficulty. 'Tis either irresistible, or has no manner of force. To talk therefore of objections and replies, of ballancing of arguments in such a question as this, is to confess, either that human reason is nothing but a play of words, or that the person himself, who talks so, has not a capacity equal to such subjects.[121]

One can certainly see their point. If the various classic paradoxes that stem from the actual parts doctrine together with infinite m-divisibility *are* indeed positive contradictions, then they can scarcely be avoided by talk of 'difficulties' and the limitation of our meagre capacities, much less by reminders of how forceful the case for infinite m-divisibility is. But, of course, all this crucially depends on whether the classic paradoxes *do* rise to positive contradictions—as the later Bayle, Berkeley, Collier, and Hume maintain—or whether they merely challenge our ability to see *how* certain things are possible, without positively showing that they are *impossible*—as Barrow, Keill, Arnauld, *et al.* maintain.

Faction (4). Affirm *both* the Actual Parts Doctrine *and* Infinite M-Divisibility; But Also Affirm that the Two are Irreconcilable—we Thus have a Fully Fledged Antinomy

This brings us to our fourth and final faction. This radical group endorses both the actual parts metaphysic and infinite m-divisibility, but then admits that the two doctrines are incompatible. We must accept both a thesis and at the same time its contradictory antithesis. Thus we have a genuine antinomy and the threat of a total crisis of reason. As Robert Boyle puts it in his 1681 *Discourse of Things Above Reason*, the problems of material structure appear to undermine 'the very foundation of our reasoning', or, in the words of Kant, the problems (along with his other three antinomies) menace us with 'the scandal of the apparent contradiction of reason with itself'.[122] It is worth quoting Boyle's confession of angst more fully:

---

[121] Hume, *Treatise*, 1. 2. 2. 6; SBN 31–2. See also G. W. Leibniz, 'Preliminary Dissertation', in his *Theodicy* (La Salle, Ill.: Open Court, 1985), 88–9 (ss. 24–5).

[122] Immanuel Kant, *Letter to Garve*, 21 Sept. 1798, quoted in D. A. Rees, 'Kant, Bayle and Indifferentism', *Philosophical Review*, 63 (1954). (Rees gives the following reference: *Werke*, Akad., Ausg., XII (1902) 255, letter 78.) Boyle, *Selected Philosophical Papers*, 235.

> [T]he truths I speak of appear not symmetrical with the rest of the body of truths, and we see not how we can at once embrace these and the rest, without admitting that grand absurdity which subverts the very foundation of our reasoning, *that contradictories may both be true*: as in the controversy about the endless divisibility of a straight line,... Upon which account I have ventured to call this... sort of things above reason *asymmetrical* or *unsociable*, of which eminent instances are afforded by those controversies (such as that of the *compositione continui*) wherein, which side soever of the question you take, you will be unable *directly* and truly to answer the objections that may be urged to show that you contradict some primitive or some other acknowledged truth.[123]

Pierre Bayle and Kant phrase the threat in a similar fashion. It seems that each opinion concerning the structure of matter in turn faces lethal objections. 'Each of these... sects,' Bayle writes, 'when they only attack, triumphs, ruins, and destroys; but in its turn, it is destroyed and sunk when it on the defensive.'[124] Kant lets this sort of martial metaphor rather run away with him in describing the threatened antinomy (along with the other three antinomies of his Dialectic):

> These subtly reasoning assertions... reveal a dialectical combat arena. There any party permitted to attack keeps the upper hand, and the party compelled to proceed merely defensively is certain to be defeated. Vigorous knights, by the same token, whether they pledge themselves to the good or to the evil cause, are sure to carry off the wreath of victory—provided they take care to have the prerogative of making the last attack and are not obliged to withstand a new onslaught by the opponent.... Perhaps, after having more exhausted than harmed each other, they will become aware of the nullity of their contest.[125]

We can usefully distinguish three subgroups in the current faction, according to whether they consider the antinomy, first, a localizable problem that does not endanger reason more generally, second, a problem that spills over and compromises our reason to a limited extent, or, third, a problem that totally humbles reason *tout court* and ushers in global scepticism.

The first subgroup allows that there is an unavoidable contradiction here, but treats it as a merely local problem that can, so to speak, be isolated behind a firewall. Now this strategy looks extremely unstable. If reason can

---

[123] Boyle, *Selected Philosophical Papers*, 235. [124] Bayle, *Dictionary*, 362.
[125] Kant, *Critique of Pure Reason*, B450–1.

lead to a contradiction over this issue, then why *wouldn't* this embarrass reason more generally? And in fact I can find no philosopher who straightforwardly belongs in this first subgroup. However, this is as good a place as any in my architectonic to mention those philosophers who think that the problems of material structure amount to full-scale contradictions, and thus conclude that matter cannot exist. They each hold that, in so far as matter exists in an external reality, it must both have actual parts and be infinitely m-divisible. But this leads to contradiction, as the various classic paradoxes show. So matter cannot exist in this alleged external realm; at best it can exist only in the mind. These philosophers, then (while not strictly belonging to this first subgroup), at least admit the following *conditional* claim. *If* matter existed in reality, it would both have actual parts and be infinitely divisible—and these two properties are incompatible. The antinomy is, if you like, 'firewalled' by denying that matter really exists. Reason's honour is upheld, but only at the cost of denying matter's external reality. The later Pierre Bayle endorses this line in his 1697 *Dictionnaire*. Faced with the unsolvable problems of material structure, we must admit that body 'can exist only *ideally*'.[126] Similarly Arthur Collier in his 1713 *Clavis Universalis*: 'Matter which is both finitely and infinitely divisible, is not at all. But this is the case of external matter. *Ergo*, there is no such thing as external matter.'[127] And William Drummond draws the same conclusion in his *Academical Questions* of 1805: the paradoxes show us that 'we can only understand extension, as sensible extension...I shall be found to have co-incided with the Bishop of Cloyne; and to have arrived by a different induction to a similar conclusion.'[128] Actually, it is not such a 'different induction' as Drummond thinks. In fact Bishop Berkeley also holds that the paradoxes can only be evaded by his idealist resolution (as we saw above in faction (2)). The later Kant should also be mentioned here: he thinks that his *critical idealism* offers the only way of avoiding the antinomy (see faction (1)).[129]

Now, one might wonder how denying that matter exists externally is supposed to help disarm the problems, if it is then maintained that matter

---

[126] Bayle, *Dictionary*, 363.   [127] Collier, *Clavis Universalis*, 50.
[128] Drummond, *Academical Questions*, 69–70
[129] This sort of argument from the paradoxes of infinite divisibility to idealism has been presented with characteristic eloquence by Jorge Luis Borges. See his 'Avatars of the Tortoise' in *Labyrinths* (New York: New Directions 1964), 208. See also 'The Perpetual Race of Achilles and the Tortoise', in his *Selected Non-Fictions* (New York: Viking Penguin 1999), 43–7.

still exists in the realm of ideas. Isn't there equally a problem of explaining the structure of bodies that exist only in this internal realm? In fact Bayle, Collier, and Drummond each fail to tell us exactly how they think idealism helps with these problems. But (to speculate) it is possible that they may have in mind Berkeley's answer: internal ideas of bodies are built up from m-indivisible *minima sensibilia* in a fashion that cannot be carried over as a coherent model for the structure of external bodies. Or perhaps they have in mind a variant of Kant's critical period solution: bodies considered as progressions of appearances have only potential parts, whereas a mind-independent 'matter-in-itself' would have to have actual parts.

The second subgroup allows that there is a fully fledged antinomy in the problems of material structure—reason *is* forced to admit a positive contradiction—and then claims that this embarrassment undermines reason's authority on *some* other issues. However, there is no claim that reason is totally undermined. A philosopher in this subgroup then endorses a certain sort of localized scepticism about the jurisdiction of reason: the paradoxes of material structure show that it cannot be trusted in some provinces, even though it may retain its authority elsewhere. This sort of localized scepticism is typically adopted in the context of Christian apologetics: the paradoxes of material structure prove insoluble, overreaching reason is thereby humbled, and its right to criticize theological claims (so the argument goes) is thereby checked. Perhaps reason can serve as a reliable guide if constrained to the mundane world of everyday life, but as the antinomy of infinite divisibility shows, it cannot reach far into the mysteries of metaphysics and theology without unravelling. Philosophers of this subgroup are thus somewhat similar to those of faction (3), such as Arnauld and Nicole, who drew pious lessons in humility from the 'difficulties' of infinite divisibility—except that the current subgroup admits a full-scale antinomy, rather than mere 'difficulties' that do not rise to positive contradictions.

While the later Henry More joins faction (3) and advances his theory of physical monads as a solution to the problems of material structure, the younger More thought that the problems amounted to an insoluble, fully fledged paradox. Whichever opinion you endorse, 'you will be wound into the most notorious Absurdities there may be'. And in the early *Antidote Against Atheism* (1653), More mobilizes the antinomy to undermine reason's right to challenge the Christian notion of *immaterial spirit*. Since reason is mired in contradiction even in the case of body—'of whose being we seem most assured'—why should we then trust its assertion that the notion of spirit is incoherent?¹³⁰ Locke himself also flirts with this peculiar strategy in the *Essay Concerning Human Understanding* (1689), defending immaterial spirit against reason's complaints by pointing—oddly enough, one might think—to the contradictions in the notion of material body.¹³¹

While More and Locke only question reason's authority to challenge the coherence of the concept of immaterial spirit, Pierre Bayle and Joseph Glanvill push this sort of local scepticism (or, better, local *fideism*) further. Bayle's handling of the problems of material structure is many-layered and spiced with irony, but his final position in the *Dictionnaire* (1697) seems to be as follows. First, he argues that, if matter exists in external reality, then there is a genuine antinomy. He therefore concludes that—if we are to be rigorous in our logic—matter does not exist externally (this is the idealist Bayle we just saw in faction (4) subgroup (1)). But then, at the close of the discussion, he confesses that, after all, he cannot truly give up his belief in external matter 'even though I feel myself completely incapable of resolving all the difficulties that have just been presented'. His final word is that there is a fully fledged antinomy, and that 'the exposition of these arguments can be of great service to religion'. Bayle quotes Arnauld and Nicole directly: the paradoxes of material structure 'check [reason's] presumption and ... keep it from being foolhardy enough to oppose its feeble light to the truths that the Church proposes on the pretext that it cannot understand them'.¹³² At least if we take all this at face value, it places Bayle squarely in the current subgroup of 'local fideists'. Lastly, Joseph Glanvill also claims in his *Vanity of Dogmatizing* (1661) that the problems of material structure rise to the level of full-scale paradox: an antinomy that he employs along with various other paradoxes to embarrass the pretensions

---

¹³⁰ More, *An Antidote Against Atheism*, in *Philosophical Writings*, 11, 12.

¹³¹ I say 'flirts' since Locke's language is somewhat evasive about whether the problems of material structure give rise to a positive contradiction or just 'difficulties': '[I]f this Notion of immaterial Spirit may have, perhaps, some difficulties in it, not easy to be explained, we have therefore no more reason to deny, or doubt the existence of such Spirits, than to deny, or doubt, the existence of Body; because the notion of Body is cumbred with some difficulties very hard, and perhaps impossible to be explained, or understood by us. For I would fain have instanced a thing in our notion of Spirit more perplexed, or nearer a Contradiction, than the very notion of Body includes in it; the divisibility *in infinitum* of any finite Extension, involving us, whether we grant or deny it, in consequences impossible to be explicated, or made in our apprehensions consistent'. John Locke, *An Essay Concerning Human Understanding*, ed. P. H. Nidditch (Oxford: OUP, 1975), II. xxiii. 31.

¹³² Bayle, *Dictionary*, 372.

of an overreaching speculative reason—'to shame confidence, and unplume *Dogmatizing*'.[133]

The position of this subgroup, in admitting an antinomy that embarrasses reason's authority only in certain restricted areas, is of course as thoroughly unstable as that of the philosopher who admits an antinomy and then would treat it as a completely isolated problem. If there *is* a genuine antinomy—a violation of the principle of non-contradiction—then *all* reasoning based on that principle is brought into question, and reason is undermined across the board. The strategy of this current subgroup, invoking the embarrassment of reason only where it helps to block rational criticism of Church-approved mysteries of faith, while at the same drawing back from a full-blown global scepticism, surely exhibits an obscurantist double standard.

The third and final subgroup of this last faction admits that the problems of material structure give rise to a genuine antinomy—a positive contradiction that reason cannot avoid—and then confesses that this result undermines reason's jurisdiction *globally*. If reason leads us to violate the principle of non-contradiction—its own most fundamental law—then it destroys *all* its own authority and 'subverts the very foundation of our reasonings' (as Boyle put the threat).[134] Reason's supposed sovereign right to govern all our judgements and inferences turns out to be self-defeating sham, and we are left with a cataclysmic epistemological collapse: a global form of scepticism about reason.

While I have suggested that this position, however calamitous, is at least much more stable than the various attempts to treat an admitted antinomy as an isolated or localizable problem (*à la* the first two subgroups), only a very few philosophers are ready to flirt with such a radically sceptical result. I just said that the later Pierre Bayle ends up endorsing a form of the *local* scepticism (or local fideism) of the previous subgroup. This is indeed where his explicit conclusion stops: the antinomy (he says) checks reason's right to challenge the Church's mysteries; but there is no express claim that it goes beyond this to undermine *all* of reason's authority. However, reading between the lines, I think it is arguable that Bayle intends the discerning reader to see that the antinomy threatens a much more global form of scepticism. Of course, it is always difficult to read through the veils of irony to Bayle's 'true' meaning. But this sort of total

---

[133] Glanvill, *Vanity of Dogmatizing*, 16, 53.   [134] Boyle, *Selected Philosophical Papers*, 235.

scepticism certainly chimes with his extreme Pyrrhonism in other sections of the *Dictionnaire*, where he is clearly and explicitly ready to cast doubt on reason across the board.[135] We have also seen that Robert Boyle is ready to toy with the idea that the problems of material structure present us with 'asymmetrical' or 'unsocial' truths that would menace *all* reason's dominion—though it should be said at once that Boyle rather leaves this conclusion dangling; he seems to simply put it to one side and move on with a shrug.[136] And finally, we should also remember that the later Kant, though not himself admitting a genuine antinomy, holds that all *transcendental realists* properly belong in this last subgroup. Rejecting transcendental idealism, they cannot avoid the antinomy (Kant says) and thus *ought* to admit the global collapse of reason if they were honest and consistent. (Of course, Collier and the other idealists of faction (4) subgroup (1) think much the same: anyone who admits the existence of external matter *ought* to realize that this leads to antinomy and the embarrassment of reason across the board.)

X. Outline of Book

The central goal of this first chapter was to introduce the reader to the Enlightenment debate on the problems of material structure and to defend a particular interpretation of that debate. I hope to have brought out the nature of these problems by way of my rational reconstruction of the paradoxes and their hierarchical interrelations. I also wanted to establish that the paradoxes are inescapably metaphysical, and that the underlying controversy turns, not on purely mathematical problems but rather on the tension between infinite m-divisibility and the actual parts metaphysic. Finally, I hope to have shown the various types of reaction to the problems of material structure that one finds in the Enlightenment literature, and to have brought out the pivotal questions that divide these various factions.

The rest of the study proceeds as follows. In Chapter 2 I present the rival actual parts and potential parts doctrines in more detail, and examine

---

[135] See, for instance, the article on 'Pyrrho' in Bayle, *Dictionary*, 194–209, esp. 199. For commentary, see Richard H. Popkin, *The History of Scepticism: From Savonarola to Bayle* (Oxford: OUP, 2003), ch. 18.

[136] Robert Boyle, *Selected Philosophical Papers*, 235–6. For commentary, see Jan W. Wojcik, *Robert Boyle and the Limits of Reason* (Cambridge: CUP, 1997), ch. 4 and 7.

the arguments that are made for and against each of these accounts. In the current chapter we saw that the actual parts metaphysic was widely thought to entail a pair of corollaries that underpin the classic paradoxes of infinite divisibility: first, that bodies have determinate numbers of parts; and, second, that they have ultimate parts. And in Chapters 3 and 4 I flesh out this claim, exploring the traditional arguments from actual parts to each of these corollaries. Chapter 3 addresses *the argument from actual parts to a determinate number of parts*, and *the argument from the definiteness of parts to ultimate parts*. Chapter 4 examines a different sort of argument from actual parts to ultimate parts: *the argument from composition*, which maintain that bodies with actual parts must resolve to metaphysical atoms on pain of vicious ontological regress. So Chapters 2, 3, and 4 examine, first, the case for the actual parts metaphysic, and, second, the case that this metaphysic leads to tension with infinite m-divisibility. I then turn in Chapter 5 to examine the other side of the conflict, the case for infinite m-divisibility, in particular, *the argument from geometry* and *the conceptual argument*. This completes the main project of the book. Along with the current introductory chapter, then, Chapters 2 to 5 jointly present a critical reconstruction and assessment of the basic conflict at the heart of the Enlightenment travails over material structure: the apparent clash between the actual parts metaphysic and the geometrization of nature.

Chapter 6 stands as something of an addendum. I have already outlined the framework of the various possible responses to the apparent conflict (in section IX of the current chapter). In this final chapter I explore in more detail one of the proposed resolutions that seems to me particularly intriguing philosophically: the field-theoretic response presaged by Henry More and developed by Boscovich and the pre-critical Kant. This approach is both interesting from a historical point of view, constituting an exceptionally early version of the field metaphysic, and raises a host of philosophical questions about the ontological status of fields, about their structure and occupancy of space, and about action at a distance.

Finally, I sum up in the Conclusion. Here I review the various results of the early modern debate over material structure, and reflect on both the historical lessons and the broader philosophical morals of the controversy.

APPENDIX
PARADOXES OF INFINITE M-DIVISIBILITY THAT DO *NOT* TURN ON ACTUAL PARTS

Much of Chapter 1 was given over to arguing that the most popular, most important, and most worrisome paradoxes of infinite m-divisibility—what I have been calling the 'classic paradoxes'—arise only within the framework of the actual parts metaphysic. They are not general problems for the notion of infinite divisibility *in abstracto*. But this does leave some paradoxes—'non-classic paradoxes'—also found in the Enlightenment literature that do *not* depend on the actual parts metaphysic. These paradoxes are motivated purely by the notion of endless divisibility, and do not turn on any particular metaphysical account of the status of the parts of matter. In this appendix I want, simply for the sake of completeness, to briskly enumerate these non-classic paradoxes. As we shall see, not only are the non-classic paradoxes much less popular in the Enlightenment literature than the classic paradoxes I have already set out in section VIII, they are also much more easily disarmed and of much less intrinsic philosophical interest.

1. The Sheer Mind-Boggling Nature of Infinite Divisibility

Some period philosophers rejected infinite divisibility simply because an endless division into smaller and smaller parts with no stopping place was too staggering—too mind-boggling—to accept. No formal contradiction is adduced, but it is simply asserted (typically with a rhetorical flourish invoking bewilderingly large numbers) that infinite divisibility beggars belief. Walter Charleton stands as a good example here, writing in his 1654 *Physiologia* as follows:

What Credulitie is there so easie, as to entertain a conceit, that one granule of sand (a thing of very small circumscription) doth contain so great a number of parts, as that it may be divided into a thousand millions of Myriads; and each of those parts be subdivided into a thousand millions of Myriads; and each of those be redivided into as many; and each of those into as many: so as that it is impossible, by multiplications of Divisions, ever to arrive at parts so extremely small, as that none can be smaller; though the subdivisions be repeated every moment, not only in an hour, a day, a month, or a year, but a thousand millions of Myriads of years?[137]

[137] Charleton, *Physiologia*, 92; see also 91, where the rhetoric is even more torrid.

A similar complaint is reported (though not endorsed) by Pierre Bayle and, here, by Jacques Rohault: 'it would from [the infinite m-divisibility of matter] follow, that a small Portion of Matter, such as a Cube, a quarter of an Inch high, might be divided into as many thin square Pieces, as would cover the whole Globe of the Earth... which, they [opponents of infinite divisibility] think, is absurd'.[138] Rohault and Bayle, along with Hobbes, make the obvious reponse.[139] This is really no argument at all, since the advocate of infinite divisibility will readily admit this sort of interminable division as unproblematic. Some *reason* needs to be given why infinite divisibility is inadmissible, not just a reminder that it is a somewhat staggering thesis.

2. If Bodies are Infinitely Divisible, Then they should Ipso Facto All be the Same Size

I have already set out this paradox, when contrasting it with classic paradox (3B) with which it is often confused. Classic paradox (3B) had it that, if infinitely divisible bodies have actual parts, then they must each have an actual infinity of ultimate parts and thus should all be the same size. By contrast, the current 'non-classic' paradox directly infers from the fact that any given number of parts can be found in an infinitely divisible body to conclude that all such bodies should be the same size as one another. This argument is clearly invalid. I refer the reader to my earlier discussion for my full reconstruction, for my critique, and for the various textual sources (see Chapter 1, section VII, paradox (3B) above).

3. If Bodies are Infinitely Divisible, they will have No Extreme Parts to Terminate their Extent

Suppose a body (or indeed any extended continuant) to be infinitely divisible. Then try to identify a last part of that body that terminates its extent. For instance, imagine a 12″ ruler with one end pointing west and one end pointing east, and then try to isolate its last, easternmost part. One might begin by

---

[138] Rohault, *System*, 34. See also Bayle, *Systême*, in *Œuvres Diverses*, iv. 303.
[139] Rohault, *System*, 34–7. Bayle, *Systême*, in *Œuvres Diverses*, iv. 303. Thomas Hobbes, *Works* i. 476–7.

pointing to the last inch-long section at the end of the ruler. But this is clearly not the *last* part, since it has subparts, some of which are to the east of others. Suppose one then point to the easternmost half inch? Again, the same problem. The easternmost hundredth inch? Thousandth inch? The same problem, ad infinitum. So it seems that, if a body is infinitely divisible, it will altogether lack terminating last parts that define its final extent, and that could be in contact with adjacent bodies. And this, according to the current objection, is absurd. This argument is endorsed by both Pierre Bayle and David Hume, each of whom suggests it was a traditional paradox of infinite divisibility well known to scholastics.[140]

One rather drastic (though apparently fairly common) response to this argument was to add extensionless points, lines, and surfaces mixed in throughout the infinitely divisible bulk of each body: as Bayle says, the paradox 'forced some of the School philosophers to suppose that nature has mixed some mathematical points in with the infinitely divisible parts to serve as connections between them and to make up the extremities of bodies'.[141] One finds this response in the *Disputationes Metaphysicae* of Fransisco Suarez, for instance (and, in an updated form, in Franz Brentano).[142] Bayle and Hume each dismiss this sort of stratagem without serious argument—it is, says Bayle, a 'subterfuge...so absurd that it does not deserve to be refuted.'[143] I am not sure that this is altogether fair, and it is at least worth noting that those proponents of infinite divisibility who admit an actual infinity of elemental parts (such as Galileo) already have point parts in their system that could readily serve as final extremities.

Whether or not Suarez's response deserves such short shrift, there is in any case a much more straightforward answer to the current paradox.[144] One could simply deny that bodies need to have final 'terminating' parts at all. Why think that the ruler needs to have an absolutely last part in the manner supposed by Bayle and Hume? This is simply asserted, but without good reason. Perhaps it is thought that we need a last part—a surface, line, or point—in order to account for the possibility of contact with adjacent bodies. Only such a last part, altogether without depth, could be in full, direct contact with another body, such that *all* of it

---

[140] 'And how could [a body] touch [another body], since all the parts that you might consider to be the last contain an infinite number of parts, and that infinite number has no part that might be the last?' Bayle, *Dictionary*, 370. See also Bayle, *Système*, in *Œuvres Diverses*, iv. 303. See also Hume, *Treatise*, 1. 2. 4. 14; SBN 43–4.

[141] Bayle, *Dictionary*, 370.

[142] On Suarez and Brentano, see Zimmerman's extremely useful account in 'Indivisible Parts', 158–60.

[143] Bayle, *Dictionary*, 370. Hume, *Treatise*, 1 2. 4. 15; SBN 44.

[144] Here I follow Dean W. Zimmerman's treatment of the current paradox. 'Indivisible Parts', esp. 160–1.

immediately touches the other body. But why think that this direct, whole-to-whole contact is required? It does not seem necessary. Why not just say that the ruler is in contact with another body if, for instance, a part of the ruler (say its easternmost inch section) and a part of the other body (say its westernmost inch section) are such that no third body, no matter how thin, could fit between them? This seems to be a perfectly natural account of what it is for to bodies to be in contact that avoids the need to posit terminating parts and whole-to-whole 'full' contact altogether. So there is no need to invoke final terminating parts to explain contact, and, absent any other need for them, we can dismiss this alleged paradox.

2

Actual Parts and Potential Parts

> Therefore we'll pass to the next, the consideration of *Bodies*, which though we see, and feel, and continually converse with; yet its constitution, and inward frame is an *America*, a yet undiscovered Region.
>
> Joseph Glanvill, The Vanity of Dogmatizing, 1661

I. Introduction

This chapter addresses the conflict between two rival theories of the internal structure of matter and the ontological status of its parts. The one theory is central to the mechanistic world-view of Galileo and Newton, and is overwhelmingly popular with partisans of the new science. The other is the classical doctrine of Aristotle, St Thomas, and the schools. Interestingly, new variants of the scholastic theory are also sponsored by Hobbes and the later Kant, who stand opposed to the mainstream of the new science on this question.

The first of our rival theories is the doctrine of actual parts, according to which the parts of bodies are so many distinct existents. These concrete parts are all already embedded in the architecture of the whole: division merely *separates* them, it does not *create* them. The whole body is an aggregate of these independently existing parts, and its structure is best described with count terms: it is a composite of *so many* distinct parts, so many independent beings. The second of our rival theories is the doctrine of potential parts, according to which the parts of bodies are merely possible or potential existents until they are generated by division. They represent ways in which the whole could be broken down, but do not exist other than as aspects of the whole until a positive act of division actualizes them as so many independent entities. Here the whole is a metaphysically simple (noncomposite) structure best described with mass terms: prior to division, it is not so many distinct things, but rather just *so much* metaphysically undifferentiated material.

I proceed as follows. Section II sets out the actual parts doctrine in detail. Here I examine the content of the doctrine and the various corollaries that were commonly thought to follow from it: for instance, that bodies cannot be substances, and that they must have an ultimate constitution of m-indivisible first parts. I also show the theory's popularity with the new philosophers of the Enlightenment and its connection to the new science. Section III sets out the potential parts doctrine and its various corollaries. I connect this theory to the scholastic tradition and examine the new variants of it found in Hobbes and the later Kant.

I then turn to assess the conflict between the system of actual parts and the system of potential parts. Section IV assesses the case for actual parts; section V, the case for potential parts.

Finally, in section VI, I sum up and pass judgement on the competing theories. My conclusion is that, for all the ferocity of the clashes between their respective champions, and notwithstanding the ultimate *de facto* historical triumph and near-universal acceptance of the actual parts account in the eighteenth century (and its continuing popularity in analytic metaphysics today), the arguments for the two theories may in fact be deadlocked in a symmetrical stalemate. The structure of this standoff suggests that there may be something bogus about the entire conflict.

II. The Actual Parts Doctrine

According to the *actual parts doctrine*, the parts into which a material body can be m-divided (that is, its logically separable parts, the parts that God could break it into) are *actual parts*. Actual parts are parts that are each a distinct being: each exists independently of the whole, independently of the other non-overlapping parts, and each exists independently of and prior to any act of division. The basic idea is that a body has actual parts in so far as it is built up from these concrete, independently existing elements. (Any given part may of course be dependent on the parts composing it, but it will be independent of the whole and independent of those parts it does not (partially or wholly) overlap.)

More formally, a material body has actual parts *if and only if*:

(1) It can be m-divided into parts, *and*
(2) those parts are each logically independent of the whole and of the other non-overlapping parts, and each exists already without any causal or logical dependence on any act of division (be it *p-division* through natural processes, *m-division* by God, or *i-division* in thought).

Thus if there are any m-indivisible parts of matter (such as More's 'physical monads' or Boscovich's force-shell atoms), these will lack actual parts because they fail to satisfy (1). (Notice also that entities like More's and Clarke's 'indiscerpible' extended spirits lack actual parts for this same reason, as does Newtonian absolute space. Such immaterial entities are not really our concern here however: we are focusing on matter.) Aristotelian substances, on the other hand, will lack actual parts because they fail to satisfy (2): they are m-divisible, but their m-divisible 'potential' parts, being identity-dependent on the whole matter–form compound, are not independent existents until they are actualized by division.

The actual parts doctrine is then asserting a thesis about the *ontological* or *metaphysical* status of material parts. The doctrine must not be confused with the view that bodies have parts that are differentiated by any heterogeneous *physical* organization or structure. For all that has been said here, bodies might be either highly complex or perfectly homogeneous in their internal *physical* architecture. In Chapter 1, section IV, I gave the example of a block of jelly that is homogeneous jelly-gloop 'all the way down': continuously solid and uniformly dense, altogether lacking any physically demarcated subparts. According to the current doctrine, the top half and bottom half of this block of jelly are each a distinct being (as are all the other parts into which the whole can be divided), notwithstanding their lack of any physical demarcation. So at the metaphysical level this physically homogeneous block of jelly is a composite of so many ontologically distinct parts. And here we can add a second, classic example (invoked in both the early modern period and in more recent discussions). On the actual parts view, it is really just a *façon de parler* to speak of Michelangelo as the creator of the *David*. This particular shapely block of marble existed long before Michelangelo laid hands on his chisel—as did its subparts and indeed numerous other perfect overlapping copies. So the Florentine did not *create* it. Rather it is just that Michelangelo's genius enabled him to bring this particular block to light, rather than, say, one of the less attractive blocks that also existed.

The actual parts doctrine can perhaps be brought out more clearly if we consider what it is denying. According to the theory, each part into which a body can be divided exists *independently* of the other parts and of the whole. So these parts are not somehow properties—aspects, features, modes—of the whole. Moreover, they are fully fledged concrete existents *prior* to any act of division. So they are not merely possible or potential existents, awaiting actualization through some positive act of division. They do not merely express possible ways in which division could *generate* parts; rather, they express an ontological structure already laid up in the whole—a structure of so many distinct, pre-existing beings that is unveiled (not created) by division.

The Actual Parts Doctrine, Arbitrary Undetached Parts, and Scattered Objects

This way of thinking about the status of material parts may well be familiar to the reader conversant in today's analytical mereology and metaphysics. Not only is the actual parts system dominant in the early modern period, it is also a forebear of a theory that continues to thrive in current metaphysics: the so-called 'doctrine of arbitrary undetached parts'. The doctrine of arbitrary undetached parts asserts that, for any given material object at a given moment in time, if R is the region of space occupied by the object at that moment, then there exists a concrete material object in each and every occupiable subregion of R at that same moment. So on this view each and every arbitrary undetached part of a given material object is itself a concrete material object and an independent being in its own right.[1]

How does the doctrine of actual parts differ from the doctrine of arbitrary undetached parts? Well, I have framed the actual parts doctrine in terms of the *divisibility* of the overall body: each part into which the whole can be m-divided exists independently of all the other parts. (And in focusing on divisibility, I have simply followed the way in which most early modern philosophers approach the issue.) On the other hand, the

---

[1] Peter van Inwagen, 'The Doctrine of Arbitrary Undetached Parts', *Pacific Philosophical Quarterly*, 62 (1981), 123–37. Richard Cartwright, 'Scattered Objects', in Keith Lehrer (ed.), *Analysis and Metaphysics: Essays in Honor of R. M. Chisholm* (Dordrecht: D. Reidel, 1975), 153–71, esp. 160–4.

doctrine of arbitrary undetached parts is framed in terms of *occupiable subregions*: for each such subregion filled by the overall body there is a part that exists independently. So the two doctrines will track each other so long as the geography of parts that can be m-divided from one another tracks the geography of occupiable subregions within the space occupied by the overall body. And many will maintain that this is the case all the way through the whole body: for each arbitrary subregion within the overall region filled by a body, it looks like there is a part that can be m-divided from the whole; and vice versa. In fact the two doctrines will only come apart if one thinks that there might be extended but m-indivisible parts of matter. For at this point partisans of the doctrine of arbitrary undetached parts are presumably still committed to saying that there are distinct (but undetachable) beings within these m-indivisible but extended atoms; but the actual parts doctrine does not imply this, since here m-divisibility has reached its limit. But whatever one makes of this subtle divergence between the two doctrines, the main point here is simply that the doctrine of arbitrary parts is at least some sort of descendant of the actual parts theory: the two doctrines are clearly closely united in spirit and represent much the same metaphysical outlook on the status of material parts.[2]

It is also worth mentioning one other connection between the early modern actual parts system and issues that are current in today's mereology and metaphysics. In many respects the debate over actual or potential parts mirrors the debate in current mereology over (what we might call) actual or potential *wholes*. Consider two or more chunks of matter that are scattered apart from one another in space. Does there *already* exist an 'actual whole' consisting of all these various disconnected chunks taken together (a so-called 'scattered object')? Or should we think in terms of a 'potential wholes' system where a whole composed of these chunks does not exist until all the various chunks are brought together to form one

---

[2] Having pointed out this difference of detail between the official versions of the two doctrines, I should stress that statements of the doctrine of arbitrary undetached parts in the current literature (from which my statement is borrowed) are generally admitted to be quite rough. In fact I think it probable that those who discuss the doctrine of arbitrary undetached parts either assume that extended but m-indivisible parts of matter are impossible (in which case the two doctrines track one another after all), or implicitly intend only to address the nature of *detachable* but undetached parts. On this latter interpretation, an advocate of arbitrary undetached parts means only to assert the existence of those arbitrary undetached parts that can be detached (i.e. m-divided from the whole), and so the doctrines will once again coincide.

continuous block, thereby actualizing this whole? I will not be exploring the parallels between actual parts and scattered objects any further here, since I am analysing the early modern debate, which focuses almost exclusively on parts. But the reader will notice that many of the arguments in the actual parts–potential parts debate will have their sister arguments in the debate over scattered objects.[3]

The Actual Parts Doctrine and the New Science

The actual parts doctrine was overwhelmingly popular with partisans of the new science of the Enlightenment. It was championed by apostles of the new philosophy during the seventeenth century and played a central role in the overthrow of the rival scholastic system. By the turn of the eighteenth century, with the success of the new science and the retreat of scholasticism, it had achieved the status of a first principle or self-evident axiom for most natural philosophers, and as that century went on it seems to have become increasingly an implicit and perhaps unconscious background presupposition. In many ways it is this theory of matter that, along with the new mechanistic model of causation and the geometrization of nature, enshrined the new world-view of Galileo and Newton. I will substantiate this fully in a moment. But first let me begin with four exceptionally explicit statements of the doctrine, each drawn from an advocate of the new philosophy.

First we have Sir Walter Charleton's 1654 tract *Physiologia*, an extended defence of Epicurean atomism and a direct influence on Newton's early metaphysics.

> Those things which can exist being actually separate; are really distinct; but Parts can exist being actually distinct, even before division. For Division doth not give them their peculiar Entity and Individuation, which is essential to them and the root of Distinction.[4]

The doctrine is also clearly set out in the *Mathematical Lectures* of Sir Isaac Barrow, the great seventeenth-century mathematician and Newton's predecessor in

---

[3] For more on actual wholes, see Cartwright's classic 'Scattered Objects', 153–71.

[4] Walter Charleton, *Physiologia Epicuro-Gassendo-Charletoniana* (1654; London and New York: Johnson Reprint Co., 1966), 109. On Charleton's influence on Newton, see J. E. McGuire and Martin Tamny, *Certain Philosophical Questions: Newton's Trinity Notebook* (Cambridge: CUP, 1983), 27, 34.

the Lucasian chair at Trinity. The *Lectures*—given at Cambridge in the 1660s and widely read when published in Latin in 1683 and in English in 1734—are the source of my metaphor of the actual part 'unveiled' by division.

[A]ll imaginable Geometrical Figures are really inherent in every Particle of Matter; I say really inherent in Fact (*actu*) and to the utmost Perfection, though not apparent to the Sense; just as the Effigies of *Caesar* lies under the unhewn Marble, and is no Thing made by the Statuary, but only is discovered and brought to Sight by his Workmanship, *i.e.* by removing the Parts of Matter which involve and overshadow it.... So if the Hand of an Angel (at least the Power of God) should think fit to polish any Particle of Matter without Vacuity, a Spherical Superfice would appear to the Eyes of a Figure exactly round; not as created anew, but as unveiled and laid open from the Disguises and Covers of its circumjacent Matter.[5]

The unveiled parts are, moreover, genuinely independent entities, for

Divisibility is the inseparable Companion of Composition.... Magnitude is composed of Parts really distinct; these may exist asunder, at least be considered separately, *i.e.* they may be really or mentally dissolved or resolved into Parts.[6]

I have already presented the next two statements of the actual parts doctrine back in Chapter 1, when setting out the quick précis account of the theory. But these exceptionally explicit statements will bear a quick review here. The first is from 1707 and Newton's lieutenant Samuel Clarke.

[S]ome of the *first* and *most obvious* Properties which we *certainly know of Matter* [include] having partes extra partes, strictly and properly speaking, that is, its consisting of such Parts as are *actually unconnected*, and are *truly distinct Beings*, and *can* (as we see by Experience) *exist separately*, and *have no dependance one upon another*.[7]

The second is from the brisk hand of Thomas Reid in 1785:

There seems to be nothing more evident than that all bodies must consist of parts, and that every part of a body is a body and a distinct being, which may exist without the other parts.... when [matter] is divided into parts, every part is a being or substance distinct from all the other parts, and was so even before the division.[8]

---

[5] Isaac Barrow, *The Usefulness of Mathematical Learning Explained and Demonstrated*, tr. John Kirkby (1734; facsimile edn.: London: Frank Cass, 1970), 76–7. Emphasis added.

[6] Ibid. 151. See also 148–9, 150, 162.

[7] Samuel Clarke, *The Works of Samuel Clarke*, 4 vols. (London: John and Paul Knapton, 1738; facsimile edn.: New York: Garland, 1978), ii. 762.

[8] Thomas Reid, *Essays on the Intellectual Powers of Man*, in *The Works of Thomas Reid*, ed. Sir William Hamilton, 2 vols. (1863; facsimile edn.: Bristol: Thoemmes Press, 1994), i. 323.

I pick out these four statements of the actual parts doctrine partly because they are so clear and unambiguous. But I also select them because they are drawn from figures that are representative of the main tradition of the new science, from the Epicurean revival of the first half of the seventeenth century to the triumph and universal acceptance of Newtonian physics in the eighteenth. The actual parts doctrine is quite orthodox in this dominant tradition in early modern physics and metaphysics, and is ratified by nearly all the new philosophers of this period. In fact the four statements of the doctrine cited above are merely the tip of an iceberg. First, the doctrine is endorsed by philosophers representative of the two great systems within the new science: the system of Descartes[9] and the Cartesians[10] on the one hand, and the system of Newton[11] and the

---

[9] 'I consider the two halves of a part of matter, however small it may be, as two complete substances.' '[I]f it [extended or corporeal substance] exists, each and every part of it, as delimited by us in our thought, is really distinct from the other parts of the same substance.' (René Descartes, *The Philosophical Writings of Descartes*, tr. and ed. John Cottingham, Anthony Kenny, Dugald Murdoch, and Robert Stoothoff, 3 vols. (Cambridge: CUP, 1991), iii. 202–3, i. 213). It is true that, for the purposes of his dynamics, Descartes holds that the parts of matter are individuated by their relative motion, such that the rupture and separation of a previously undifferentiated portion of matter creates two distinct bodies from one. (Descartes, *Philosophical Writings*, i. 233. Also see Dennis Des Chene's *Physiologia: Natural Philosophy in Late Aristotelian and Cartesian Thought* (Ithaca, NY: Cornell University Press, 1996), 370–2, for a clear account of this strain in Descartes's thinking.) And this may seem less like an actual parts account and more like a potential parts analysis where division creates rather than unveils parts. But this account applies merely to dynamics. At the metaphysical level—the level of individuation into 'really distinct' substances or independent beings, rather than the merely physically unified beings that concern dynamics—Descartes consistently maintains an actual parts account.

[10] See e.g. the leading Cartesian popularizers Jacques Rohault and Antoine Le Grand. Jacques Rohault, *System of Natural Philosophy*, tr. John Clarke (1723; facsimile edn.: New York: Johnson Reprint Co., 1969), 31. Antoine Le Grand, *An Entire Body of Philosophy*, tr. Richard Blome, 2 vols. (London, 1694), i. 99, 97.

[11] I cannot find an explicit statement of the doctrine in the mature Newton's published writings, though it is strongly suggested in a draft written around the period of the *Principia Mathematica* 2nd edn. See J. E. McGuire, 'Newton on Place, Time and Space: An Unpublished Source', *British Journal for the History of Science*, 11 (1978), 114–29, 117. Compare also the early 'De Gravitatione et Aequiponidio Fluidorum', in Isaac Newton, *Unpublished Scientific Papers of Isaac Newton*, tr. and ed. A. R. Hall and M. B. Hall (Cambridge: CUP, 1962), 132–3. The doctrine is clearly stated in Clarke's 1716 correspondence with Leibniz, which was famously written under Newton's close guidance: 'Parts, in the corporeal sense of the word, are separable, compounded, ununited, independent of and moveable from, each other.' (Clarke's Fourth Reply to Leibniz, paras. 11–12, in Samuel Clarke and G. W. Leibniz, *The Clarke–Leibniz Correspondence*, ed. H. G. Alexander (Manchester: Manchester University Press, 1956), 48.) And Newton deploys an argument that implicitly assumes an actual parts account in his early notebooks

Newtonians[12] on the other. Second, the overwhelming majority of philosophers outside of these two main factions also accepted the actual parts account—including Galileo,[13] Gassendi,[14] Cambridge Platonists like Henry More[15] and Ralph Cudworth,[16] Pierre Bayle,[17] Berkeley,[18] Christian Wolff,[19] Hume,[20] Roger Boscovich,[21] Leonhard Euler[22], and the pre-critical Kant.[23]

---

from the 1660s: Isaac Newton, 'Certain Philosophical Questions', tr. and ed. J. E. McGuire and Martin Tamny, in McGuire and Tamny, *Certain Philosophical Questions*, 330–48C, 341.

[12] We have already seen an endorsement of the doctrine from Newton's great advocate Samuel Clarke in the main body of the text and (in n. 11) from the correspondence with Leibniz. See also Samuel Clarke, *Works*, iii. 761, 790, 850, 891, and Clarke, *A Demonstration of the Being and Attributes of God and Other Writings*, ed. Ezio Vailati (Cambridge: CUP, 1998), 116. Compare also the Newtonian theologian William Wollaston, *The Religion of Nature Delineated* (1724; facsimile edn.: Delmar, NY: Scholars' Facsimiles and Reprints, 1974), 74.

[13] Galileo seems to be implicitly committed to the actual parts doctrine when he argues that, if something is divisible ad infinitum, then—prior to division—it must already contain an infinite number of parts. See Galileo Galilei, *Dialogues on Two New Sciences*, tr. Henry Crew and Alfonos de Salvio (New York: Macmillan, 1914), 34. On Galileo's commitment to actual parts, see my discussion in Ch. 3 s. VI.

[14] Gassendi endorses the same argument as Galileo (see n. 13 above) and thus seems to be implicitly committed to actual parts. Lynn Suminda Joy, *Gassendi the Atomist* (Cambridge: CUP, 1987), 151–2, 155–6.

[15] Henry More endorses the same argument as Galileo (see n. 13 above) and thus seems to be implicitly committed to actual parts. Henry More, *An Antidote Against Atheism*, in *Philosophical Writings of Henry More*, ed. F. I. Mackinnon (New York and Oxford: OUP, 1925), 11.

[16] Ralph Cudworth, *True Intellectual System*, 3 vols. (London: Thomas Tegg, 1845); iii. 390, 393, 394.

[17] Pierre Bayle, *Historical and Critical Dictionary*, tr. and ed. Richard H. Popkin (Indianapolis: Hackett, 1991), 138, 302, 304, 305, 306, 338, 360. *Système Abrégé*, in Pierre Bayle, *Œuvres Diverses*, 5 vols. (1731; facsimile edn.: Hildesheim: Georg Olms, 1968), iv. 297, 299.

[18] George Berkeley endorses the same argument as Galileo (see n. 13 above) and thus seems to be implicitly committed to actual parts. *The Works of George Berkeley*, ed. A. A. Luce and T. E. Jessop, 9 vols. (London: Thomas Nelson & Sons, 1948–9), i. 42, ii. 283; see also v. 160.

[19] Christian Wolff, *Cosmologia Generalis*, s. 119, *Ontologia*, s. 531, in Wolff, *Gesammelte Werke*, ed. Joannes Ecole (Hildesheim: Georg Olms Verlag, 1962).

[20] On Hume's commitment to actual parts, see my discussion in Ch. 1, s. VI.

[21] Roger Boscovich, *A Theory of Natural Philosophy*, tr. J. M. Child (Cambridge, Mass.: MIT Press, 1966), ss. 83–4.

[22] Leonhard Euler, *Letters of Euler on Different Subjects in Natural Philosophy* (New York: Arno Press, 1975), 59.

[23] Immanuel Kant, *Physical Monadology*, in *The Works of Immanuel Kant: Theoretical Philosophy 1755–70*, tr. and ed. David Walford and Ralf Meerbone (Cambridge: CUP, 1992), 1: 473–87, 1: 477. In the later, critical period Kant will back off from the actual parts doctrine and endorse the potential parts account instead: see Kant's *Critique of Pure Reason*, tr. Werner S. Pluhar (Indianapolis: Hackett, 1996), B333, B552, and *Metaphysical Foundations of Natural Science*, in Immanuel Kant, *Philosophy of Material Nature*, tr. and ed. J. W. Ellington (Indianapolis: Hackett, 1985), 54.

(Leibniz also endorses a position that is very close to the actual parts account.²⁴) Now, I certainly do not wish to overstate the similarities among this diverse array of philosophers or to underestimate their many deep differences concerning fundamental issues in the metaphysics of matter. But these profound differences make it all the more interesting to note that these *lumières* of Enlightenment natural philosophy are indeed one and all united in their endorsement of the actual parts doctrine.

Four Widely Accepted Corollaries of the Actual Parts Doctrine

The actual parts doctrine was traditionally thought to entail the following four propositions as corollaries. (Not every partisan of the actual parts doctrine accepted each of these supposed entailments—though each inference certainly enjoyed wide and significant support.)

1. *Bodies are compounds*

The actual parts doctrine states that the parts of bodies are each fully fledged distinct entities. This implies that the whole gross extended body is a compound or composite, a structured aggregate of these pre-existing, independently existing parts. Since each actual part is a distinct entity, the whole must be conceptualized as a composite structure, a compound aggregated from ontologically prior concrete elements. Those who endorse the actual parts doctrine generally admit this first corollary.²⁵

(Notice that this corollary connects the actual parts system to a popular and well-known early modern argument against materialism: since any

---

[24] Leibniz certainly accepts that bodies have actual parts: 'A body is not a substance but an aggregate of substances, since it is always further divisible, and any given part always has another part, to infinity.' ('Notes on Some Comments by Michel Fardella' (1690), in Gottfried Wilhelm Leibniz, *Philosophical Essays*, tr. and ed. Roger Ariew and Daniel Garber (Indianapolis: Hackett, 1989), 103.) But he does not quite endorse the actual parts doctrine. For Leibniz each actual part is either the body of a living organism or a composite of such organic bodies, and so what shapes and sizes of actual parts one finds will depend on the shapes and sizes of these physically organized living structures. So Leibniz agrees that material bodies are composites and that there are actual parts all the way down—but he does not accept that there are *arbitrary* actual parts, actual parts answering to every possible m-division.

[25] For instance, when Pierre Bayle writes that 'The idea of matter [is] that of a composite being, that of a collection of several substances' (Bayle, *Dictionary*, 306). And Samuel Clarke writes that matter is 'a Being that consists in a Multitude of separable and distinct Parts' (*Works*, iii. 790). See also the quotation from Leibniz in n. 24.

material body is a collection or aggregate rather than a unified simple being, no purely material system can support the unity of consciousness; therefore the mind must be identified with a simple immaterial substance rather than with any material body. This famous argument presupposes an actual parts framework.[26])

2. Bodies are not substances

A second important corollary follows from this first point. Since extended entities constructed from actual parts are aggregates, they fail to meet the traditional criteria for substancehood in two ways. First, as aggregates, their existence is a derivative one, depending on the ontologically prior existence of their parts. (Compare an army: a paradigmatic aggregate whose existence is derivative on the prior existence of its members.) In this sense, bodies are not substances but rather modes of their parts. Second, as aggregates, bodies lack the unity typically required for substancehood. Thus those who follow through the actual parts view to this corollary will deny that they are substances in full metaphysical rigour: at best they are collections of substances—in Leibniz's phrase '*substantiata*', 'of-substances'.[27]

While corollary (1) is accepted by all philosophers who endorse the actual parts doctrine, this second corollary is somewhat less popular. (Clearly whether or not one accepts this second corollary will depend on how demanding a set of criteria for substancehood one endorses.) But the inference is certainly prominent in the writings of some figures.[28] This line of reasoning is famously pressed furthest in the rational cosmology of

---

[26] For a detailed historical account of this argument see Ben Lazare Mijuskovic, *The Achilles of Rationalist Arguments: The Simplicity, Unity, and Identity of Thought and Soul from the Cambridge Platonists to Kant* (The Hague: Martinus Nijhoff, 1974).

[27] Leibniz, *Philosophical Essays*, 200, 204, 205, 227.

[28] The Cambridge Platonist Ralph Cudworth, for instance, writes that the nature of corporeal substance 'is nothing but *aliud extra aliud*, "one thing without another," and therefore perfect alterity, disunity, and divisibility So that no extensum whatsoever, of any sensible bigness, is truly and really one substance, but a multitude or heap of substances, as many as there are parts, into which it is divisible' (Cudworth, *True Intellectual System*, iii. 393). Similarly, Samuel Clarke writes that 'The meaning of "parts," in questions of this nature, is "separable, compounded, un-united parts," such as are the parts of matter which, for that reason, is always a *compound*, not a simple substance No matter is *one* substance, but a *heap* of substances.' The Answer to a Sixth Letter, in *Demonstration*, 116.

Leibniz: bodies are aggregates of actual parts, hence dependent entities lacking unity, hence fundamentally phenomenal.[29]

3. *Bodies have a determinate number of parts*

A third widely accepted corollary of the actual parts doctrine is that bodies have a determinate number of parts. (This corollary was previously touched upon in Chapter 1, section VII; see also Chapter 3, section II.) According to the actual parts doctrine, *all* the parts into which a body can be m-divided are given in their totality prior to any act of division. There is nothing indefinite or open-ended about the number of parts: any part that can be found in the whole is there from the start; no additional parts are subsequently generated or created. The number of parts is therefore determinate. Notice that this means that, if bodies are m-divisible ad infinitum, they must then have an *actual infinity* of parts (a completely given greater-than-finite number of parts) and not merely a potential (open-ended, ever-increasable, and indeterminate) infinity. For many thinkers of our period, this is a paradoxical conclusion (as we saw in Chapter 1, section VIII).

4. *Bodies have ultimate parts*

This last corollary (which was previously touched upon in Chapter 1, section VII) is argued for in two ways. First, many maintain that it follows from corollary (1) (the doctrine that bodies are composites) and the principle that composites are ontological parasites, presupposing a base of simple first parts. The argument is most familiar from Leibniz: it is his famous demonstration of monads from the opening paragraphs of *Principles of Nature and Grace* and *The Monadology*. But it is also found throughout much of rest of the Enlightenment literature. I document and discuss this *argument from composition to ultimate parts* in detail in Chapter 4 below.

The second argument from the actual parts doctrine to the existence of ultimate parts can also be found in the Enlightenment literature. The reasoning is that ultimate parts follow by way of corollary (3), the proposition that a body with actual parts has a determinate number of parts. *All* the parts into which such a body can be m-divided are completely given,

---

[29] 'Matter is an aggregate, *not a substance but a substantiatum* as would be an army or a flock; and insofar as it is considered as making up *one* thing, it is a phenomenon, very real, in fact, but a thing whose *unity* is constructed in our conception.' Leibniz, letter to Samuel Masson, 1716, in *Philosophical Essays*, 227. For an excellent discussion of Leibniz on this point, see R. C. Sleigh, *Leibniz and Arnauld* (New Haven: Yale University Press, 1990), ch. 5.

extant, and concrete, prior to any act of division. There is, then, no indeterminacy in the structure of the body, no open-ended series of ever-smaller parts coming successively into existence. And since *all* the parts are given, it follows (according to the current argument) that smallest, ultimate parts must be given: m-indivisible metaphysical atoms. I document and discuss this *argument from the definiteness of parts to ultimate parts* fully in Chapter 3.

(Notice that this corollary will split the physical realm into two sharply distinct domains. On the one hand we have the world of ultimate parts—the fundamentally real substances underlying the sensible world of gross bodies. And, on the other, we have the derivative, secondary realm of m-divisible bodies—aggregates of substances—an entirely different class of entity, since they are merely upshots of the structural relations among their (ultimate) parts.[30])

III. The Potential Parts Doctrine

According to the rival *doctrine of potential parts*, the parts into which a body can be m-divided are not distinct existents prior to their being actualized by some operation of division. Rather, division *creates* these parts as so many freshly minted beings—it does not simply unveil so many pre-existent things. Prior to the act of division that actualizes or creates the parts, the whole is best described as containing *possible* or *potential* parts. But such 'potential parts' simply represent ways in which the whole could be broken down, and talk of potential parts is really just talk about the modal properties of the whole original. Suppose, for instance, that a body is divisible into (at least) four parts. The potential parts theorist may then concede that it contains four potential parts. But, properly understood, this is not to concede the existence of four shadowy pseudo-entities. Rather it is just to say of the whole—the one thing that is there to be counted—that it has the power or the ability to be divided into four distinct parts, thereby creating four new beings. (Of course, these parts are not created *ex nihilo*. The matter from which they are made certainly existed beforehand. But in so far as we are counting distinct, independent

---

[30] This point is explored thoroughly in work of Ivor Leclerc. For instance, see his 'The Problem of the Physical Existent', *International Philosophical Quarterly*, 9 (1969), 48; *The Nature of Physical Existence* (London: George Allen & Unwin, 1972), 246, 282–3; *The Philosophy of Nature* (Washington, DC: Catholic University of America Press, 1986), 61.

entities, division creates several new beings from one.) On this view, then, the whole is not a composite or aggregate structure, a construction from so many distinct actual parts. Nor then is the whole an ontologically derivative entity, depending for its existence on the ontologically prior existence of component parts. It has no particular inherent structure of ontologically distinct parts prior to division. Rather it is a simple, noncomposite entity: one metaphysically undifferentiated whole, lacking any embedded real parts.[31]

As with the rival actual parts doctrine, it is important to appreciate that this is a *metaphysical* thesis about the ontological status of the parts of material bodies, not a thesis about their internal *physical* architecture. For all that the potential parts doctrine asserts, bodies might be either completely homogeneous or highly complex in terms of their internal physical organization. (Here we can avoid a common mistake. The actual parts–potential parts controversy is quite orthogonal to questions about the actual physical microstructure of material bodies, at least for all that has been said so far. For instance, it has no direct logical bearing on the question of whether bodies resolve to physically unsplittable parts in the manner of Boyle, or whether they are Cartesian plenum substances, physically divisible ad infinitum. Either sort of physical structure could be thought of in terms of actual or potential parts.)

Now, the potential parts theorist will of course concede that there are *some* composite beings. To give the classic example, the potential parts theorist will presumably allow that a pile of sticks is a composite or

---

[31] For a recent version of the potential parts approach, see Roberto Casati and Achille C. Varzi, *Parts and Places: The Structure of Spatial Representation* (Cambridge, Mass.: MIT Press, 1999), 99–109. Casati and Varzi offer a formalization and defence of the attempt to conceptualize potential parts as modal features of the undivided whole. Compare also William Charlton's defence of the potential parts system in his 'Aristotle's Potential Infinities', in Lindsay Johnson (ed.), *Aristotle's Physics: A Collection of Essays* (Oxford: OUP, 1991), 129–49. For an example of the attempt to translate away talk about (and hence ontological commitment to) undetached parts, see the 17th-century potential parts theorist Kenelm Digby: '[I]f (for example) a rodde be layed before us, and halfe of it be hid from our sight, and the other half appeare; it is not one part or thing that shew it selfe, and an other part or thing that doth not shew it selfe; but it is the same rodde or thing, which sheweth itself according to the possibility of being one new thing, but doth not shew itselfe according to the possibility of being the other of the two thinges, it may be made by division.' Kenelm Digby, *Two Treatises, in the one of which the Nature of Bodies is expounded; in the other, the Nature of Man's Soule; is looked into* (1644; facsimile edn.: Stuttgart: Friedrich Fromman Verlag, 1970), 13. For a similar recent attempt, see Ned Markosian, 'Simples', *Australasian Journal of Philosophy*, 76 (1998), 213–28, esp. 223–5.

aggregate entity. If one wants to call this sort of entity a material *body*, then the potential parts theorist will qualify his view and grant that there are indeed some composite material bodies. And if one wants to talk of individual sticks or subgroups of sticks within the pile as *parts* of the whole pile, then the potential parts theorist will allow that these are indeed actual parts—parts that are each a distinct concrete entity, independent of the whole pile and all the other non-overlapping parts. But the core thesis of the potential parts system is that at some level of resolution one will encounter extended and (m-)divisible material bodies that—contra the rival actual parts theory—are genuinely simple and noncomposite beings, their m-divisibility notwithstanding. It is these simple material bodies that merely have potential parts.

The potential parts theorist then owes us an account of what these m-divisible but noncomposite simple material bodies are, and an account of when one arrives at the level of resolution into such individual units. They owe us a reason for thinking of these m-divisible bodies as simple, non-composite entities, while other m-divisible beings (such as the pile of sticks) are clearly composite. And here there are numerous possible versions and subversions of the potential parts system. For instance, Aristotelians and scholastics will appeal to substantial forms—paradigmatically, the form of an organism—as the unifying principle behind the simplicity of their divisible but noncomposite substances. The basic individuals of this system are then those portions of matter that are structured by a substantial form, and these material bodies will have merely potential, not actual, parts. Other potential parts theorists will appeal to p-indivisibility as the principle of simplicity. Those bodies that are unsplittable by any natural force (such as Newton's and Boyle's extended corpuscles) will then count as the basic noncomposite units, their m-divisibility by divine power notwithstanding. Others might invoke internal qualitative homogeneity as providing the standard for demarcating simples. (In these last two cases, my comment above about the potential parts system being orthogonal to issues of physical microstructure would have to be qualified accordingly.) Other potential parts advocates might appeal to some sort of principle of maximal continuity as the criterion of simplicity, such that simple bodies are those that occupy the largest continuous matter-filled regions of space around.[32]

---

[32] For a recent defence of the maximal continuity account (and an overview of other approaches) see Markosian, 'Simples'. For an early modern statement of this approach, see

Or again, other potential parts theorists might deploy some sort of observer-relative or mind-dependent standard of simplicity. We shall see historical examples of some of these versions of the potential parts system in the following subsection. But first let us begin with some more general statements of the potential parts doctrine from the early modern texts.

Sir Kenelm Digby gives a clear endorsement of the doctrine in his widely read quasi-Aristotelian system of natural philosophy, the *Two Treatises* of 1644. (This quotation appeared in Chapter 1's précis account of the potential parts doctrine, but it will stand review.) Any given body, Digby writes,

> is but one *whole* that may indeed be cutt into so many severall partes: but those partes are not really there, till by division they are parcelled out: and then, the whole (out of which they are made) ceaseth to be any longer; and the parts succeed in lieu of it; and are, every one of them, a new *whole*.
>
> [T]he partes which are considered in Quantity, are not diverse things: but are onely a vertue or power to be diverse thinges.[33]

Thus far I have not specified what particular sort of division is required to actualize the parts of a body. Clearly this will depend on which account is given of the criterion of simplicity and individuation, and thus on which of the various permutations of the potential parts system outlined above one adopts. But to tie this to some explicit period statements of the potential parts account, here we can distinguish three possible views. One might hold (i) that the parts need to be actually broken apart and separated one from another—either through natural processes (p-division) or God's power (m-division). Or one might hold (ii) that the parts need be distinguished in thought in order for them to be actualized as distinct entities—that is, that they be i-divided in some mental representation. Or, finally, one might hold (iii) that either of these forms of division—actual separation through p- or m-division, or mental discrimination through i-division—is sufficient to create the parts. One can see these distinct ver-

---

Anthony Collins, *A Letter to the learned Mr. Henry Dodwell, containing some remarks on a (pretended) Demonstration of the Natural Immortality of the Soul* (1709), in Samuel Clarke, *Works*, iii. 749–53, 751. (Collins's pamphlet is reprinted in its entirety here.) Compare also John Jackson, *A Dissertation Concerning Matter and Spirit* (1735; facsimile edn.: Bristol: Thoemmes Press, 1994), 10.

[33] Digby, *Two Treatises*, 10, 13. Kenelm Digby—along with his mentor Thomas White—develops a two-sided system of natural philosophy that, in part, employs the corpuscularian-mechanistic methods of the new science and, in part, invokes the scholastic Aristotelian modes of explanation. On the present issue—the structure of physical continua—the *Two Treatises* is strictly Aristotelian.

sions of the potential parts doctrine distinguished in Galileo's and Walter Charleton's accounts of the view. (Neither of these two accept the view— they are merely reporting it in order to attack it.) First, we have the character Simplicio, the mouthpiece of orthodox scholasticism and patsy of Galileo's 1638 *Dialogues Concerning Two New Sciences*:

> Parts cannot be said to be in a body which is *not yet divided* or at *least marked out*; if this is not done we say that they exist potentially.[34]

Similarly, in his hostile and sharply polemical *Physiologia* of 1654, Walter Charleton distinguishes between different ways a potential parts theorist— here the scholastic Albertinus—might claim that parts are actualized. A physical continuum might be said to be composed

> *of a Simple Entity, before Division*, Indistinct; as not a few of our Modern Metaphysicians have dreamt, among whom *Albertinus* was a Grand Master. Who, that he might palliate the Difficulty of the Distinction of Parts, that threatened an easie subversion of his phantastick position; would needs have it that all Distinction doth depend *ab Extrinseco*, i.e ariseth only from mental Designation [i.e. i-division], or actual Division [i.e. p- or m-division].[35]

The Potential Parts Doctrine in Scholasticism, Hobbes, and the Later Kant

Scholasticism

The potential parts doctrine stems originally from Aristotle: it is, in fact, a particular application of his famous view that the parts of continua are posterior to the whole and require some positive operation to actualize, here brought to bear on to the specific case of material bodies. Endorsed by St Thomas, the doctrine is central to orthodox scholastic natural philosophy. The basic unit of the material realm, according to this system, is an Aristotelian substance—a quantity of matter organized by a particular form (paradigmatically, an organic form). Such a substance is a genuinely unified whole, not a composite or aggregate of the parts into which it may be divided. Prior to division, these parts do not exist independently: since their identity is determined by their functional role in the whole substance, they are merely aspects or features of that whole. Only when division is carried through are the parts actualized as distinct entities. The standard

---

[34] Galileo, *Two New Sciences*, 34. Emphasis added.   [35] Charleton, *Physiologia* 108.

scholastic view is that a positive separation of parts (through p- or m-division) is necessary for their actualization, not merely a mental discrimination (i-division). Without such an actual rupture, the parts retain their functional role determined by the organizing substantial form, and thus remain aspects or features of that whole.[36]

In addition to being sponsored by orthodox scholasticism, the potential parts doctrine also finds support from one or two of the new philosophers who are—on this question—outside the mainstream of the new science. I have said that the actual parts doctrine is a central pillar of the new science and, as that system triumphs over scholasticism, the doctrine becomes increasingly axiomatic for Enlightenment philosophers. This is indeed true of the mainstream of the new science from Galileo and the neo-Epicureans to Newton and the eighteenth-century Newtonians, as I have shown above. Nonetheless, a small group of partisans of the new science stands outside of the mainstream on this question and endorses the potential parts doctrine. This group of potential parts apologists within the tradition of the new philosophy includes two of the great heavyweights: Hobbes and the later Kant. I take each in turn.

Hobbes

As I just noted, in the main tradition of scholastic natural philosophy the creation of parts as so many distinct beings requires an actual rupture and separation of parts one from another (that is, an act of either p- or m-division); mere mental discrimination (i-division) is insufficient to actualize parts. For Hobbes, on the other hand, parts are actualized by *i-division*, discrimination in mental representation. This new version of the potential parts theory has extremely radical consequences for Hobbes's metaphysics, since it relativizes the individuation of particular material entities to observers. It constitutes a far-reaching system of anti-realism that has not been sufficiently appreciated in the secondary literature.

---

[36] The classic scholastic account of the parts of organisms is clearly presented in Digby's *Two Treatises*. When separated from the whole organism, parts lose their functional role and thus their identity determined by that role: they become an entirely new thing. '[I]f you sever any of these partes from the whole body; the hand can no more hold; nor the eye see; nor the foote walke; which are the powers that essentially constitute them to be what they are: and therefore they are no longer a hand, an eye, or a foote.' Digby, *Two Treatises*, 13. For more on the Aristotelian–scholastic version of the potential parts doctrine, see Theodore Scaltsas, *Substance and Universals in Aristotle's Metaphysics* (Ithaca, NY: Cornell University Press, 1994), 83–7.

Actual Parts and Potential Parts / 97

The main sources for Hobbes's version of the potential parts doctrine are the classic 1656 treatise *De Corpore*, and the manuscript text *Anti-White, or Thomas White's 'De Mundo' Examined*, composed during 1642–5 (and only recently published). An anti-realist version of the potential parts doctrine is found consistently in both texts. First, parts are created by division, not merely unveiled: 'it is manifest, that nothing has parts till it be divided; and when a thing is divided, the parts are only so many as the division makes them'.[37] This is a clear statement of the potential parts doctrine. Second, Hobbes's account of the generation of parts by division sits squarely alongside his *per mentem*, observer-relative definitions of 'division', 'part', 'whole', 'one', and 'compound'. Each of these concepts is defined relative to some frame of mental consideration: something is a part only if it is thought of in such and such a way, and likewise for the concepts of whole, one, and compound.[38] The division that creates parts is likewise a mental i-division, as we can see first in the case of the division of space and time, and then of body.

[T]o *make parts*, or to *part* or DIVIDE space or *time*, is nothing but to consider one and another within the same; so that if any man divide space or time... his first conception is of that which is to be divided, then of some part of it, and again of some other part of it, and so forwards as long as he goes on in dividing.

A body, and the magnitude, and the place thereof, are divided by one and the same act of the mind... it cannot be done but by the mind.[39]

We are said to divide something when we consider first the thing and then something smaller within it.[40]

The division that creates parts is thus i-division: discrimination in mental representation. Hobbes is quite clear that it is not an actual separation that individuates parts, and goes out of his way to distance himself from this traditional scholastic view:

There are people who understand 'division' only as 'the separation and dispersal of a continuum,' in which case a knife, or a sword, or some other instrument is

---

[37] Thomas Hobbes, *De Corpore*, ch. 7.9, in *The Collected Works of Thomas Hobbes*, ed. W. Molesworth, 11 vols. (London, 1839–45; facsimile edn.: London: Routledge Thoemmes Press, 1992), i. 97. See also *Leviathan*, in *Works*, iii. 677, where division again 'makes' parts.

[38] Hobbes, *De Corpore*, in *Works*, i. 95–6.

[39] Ibid. i. 95–6, 108.

[40] Hobbes, *Thomas White's 'De Mundo' Examined*, tr. Harold Whitmore Jones (London: Bradford University Press, 1976), 29.

necessary to the nature of division ...—as if half a marble column is not part of it unless the column is broken.⁴¹

[B]y *division*, I do not mean the severing or pulling asunder of one space or time from another (for does any man think that one hemisphere may be separated from the other hemisphere, or the first hour from the second?) but diversity of consideration; so that division is not made by the operation of the hands but of the mind.⁴²

In this manner, Hobbes retains the scholastic doctrine of potential parts, but overhauls their account of actualization. Parts are now created by 'diversity of consideration' or discrimination in thought: i-division.

The upshot of this new version of the potential parts doctrine is an account of the material realm that is anti-realist in a way that has not been appreciated in most of the secondary literature. Far from having nothing innovative to say concerning individuation and the status of parts,⁴³ Hobbes's theory amounts to a radical and far-reaching anti-realism. Parts are actualized by a mental operation of i-division. How many distinct entities there are depends on the observer's conventional attitudes. Whether or not something is a compound—or a part, or a whole—depends on how it is conceived. Indeed, what things count as the units or basic individuals of the material realm turns again on how the observer represents it. Particular, determinate material beings are in this sense mind-dependent. This sort of anti-realism permeates Hobbes's ontology of material nature.⁴⁴

(It is interesting to draw a contrast here between Hobbes on the one hand and Leibniz on the other. Famously, for Leibniz, reality is fundamentally

---

⁴¹ Hobbes, *Thomas White's 'De Mundo' Examined*, 29; see also 63–4.

⁴² Hobbes, *De Corpore*, ch. 7.5, in *Works*, i. 96.

⁴³ The judgement, for instance, of Frithiof Brandt's classic *Thomas Hobbes's Mechanical Conception of Nature* (Copenhagen: Levin Munksgaard, 1928), 258.

⁴⁴ Compare Gary S. Herbert, 'Hobbes's Phenomenology of Space', *Journal of the History of Ideas*, 48 (1987), 709–17, esp. 714, 716–17. Herbert reaches the same conclusion that I have about Hobbes's anti-realism—though it is interesting to note that, where I reach this conclusion by focusing on Hobbes's account of parts, wholes, and division, Herbert reaches it by way of an examination of his distinction between 'real' and 'imaginary' space. Hobbes's anti-realism on this issue may be presaged in Stoic natural philosophy. According to Dirk Baltzly, there is some tentative evidence that certain Stoics adopted the view that parts are actualized by i-division, and employed this doctrine to block the actual parts-motivated Zeno paradoxes. Dirk Baltzly, 'Who are the Mysterious Dogmatists of *Adversus Mathematicus* ix 352?', *Ancient Philosophy*, 18 (1998), 145–70.

discrete, whereas continuous entities are necessarily ideal.⁴⁵ This can now be contrasted with Hobbes, who holds that all discrete things—all ceterminate particular entities—are mind-dependent, whereas matter prior to this individuation in thought is fundamentally continuous.)

Kant

The later, critical period Kant stands as our second major Enlightenment proponent of the potential parts doctrine. Although the earlier Kant of the 1756 *Physical Monadology* endorses the actual parts account of matter,⁴⁶ in the 1781/7 *Critique of Pure Reason* and the 1786 *Metaphysical Foundations of Natural Science*, Kant backs off from this view and embraces the potential parts system instead.

In this later period Kant indeed maintains that the actual parts analysis of matter is inevitable for the transcendental realist (i.e. for one who takes the objects of our knowledge to be things-in-themselves rather than appearances structured by the forms of sensibility and the categories of the understanding). Given transcendental realism, 'a whole must already contain within itself all the parts in their entirety into which it can be divided'.⁴⁷ From this it follows (Kant claims) that the transcendental realist must admit that bodies are constituted from simple first elements, and thus that they run into contradiction when this result is set against the geometrical proofs of infinite m-divisibility. This constitutes a *reductio* on their system. (It is the 'Second Antinomy' of the four conflicts that plague transcendental realism in the *Critique of Pure Reason*.)

On the other hand, this later Kant claims that the potential parts doctrine is inevitable for the transcendental idealist (i.e. for one who takes the objects of knowledge to be appearances structured by the forms of sensibility and the categories of the understanding rather than things-in-themselves). Given transcendental idealism, bodies are (suitably structured) progressions of appearances. And in these progressions, the intuition of the whole body is presented prior to the intuition of the parts that one can generate from it via a successive process of division and subdivision. Parts are only given, Kant says, 'in the progressing decomposition... that first makes the series [of parts generated by division and subdivision] actual'.⁴⁸

---

⁴⁵ Leibniz, *Philosophical Essays*, 185.   ⁴⁶ See n. 23 above.
⁴⁷ Kant, *Metaphysical Foundations*, in Kant. *Philosophy of Material of Nature*, 53; see also 55 and the *Critique of Pure Reason*, B440 and B462–70.
⁴⁸ Kant, *Critique of Pure Reason* B552.

100 / Actual Parts and Potential Parts

So the parts are only given as one actualizes them through division, and we thus have a transcendental idealist version of the potential parts doctrine.

All this comes out clearest when Kant discusses the case of infinitely divisible matter. Since the parts of any given body need to be actualized by a positive operation of division before they exist as independent entities, their total number at any time will be indexed to how far division has proceeded. The total collection of parts in an infinitely divisible body is thus only ever *potentially* infinite (ever increasable but always actually finite):

[W]ith regard to what is actual only by its being given in representation, there is not more given than is met with in the representation, i.e., as far as the progression of the representation reaches. Therefore, one can only say of appearances, whose division goes on to infinity, that there are as many parts of the appearance as we give, i.e., as far as we want to divide. For the parts insofar as they belong to the existence of an appearance exist only in thought, namely, in the division itself. Now, the division indeed goes on to infinity, but it is never given as infinite; and hence it does not follow that the divisible contains within itself an infinite number of parts in themselves, that are given outside our representation, merely because the division goes on to infinity.... Any division of the object (which is itself unknown) can never be completed and hence can never be entirely given. Therefore, any division of the representation proves no actual infinite multitude to be in the object.[49]

[W]e are by no means permitted to say of a whole which is divisible to infinity that it *consists of infinitely many parts*. For though the intuition of the whole contains all the parts, it yet does *not* contain the *whole division*; this division consists only in the progressing decomposition, or in the regression itself that first makes the series actual.... [N]ot contained therein is the whole *series of the division*, which is infinite successively and never *whole* and hence can exhibit no infinite multitude of parts and no gathering together of such a multitude in a whole.[50]

This is simply one application of Kant's broader claim in the critical period that, given transcendental idealism, the 'conditions' of any 'conditioned' intuition are only ever presented in incompleteable, open-ended series of empirical progressions. (In the current type of case the conditions are the parts and the conditioned is the whole. In another type of case, the conditions are larger enclosing regions of space around a conditioned given region; in yet another, the conditions are the prior causes of a conditioned

---

[49] Kant, *Metaphysical Foundations of Natural Science*, 54.
[50] Kant, *Critique of Pure Reason*, B552; see also B333.

given event.) On this view, we can never encounter an unconditioned condition (such as an elemental first part, the entirety of space, or an uncaused cause). Nor can we ever encounter the completed totality of any progression of conditions (such as an entire series of parts within parts, the full sequence of spaces outside of spaces, or the complete chain of causes).[51]

Kant is not particularly clear about whether he thinks an actual separation (p- or m-division) or just a diversity of consideration (i-division) is necessary to actualize the parts of matter as distinct entities. Is he thinking of the 'progressing decomposition...that first makes the series actual' as a physical or a mental operation? There are perhaps hints in the above quotations that Kant is thinking in terms of i-division rather than the actual rupturing of p- or m-division. (For instance: 'the parts insofar as they belong to the existence of an appearance exist only in thought, namely, in the division itself'). But in the context of transcendental idealism such hints are hardly unambiguous. I think it is probably safest to simply say that Kant was less interested in this issue than he was in the more important claim that *some* sort of division or other is necessary to actualize the parts of matter as distinct entities. For the later Kant, matter must be understood in terms of the potential parts metaphysic, not the actual parts doctrine.

Two Corollaries of the Potential Parts Doctrine

Whether one holds with the scholastics that rupture and separation actualizes parts, or with Hobbes that parts are actualized by mental discrimination, two important corollaries follow from the potential parts account. In fact I already touched on each of these points when examining Kant's version of the potential parts doctrine, but they are of fundamental importance to the problems of material structure and it will be worthwhile setting them out in a little more detail.

1. *Bodies have an indeterminate number of parts*

According to the potential parts doctrine, the parts of a body are only actualized as distinct entities once an operation of division is carried out (be it p-, m-, or i-division, depending on which particular version of the doctrine is adopted). The parts 'are not really there, till by division they are

[51] Ibid. B525–43.

parcelled out', as Digby puts it.[52] It follows that how many distinct parts there are in existence at any given time depends on how far forth the process of division and subdivision has been carried. In this sense the number of parts is indeterminate. Granted, at any one moment there will only be a determinate number of parts in existence. But more can be generated, and the number of parts will always depend on how far division is carried. So, in contrast to the actual parts doctrine, according to which all the parts are really inherent as distinct existents in the structure of the whole from the start, here it follows that the number of parts in a body is (in this sense) indeterminate. (Notice that this is perfectly compatible with there being a given maximum number of parts into which a body can be divided. If a given body is only finitely m-divisible, say into a hundred parts, then obviously only a hundred parts can ever be actualized from that body. But this is quite compatible with the present corollary, which simply states that the number of parts in existence at any given time is indexed to how far the whole has been divided.)

2. *If infinitely divisible, bodies have only a potential infinity of parts*

This second corollary runs as follows. It is an extremely popular view (though often a tacit one) that a single act of division can, in the one stroke, only divide a body into a finite number of parts.[53] If this is granted, then it follows that, however many such acts of division are carried out in succession, only a finite number of parts will ever be actualized. One can successively divide and subdivide as long as one likes, but at any stage of the process only a finite number of parts will have been generated. Thus, if it turns out that matter is divisible ad infinitum, this will imply merely a *potential infinity* of parts: an ever-increasable but always actually finite number of parts.

This result should be contrasted with corollary (3) of the actual parts doctrine. Recall that, given the actual parts doctrine, if a body is divisible ad infinitum, then there must be an actual infinity of distinct parts already embedded in its structure. Of course, just as the potential parts theorist argues that only a potential infinity of parts will ever be *actualized* by a successive process of division and subdivision, so too the actual parts

---

[52] Digby, *Two Treatises*, 10.
[53] I discuss the rival view—that a single act of division can split a body into a greater than finite number of parts—in Ch. 3, s. V.

theorist would claim that only a potential infinity of parts will ever be *unveiled*. But, nonetheless, if that process can continue ad infinitum, then there must already be an actual infinity of pre-existent parts laid up in the whole. In contrast, the potential parts doctrine neatly allows one to grant that matter is infinitely divisible without admitting an actual infinity of parts and the paradoxes that supposedly follow such an admission. This is one of the central virtues Aristotle and his scholastic followers claim for the doctrine of potential parts. The *locus classicus* here is the *Physics*, where the doctrine is mobilized to block Zeno's paradoxes of continua.[54] The same point is stressed by those Moderns who endorse the potential parts doctrine—for instance, when Hobbes answers Zeno in *De Corpore*[55] and when Kant deploys a potential parts account to disarm the Second Antinomy.[56]

IV. The Case for Actual Parts

In this section I present a survey of the classic arguments for the actual parts doctrine. Given the overwhelming popularity of the doctrine with the new philosophers of the Enlightenment, explicit arguments for this theory are surprisingly rare in the period literature. In the early and mid-seventeenth century, it is true, the continuing prominence of the rival scholastic system does motivate some philosophers affiliated to the new science to argue actively for actual parts. But even in this early period, and

---

[54] Aristotle, *Physics*, tr. R. P. Hardie and R. K. Gaye, in *The Complete Works of Aristotle*, ed. Jonathan Barnes, 2 vols. (Princeton: Princeton University Press, 1984), i. 315–446 (esp. books 3, 6, and 8). For useful commentary, see David Bostock, 'Aristotle, Zeno, and the Potential Infinite', *Proceedings of the Aristotelian Society* (1972–3), 37–51. Jonathan Lear, 'Aristotelian Infinity', *Proceedings of the Aristotelian Society* (1979–80), 188–210. Fred D. Miller, Jr., 'Aristotle against the Atomists', in Norman Kretzman (ed.), *Infinity and Continuity in Ancient and Medieval Thought* (Ithaca, NY: Cornell University Press, 1982), 87–111. Maurice Clavelin, *The Natural Philosophy of Galileo* (Cambridge, Mass.: MIT Press 1974), 39–45.

[55] '[T]he force of that famous argument of Zeno against motion, consisted in this proposition, *whatsoever may be divided into parts, infinite in number, the same is infinite*; which he without doubt, thought to be true, yet nevertheless is false. For to be divided into infinite parts, is nothing else but to be divided into as many parts as a man will. But it is not necessary that a line should have parts infinite in number, or be infinite, because I can divide it and subdivide it as often as I please; for however many parts soever I make, yet their number is finite.' (Hobbes, *De Corpore*, in *Works*, i. 63–4.)

[56] Kant, *Critique of Pure Reason*, B551–4; see also Kant, *Metaphysical Foundations*, 53–4.

certainly by the close of the seventeenth century, the actual parts doctrine is often treated as a first principle rather than as a contentious theory requiring argument. And in the eighteenth century, with the triumph of the new science and scholastic natural philosophy decisively routed, the doctrine is typically treated as undeniable fact—a banal truism that is not only assumed without argument, but also often simply presupposed without explicit statement. Leibniz, Wolff, Berkeley, Hume, and the pre-critical Kant are all examples of leading philosophers of this later period who either treat the doctrine as a first principle or merely offer the barest of throwaway arguments for it. It is then to the earlier figures we must turn for explicit and properly developed arguments for actual parts.

Three key philosophers should be mentioned here among the few actual parts theorists who explicitly argue for the doctrine: Galileo Galilei, Walter Charleton, and Pierre Bayle. Galileo deserves credit as the first to seriously popularize the case for actual parts in the modern period. His last great work, the 1638 *Dialogues Concerning Two New Sciences*, introduces the actual parts alternative to scholastic orthodoxy and can be regarded as the first widely read text of the new science to defend the doctrine. On the first day of the *Dialogues* the discussion centres on the rival actual parts and potential parts theories and their application to the traditional problem of Aristotle's wheel. Salviati, Galileo's spokesman for the new science, argues for the actual parts theory and from thence to a system of ultimate first parts; Simplicio, the beleaguered spokesman of traditional scholasticism, offers a half-hearted defence of the received potential parts doctrine.[57]

Walter Charleton and Pierre Bayle are by no means the grandest of the various Enlightenment philosophers to endorse the actual parts doctrine. However, it is these two who go the furthest in self-consciously and explicitly presenting arguments for the doctrine, rather than simply treating it as a first principle. Accordingly I will be focusing on their texts in my examination of the case for actual parts. A brief introduction to each of these less well-known philosophers may not be out of place here.

Sir Walter Charleton (1620–1707), a prominent figure in the atomist revival in seventeenth-century natural philosophy, was an original member of the Royal Society and personal physician to Charles I. His *magnum opus* is the 1654 *Physiologia Epicuro-Gassendo-Charletoniana, or a Fabrick of Science Natural, Upon the Hypothesis of Atoms*, which—as the title confesses—is a defence of

---

[57] I defend this interpretation in Ch. 3, s. VI.

the Epicurean system of atoms and the void. As the title also acknowledges, it is deeply indebted to Pierre Gassendi. Large parts of the work are simply free translations lifted from Gassendi's 1649 *Animadversiones*, and Charleton also appears to have had access to manuscript versions of Gassendi's *Syntagma Philosophicum* (which was eventually published in 1658, four years after Charleton's *Physiologia*). However, Charleton certainly expands liberally on Gassendi's work, adding original arguments of his own. Moreover, he goes much further than Gassendi in the radical Epicureanism of his system: while Gassendi is merely concerned to argue the existence of p-indivisible atoms in the infinitely f-divisible void,[58] Charleton maintains that *all* physical quantities—including space and body—must resolve to f- and m-indivisible *minima*. The *Physiologia* is truly a great read: an almost Shandean concoction of verbose and mock-imperious declamations of the new system and witheringly sharp-tongued put downs of the received scholastic potential parts account. It had a direct influence on the thought of Newton, whose early Trinity notebooks repeatedly evince arguments drawn word-for-word from Charleton's text. The Newton of this early period follows Charleton in positing f- and m-indivisible *minima* of space and matter.

Pierre Bayle (1647–1705) is perhaps a more familiar figure. As editor of the *Nouvelles de la République des Lettres*, as lecturer at Sedan and Rotterdam, and—of course—as the gossipy ur-*philosophe* behind the sprawling and subversive volumes of the *Dictionnaire historique et critique*, Bayle developed a wealth of sceptical arguments that Voltaire described as 'the arsenal of the Enlightenment'. Rarely shy of hyperbole, Voltaire goes on to describe 'immortal Bayle, the honour of human nature' as the 'greatest dialectician who ever lived'.[59] And in Bayle's armoury of arguments we find by far the most systematic and comprehensive defence of the actual parts doctrine I have discovered in the period literature. For our purposes, the texts of interest are, first, the *Système Abrégé de Philosophie* (1675–7) and, second, the articles on Leucippus, Spinoza, and Zeno of Elea in the *Dictionnaire* (1697).

---

[58] See, for instance, E. J. Dijksterhuis *The Mechanization of the World Picture* (Oxford: OUP, 1964), 426. McGuire and Tamny, *Certain Philosophical Questions*, 60. (McGuire and Tamny give the following reference: Gassendi, *Opera*, Liber III, Sectio I, Cap. V, 256–66.)

[59] Voltaire, *Œuvres Complète*, ed. Louis Moland (Paris: Garnier Frères, 1877–80), xiv. 546. Quotation and reference from Robert C Bartlett, introduction to Pierre Bayle, *Various Thoughts on the Occasion of a Comet* (Albany, NY: State University of New York Press, 1999), pp. xxiii–xlvii, p. xxiii.

106 / Actual Parts and Potential Parts

The pithy *Système* was drawn up from Bayle's lectures at Sedan in the 1670s—'à l'usage des Etudiens', as the title-page has it. It divides into four sections: 'la logique', 'la morale', 'la physique' and 'la metaphysique'. The division covering physics presents a more or less orthodox Cartesian system of natural philosophy and can be seen as a text set squarely in that framework.[60] The later *Dictionnaire* is of course anything but orthodox, and, by comparison with the terse syllogisms of the *Système*, the three articles in the *Dictionnaire* are more playful, ironic, and discursive, in keeping with the almost Rabelaisian tenor of this vast work. However, buried in the footnotes of these articles we do find the case for actual parts explicitly developed and prosecuted further.

Argument A1. How could M-Division Proceed Unless there were Distinct Parts Prior to Division?

According to this first argument, m-division (that is, the actual rupture and separation of parts, by God's power if necessary) is only possible if those parts exist as distinct entities in the first place. Our concept of such a division presupposes the prior existence of parts that are *already* logically distinct from one another. So the modal property m-divisibility entails the existence of actual parts. Charleton puts the argument as follows. (He is speaking here of a *physical* continuum.)

[A]s those parts, which are deduced from a Continuum, must be præexistent therein before deduction (else whence are they deduceable?) so also must those, which yet remain deduceable, be actually existent therein, otherwise they are not deduceable from it.[61]

The parts into which a body can be m-divided must already be laid up in it, 'else whence are they deduceable?' In Charleton's *Physiologia*, this section of reasoning occurs as the initial step of his larger argument to show that infinite m-divisibility entails the prior existence of an actual infinity of distinct parts. In short, here he is establishing that, if a body can be

---

[60] This then raises the possibility that Bayle's arguments for actual parts set out in this division were in fact quite traditional in Cartesian circles, notwithstanding the lack of explicit textual statements of any such arguments in the published writings of major figures such as Descartes, Arnauld, Malebranche, Rohault, or Le Grand. Perhaps it is just that Bayle was the only one to publish arguments that were implicit but never quite stated in the broader milieu of Cartesian natural philosophy.

[61] Charleton, *Physiologia*, 93.

m-divided into *x* parts, it must already contain *x* pre-existent parts. Parts are unveiled by division, not created. The next step will then be to apply this to the case of infinite divisibility—thereby finding '*Aristotle's subterfuge of Infinitude Potential...openly Collusive*':[62] given the argument above, infinite m-divisibility implies the pre-existence of an actual infinity of parts, not merely an open-ended, ever-increasable potential infinite.

This same argument from infinite m-divisibility to an actual infinity of pre-existent parts is found in Galileo, Henry More, the younger Newton, Bayle, Berkeley, and Hume. It is of course a classic Epicurean argument, handed down to the modern era by way of Lucretius.[63] In More, Newton, and Hume the initial step is made quite explicit. Thus More:

[N]*othing can be divisible into parts it has not*; therefore if a body be divisible into infinite parts, it has infinite extended parts.[64]

Newton:

I shall use one argument to show that it [matter] cannot be divisible *in infinitum*, and it is this: *nothing can be divided into more parts than it can be possibly be constituted of*.[65]

And Hume:

'Tis also obvious that whatever is capable of being divided *in infinitum*, must consist of an infinite number of parts, and that 'tis impossible to set any bounds to the number of parts without setting bounds at the same time to the division.[66]

In Galileo, Bayle, and Berkeley, on the other hand, the initial step is not explicitly spelt out. These philosophers each argue directly for the conclusion that infinite m-divisibility implies an actual infinity of parts without openly stating the general principle that m-divisibility into *x* parts implies

---

[62] Ibid. 92–3.
[63] Lucretius, *De Rerum Natura*, ed. and tr. Cyril Bailey, 3 vols. (Oxford: Clarendon 1947), i. 615–18. Compare also a clear 14th-cent. statement of the argument in Nicholas of Autrecourt, *The Universal Treatise* (Milwaukee, Wis.: Marquette University Press, 1971), 78.
[64] More, *Antidote Against Atheism*, in *Philosophical Writings*, 11. Emphasis added.
[65] Newton, 'Certain Philosophical Questions', 341. Emphasis added.
[66] David Hume, *A Treatise of Human Nature*, ed. David Fate Norton and Mary J. Norton (Oxford and New York: OUP, 2000), 1. 2. 1. 2. References abbreviated 'SBN' give the corresponding page numbers in David Hume, *A Treatise of Human Nature*, ed. L. A. Selby-Bigge, 2nd edn. with text revised and notes by P. H. Nidditch (Oxford: OUP, 1978). Here the SBN reference is 26–7. See also *Treatise*, 1. 2. 2. 2; SBN 29.

the prior existence of those x parts.[67] But it is clearly the same argument they are employing, and it clearly involves a commitment—albeit tacit—to the same general principle. So each of these thinkers is likewise committed to our current argument: m-divisibility implies the prior existence of distinct parts, 'else whence are they deduceable?'

The argument is, of course, question-begging. It turns on the central principle that m-division is only possible if distinct parts already exist, and that, if a body is m-divisible x times, then it must already contain x parts. But this is of course exactly what the potential parts theorist denies, and no additional reason has been adduced for thinking their account wrong. The potential parts theorist will simply say: 'M-divisibility merely presupposes an extended entity whose parts—prior to division—are not independent of the whole but rather aspects or features of it. An act of m-division "actualizes" these parts as newly independent existents. "Whence are they deduceable?" *As distinct existents*, they are generated or created by division, not deduced or unveiled by it.'

Argument A2. Whatever Occupies a Distinct Place is a Distinct Thing

A second argument runs as follows. Bodies are extended in space and so have spatially distinct parts: for instance, any given body will have a left half and a right half, each of which occupies a different region of space. But these spatially distinct parts must be actual parts, for whatever occupies a distinct place is a distinct thing.

This argument is reported again and again by Henry More. His three main philosophical works—the 1653 *Antidote against Atheism*, 1659's *The Immortality of the Soul*, and the later 1671 *Enchiridium Metaphysicum*—each discuss it in turn. In each of these texts More is reporting the reasoning in order to respond to it. Of course, as a partisan of both extended m-indivisible 'physical monads' and extended m-indivisible penetrable spirits, he is a stalwart opponent of the argument. In all three texts More's discussion of the argument takes place in the context of his defence of the possibility of extended, penetrable, and 'indiscerpible' (that is, m-indivisible) spirits against the charge that whatever is extended has actual parts and so is

---

[67] Galileo, *Two New Sciences*, 34. Bayle, *Dictionary*, 356, and *Œuvres Diverses*, iv. 299. Berkeley, *Works*, i. 42, 283. See also my discussion of this form of argument in Ch. 3, s. II.

m-divisible. But while More's driving interest here is the notion of spirit, the debate is conducted in terms of the more general question of whether every entity that is extended is *ipso facto* composed of actual parts. So the reasoning here applies equally to bodies as to spirits, as More himself notes.[68] The argument More reports is as follows:

it is objected,... that *Extension* cannot be imagined without *diversity of parts*, nor *diversity of parts* without a *possibility* of *division*, or separation of them; because *diversity of parts* in any Substance supposes *diversity of substances*, and *diversity of Substances* supposes *independency of one another* ...[69]

[T]here could be no extension which has not parts real and properly so called into which it may be actually divided, *viz.* for this reason, that only is extended which has *partes extra partes*, which being *substantial*, may be separated one from another, and thus separate subsist.[70]

While More merely reports the argument, Isaac Barrow endorses it. Here is Barrow in his *mathematical lectures*:

The next Affection of Magnitude, which follows its *Extension*, is its *Composition*, i.e. that Magnitude contains within it or consists of different Things. For because it is extended, and its Extremes are distant from one another, therefore the *Whole* is not in the same indivisible *Place*; whence different Things may be assigned in it; and consequently it is compounded of different Things, i.e. of *Parts* ...[71]

Bayle also endorses the argument—in fact, he presses it over and over again. His clearest statement of the reasoning is the following crisp syllogism from the *Système*:

Ce qui est étendu occupe en même tems plusieurs lieux. Or une même chose ne sauroit en même tems occuper plusieurs lieux. Donc ce qui est étendu n'est pas une même chose, ou pour m'exprimer autrement, ce qui est étendu n'est pas une entité unique. Ainsi dans une être étendu il y a plusieurs entitez ou parties distinctes.[72]

---

[68] More, *Enchiridium Metaphysicum*, in *Philosophical Writings*, 217, 219.

[69] More, *Antidote against Atheism*, appendix III, in *A Collection of Several Philosophical Writings* (London: 1712), 186.

[70] More, *Enchiridium Metaphysicum*, in *Philosophical Writings*, 214; see also 212 and 216.

[71] Barrow, *Usefulness*, 148–9.

[72] i.e. 'That which is extended occupies several places at the same time. But the same thing cannot occupy several places at the same time. Therefore that which is extended is not one same thing, or to express myself differently, is not one unique thing. Thus in an extended entity there are several entities or distinct parts.' (Bayle, *Système*, in *Œuvres Diverses*, iv. 297; the

Our final token of the argument comes a century later in the *Critique of Pure Reason*. Kant is here reporting the reasoning rather than endorsing it. (The argument comes in the Second Antinomy, and as such is attributed to the transcendental realist.) Kant's statement of the argument is terse and to the point:

[A]nything real that occupies a space comprises a manifold [of elements] outside one another and hence is composite.[73]

Versions of the argument can also be found in the recent literature on arbitrary undetached parts.[74]

But, again, this argument will not convert anyone. The potential parts advocate will simply reject the argument as begging the central question. The argument's crucial assumption—that whatever is f-divisible necessarily has actual parts, parts that each exist independently prior to division—is exactly what the potential parts doctrine denies. No new reason has been given for accepting this claim, so the argument is a *petitio*. I think that Bayle at least implicitly accepts this point: while he sometimes presses the current argument as if it were a free-standing demonstration of the actual parts doctrine, elsewhere he sees the need to add additional support to the central premise on pain of begging the question. We shall see this additional reasoning under argument (A3) following.

Argument A3. Incompatible Properties Require Distinct Subjects

This next argument is again an attempt to prove that bodies have actual parts from the fact that they are extended. But rather than conclude this

argument is stated again on 299.) The same argument occurs in the 'Spinoza' article and, here, in the 'Zeno of Elea' article in the *Historical and Critical Dictionary*: 'Every extension, no matter how small it may be, has a left side and a right side, an upper and a lower side. Therefore it is a collection of distinct bodies. I can deny concerning the right side what I affirm about the left side. *These two sides are not in the same place. A body cannot be in two places at the same time, and consequently every extension that occupies several parts of space contains several bodies.*' Bayle, *Dictionary*, 360; see also 302. Emphasis added—in order to discriminate the statement of our current argument (A2) from Bayle's interwoven endorsement of argument (A3).

[73] Kant, *Critique of Pure Reason*, B464.
[74] Dean W. Zimmerman, 'Could Extended Objects Be Made Out Of Simple Parts? An Argument for "Atomless Gunk"', *Philosophy and Phenomenological Research*, 56 (1996), 1–29: 8. Compare also Phillip Cummins, 'Bayle, Leibniz, Hume and Reid on Extension, Composites and Simples', *History of Philosophy Quarterly*, 7 (1990), 299–313, 304.

directly (and thereby beg the question), this time an additional appeal to the impossibility of a single individual supporting incompatible qualities bolsters the inference.

The argument runs as follows. Nothing can have two incompatible qualities at the same time, unless it is a composite of distinct elements that are each the true subjects of those qualities. For instance, nothing can be both black and white, unless it is composed of two or more distinct individuals, one of which is black and one of which is white. So no truly individual, noncomposite entity can have incompatible qualities at the same time. But every extended entity can have incompatible qualities at the same time (since its spatially distinct parts might support incompatible qualities: one half might be black, the other white). So every extended entity is a composite of distinct elements; that is to say, every extended entity has actual parts. Now, all bodies are extended, so all bodies have actual parts.[75]

This is Bayle's favourite argument for the actual parts system. He returns to it again and again, prosecuting it, in turn, against partisans of m-indivisible extended microbodies, potential parts theorists, and Spinozistic monists. Our first statement is an early version drawn from the *Système*, where Bayle mobilizes it against m-indivisible extended material atoms:

Les Etres dont on peut assurer avec vérité des choses contradictoires, sont réellement distinguez les uns des autres Or dans un Atome il y a des Etres desquels on peut assurer avec vérité des choses contradictoires. Donc il y a dans un Atom quelques êtres distinguez réellement.[76]

In the later *Dictionnaire* the argument occurs in the 'Zeno of Elea' article,[77] and here in the 'Spinoza' article, where it forms the basis of a notoriously ferocious polemic against Spinozistic monism:

---

[75] Notice that argument (A2) could be interpreted as a version of argument (A3): a version that happens to focus on the differing properties of spatial location that each subpart has. But, as Theodore Sider notes (*Four-Dimensionalism* (Oxford: OUP, 2001), 89), the general form of argument (A3) can seem more compelling when it is applied to intrinsic properties rather than extrinsic properties like spatial location.

[76] i.e. 'Those beings of which one can correctly affirm contradictory things, are really distinct one from another. But in an atom there are beings of which one can affirm contradictory things. Therefore in an atom there are beings which are really distinct.' Bayle, *Système*, in *Œuvres Diverses*, iv. 297–8; see also 299.

[77] Bayle, 'Zeno of Elea', in *Dictionary*, 360.

> [E]xtension is composed of parts which are each a particular substance.... The Scholastics have perfectly well succeeded in showing us the characteristics and the infallible signs of distinction. When one can affirm of one thing, they tell us, what one cannot affirm of another, they are distinct... Applying these characteristics to the twelve inches of a foot of extension, we will find a real distinction between them. I can affirm of the fifth that it is contiguous with the sixth, and I can deny this of the first, the second, and so on....
> Incompatible modalities require distinct subjects.[78]

Bayle is extremely confident in this demonstration of the actual parts doctrine and uses it to ground his sharp-tongued dismissal of Spinozistic monism. Spinoza 'has to teach that extension is a simple being, as exempt from composition as the mathematical points. But is it not a joke to maintain this?' Spinoza's simple substance supports contradictory qualities, but '[s]imply to report such things is to refute them, is to show the contradictions clearly; for it is obvious either that nothing is impossible, not even that two plus two equals twelve, or that there are in the universe as many substances as subjects that cannot be designated in the same way at the same time'. All in all, '[a] man of good sense would prefer to break the ground with his teeth and his nails than to cultivate as shocking and absurd a hypothesis as this'.[79]

Notwithstanding Bayle's ebullient faith in the argument, I think we can show that it is a failure. Consider again his claim that 'there are in the universe as many substances as subjects that cannot be designated in the same way at the same time'. This principle—which is clearly at the core of his argument—effectively identifies *substances* with *subjects*. Bayle asserts that the principle is incontestable: to reject it would be as unconscionable as asserting that two and two make twelve. However, it is just this identification that seems, on inspection, much too quick.[80]

As is well known, there are at least two rival definitions of the term 'substance' running through the literature of our period. First, there is the definition of a substance as that in which properties ultimately inhere—a *subject* of qualities that is not itself also a quality. Second, there is the

---

[78] Bayle, 'Spinoza', in *Dictionary*, 306. Compare also Edmund Law, *An Enquiry into the Ideas of Space, Time, Immensity and Eternity* (Cambridge, 1734; facsimile edn.: New York: Garland, 1976), 106–7.

[79] Bayle, 'Spinoza', in *Dictionary*, 304–5, 310–11, 312.

[80] This point has been well made in Tad Schmaltz, 'Spinoza on the Vacuum', *Archiv für Geschichte der Philosophie*, 81 (1999), 174–205, esp. 177–9.

definition of substance as that which is *independent* or *self-sufficient*, not requiring anything else in order to exist.[81] Now, all created beings are dependent for their existence on God. But this dependence is typically bracketed off as obvious and uninteresting, leaving us with a more useful distinction between those created beings that are dependent on other created beings, and those created beings that are not so dependent. These latter relatively independent beings may loosely be called substances (in the second sense of the term, denoting independence or self-sufficiency) even though they are of course dependent on God.[82]

The problem with Bayle's argument is that it proceeds from premises about substances in the first sense (subjects of properties) to a conclusion about substances in the second sense (independent beings). It is true, no doubt, that if something can support contradictory qualities at the same time, then it must have different parts in which these respective qualities inhere. Something can only be black and white at the same time if (say) half of it is black and the other half white. And we may well have to talk about these two halves as different subjects if we want to avoid the embarrassment of one subject supporting contradictory qualities. But unless one slides (as Bayle, in effect, does) from talk about 'substances' in the sense of subjects to 'substances' in the sense of independent beings, nothing here shows that the black half and white half are necessarily logically independent of one another. Perhaps, for all that has been said here, they are the logically inseparable halves of a single m-indivisible extended microbody. Or perhaps they are the two sides of a potential parts theorist's undivided whole: *aspects* or *features* of that whole that are identity-dependent upon it prior to actualization by division.

There is another way of presenting the same underlying complaint against Bayle's argument. Bayle confesses his whole argument to turn on a fundamental axiom he quotes from Coimbra's scholastic logic. 'If there is anything certain and incontestable in human knowledge', he writes, 'it is this principle ... two opposite terms cannot be truly affirmed of the same subject, in the same respect, and at the same time'.[83] Now, I have already

---

[81] The *locus classicus* of the first definition is Descartes, 'Second Replies', in *Philosophical Writings of Descartes*, ii. 114. For the second definition, see Descartes, *Principles*, part 1, s. 55, in *Philosophical Writings of Descartes*, i. 211. Descartes seems to assume that whatever fits one of these definitions will necessarily fit the other.

[82] See, again, Descartes, *Principles*, part 1, s. 55, in *Philosophical Writings of Descartes*, i. 211.

[83] Bayle, 'Spinoza', in *Dictionary*, 309. Bayle cites 'the logic of Coimbra, In cap. X Aristotelis de Praedicamentis'.

pointed out that this principle tells us something only about substances in the first sense (subjects of properties), and not about substances in the second sense (independent beings). This was my first point. But even if we allow this to pass, the underlying complaint can still be brought out if we compare Bayle and Coimbra's principle with the more felicitous version from which it is ultimately derived. Plato's famous version of the principle of individuation in *Republic* 4 runs as follows: 'the same thing will not be willing to do or undergo opposites *in the same part of itself*, in relation to the same thing, at the same time'.[84] As my italicization indicates, this version of the principle has an additional qualification: it rules out only the possibility that 'one thing' could have incompatible qualities at the same time, in the same respect, and *in the same part of itself*. And this principle is clearly preferable to Bayle's and Coimbra's, since it does not assume that if one thing supports incompatible qualities in different parts, then those parts are necessarily ontologically independent of one another, thereby begging the question against the opponent of actual parts. So those who reject the actual parts doctrine will simply reject Bayle and Coimbra's principle of individuation in favour of Plato's more nuanced version.

Argument A4. The Parts into which a Body is M-Divisible can Exist Separately; Therefore they are Actual Parts

Our last argument for actual parts proceeds as follows. Since the parts that result from m-division can exist separately from one another, they are certainly so many distinct existents after that division. Moreover, even though they have been divided one from another, they remain the self-same entities they were prior to division: division merely splits them apart; it does not radically alter their identity. So they were distinct existents even before division. That is to say, they were—and are—actual parts.

Once again, it is Walter Charleton's *Physiologia* that provides us with an early and particularly clear statement of this reasoning:

Those things which can exist being actually separate; are really distinct: but Parts can exist being actually separate; therefore they are really distinct, even before

---

[84] Plato, *Republic*, tr. G. M. A. Grube (Indianapolis: Hackett, 1992), marginal reference 436b. Emphasis added.

division. For Division doth not give them their peculiar Entity and Individuation, which is essential to them and the root of Distinction.[85]

It is also found in Bayle's *Système*:

Ainsi les parties de Continu deviennent-elles distinguées par la division, c'est une preuve qu'elles étoient distinctes auparavant, & au contraire étoient-elles identifiées avant la division, elles doivent conserver leur identité après cette division, ce qui est manifestement faux ...[86]

Lastly, the same reasoning also occurs in Newton's early Trinity notebooks. The language of the notebooks clearly shows the direct influence of Charleton, but Newton does add a new twist to the argument. It is no longer the mere fact that parts can exist separately that establishes the actual parts doctrine: now it is also the fact that they can be separated *and then reunited* to form the same whole original. First the rehearsal of Charleton's argument:

Those things that can exist being actually separate are really distinct; but such are the parts of matter.[87]

Then the new appeal to the possibility of *reconstructing* a divided body: contra the potential parts theorist, matter is not built up from

a simple entity before division indistinct. For this must be a union of the parts into which a body is divisible, since those parts may again be reunited and become one body again, as they were before at the creation.[88]

However, either in its original form or with Newton's additional twist, the current argument once again simply begs the question against the potential parts theorist. The potential parts theorist will admit (of course) that, once divided, the parts of a body are indeed distinct existents. They will also admit that those distinct parts can be reunited to reconstitute the original body. But they will deny that these distinct parts, divided one

---

[85] Charleton, *Physiologia*, 108.
[86] i.e. 'The parts of a continuum come to be distinguished by division, which is a proof that they were distinct beforehand, and if, on the contrary, they had been identical [as opposed to distinct] before the division, they should keep their identity after the division, which is manifestly false.' Bayle, *Système*, in *Œuvres Diverses*, iv. 299; see also Bayle, 'Spinoza', in *Dictionary*, 305–6. And compare Nicholas of Autrecourt, *The Universal Treatise* (Milwaukee, Wis.: Marquette University Press, 1971), 78.
[87] Newton, 'Certain Philosophical Questions', 339.
[88] Ibid. 337.

from another, are the 'self-same' entities as the parts of the whole body prior to division, or post-reconstitution. There is in fact a radical difference between the parts of an undivided (or reconstituted) whole and the parts that result from division. The parts of an undivided whole are identity-dependent on that whole. They are merely aspects or features of it—they represent potential or possible ways it could be broken down: as Digby puts it, 'the partes which are considered in Quantity, are not diverse thinges: but are onely a vertue or power to be diverse thinges'. Only when some positive operation of division actualizes these possibilities do we have parts as independent existents. All this is but to reiterate the content of the potential parts doctrine—but what other reply could there be to such a manifest *petitio*?[89]

Now, an actual parts advocate may feel that this response—while perhaps correct in its identification of a formal *petitio*—does not really plumb the depths of Charleton's, Bayle's, and Newton's underlying complaint. Certainly they each assume that the parts post-division are identical with the parts pre-division (or post-reconstitution), and this is just what the potential parts doctrine denies. But their deeper point is surely this: the view that an operation of division could radically alter the identity of some portion of matter, turning it from a dependent feature of a larger whole into an 'actualized' independent existent, is so much occult scholastic nonsense. First, the view seems wholly unmotivated. It may, perhaps, be plausible in the case of the parts of an *organism*, since here the part's integration in the whole may give it a new identity in virtue of its functional role within that organism. But at least in the case of *unorganized* bodies—homogeneous lumps of metal, rocks, pools of water—the view that the identity of parts radically changes when they are divided one from another seems an *ad hoc* conjuring trick.[90] Second, the view that the identity of parts changes with division requires that material entities be individuated by something external to their own intrinsic properties. As Charleton snarls,

---

[89] This comes out particularly clearly in a recent statement of the argument in the current literature. 'Suppose [a] desk is divided into halves...Since the halves are spatially separated, they are distinct objects. These objects were presumably not created by division, so they must have existed before division. But surely it does not matter whether the division *actually* occurs, so anything *potentially* divisible must actually have parts.' Sider, *Four-Dimensionalism*, 90. As Sider clearly notes, the argument depends on the assumption that division does not create parts—and of course this is exactly what the potential parts theorist denies.

[90] For a recent statement of this sort of attack on the potential parts account, see Andrew Pyle, *Atomism and its Critics* (Bristol: Thoemmes Press, 1995), 391.

the potential parts theorist 'would needs have it that all Distinction doth depend *ab Extrinseco*, i.e. ariseth only from *mental Designation*, or actual Division'. But—'O the Vanity of affected subtilty!'—this is absurd, for—as we saw above—'Division doth not give [parts] their peculiar Entity and Individuation, which is essential to them and the root of Distinction.' In fact, individuation cannot depend on anything external to the individuated entity, as is 'verified by that serene Axiome, *Per idem res distinguitur ab omni alia, per quod constituitur in suo esse*'.[91] In much the same vein, Bayle adds to his barrage of objections to potential parts the claim that 'il est certain que chaque être est distinct de tout autre par soi même, & non par quelque chose que lui arrive de dehors, tel qu'est division'.[92] In short, the identity of a given being is determined by its own intrinsic properties, not by anything external to it. So the potential parts theorist's alchemical individuation through division is a nonsense.

To deal with this latter objection first: the 'serene Axiome' notwithstanding, this is—once again—so much question begging. Certainly the potential parts doctrine does entail that portions of matter are reidentified as they are divided from or reintegrated into larger bodies, and that, therefore, the individuation of particular entities can depend on factors '*ab Extrinseco*'. This is just to spell out their doctrine of the actualization of parts through division. But what of it? Charleton and Bayle assert that this is absurd, but no additional reason has been given, thus far, in support of this claim.

As for the former charge that the potential parts doctrine is unnatural or *ad hoc* in its claim that the identity of parts changes radically with division: this objection is certainly to the point and captures, I think, much of the deep intuitive force behind the actual parts doctrine. For the actual parts theorist, the view that parts metamorphose when divided from the whole, shifting ontological category from mere *possibilia* to become actual physical *concreta*, is entirely *ad hoc*. After all, whether a particular part is attached or unattached seems to have no bearing on its intrinsic nature of properties. One can still trip over half a brick, whether or not it is attached to the whole. So what is the motivation for claiming that it is a

---

[91] i.e. 'A being is distinguished from everything else by the same feature which makes it itself.' Charleton, *Physiologia*, 108.

[92] i.e. 'it is certain that each being is distinct from all the others through its own self, and not through something that comes to it from without, such as division'. Bayle, *Système*, in *Œuvres Diverses*, iv. 299.

dependent feature or mere possibility in the one case but a distinct being and concrete actuality in the other?

How telling is this sort of attack on the potential parts system? Well, whether the potential parts theorist's account of actualization and individuation is well motivated or *ad hoc* will ultimately depend on their overall case for the doctrine. As we shall see, potential parts advocates typically argue that their doctrine of actualization through division, far from being unnatural in any way, is in fact implicit in our pretheoretical concept of division and everyday scheme of individuation. (I examine the charge that the potential parts doctrine is *ad hoc* further in my discussion of argument (P1), in section V below. The complaint is unfair, or so I argue.)

V. The Case for Potential Parts

Although we shall see arguments in this section credited to a variety of figures—including Hobbes, Spinoza, Anthony Collins, and the critical period Kant—I will be concentrating, in the main, on the case for potential parts presented in Sir Kenelm Digby's *Two Treatises*. Digby is certainly not as well known as these other figures, but his discussion is much more explicit and protracted than the often oblique arguments found in the texts of these other philosophers. Moreover, Digby's case for potential parts is a particularly clear statement of the classic Aristotelian–scholastic arguments, and as such can serve as an archetype of the dominant system of Oxford and the Sorbonne that advocates of actual parts were rebelling against.

Sir Kenelm Digby (1603-55) was a prominent member of the Royal Society and a pupil of Sir Thomas White (the target of Hobbes's *Anti-White*). Along with White he forged a system of natural philosophy that was an attempt to synthesize the new science with the traditional Aristotelian system. However, on our specific question of the structure of physical continua and the status of the parts of bodies, Digby fiercely championed the orthodox Aristotelian potential parts account. His *magnum opus* is the 1644 *Two Treatises, In the one of which the Nature of Bodies is expounded; in the other, the Nature of Man's Soule; is looked into.* In the preliminary stages of the first *Treatise*, Digby sets out the traditional version of the potential parts theory and presents both arguments for the doctrine and pre-emptive strikes on the rival actual parts account. This section of the first *Treatise* was in all likelihood written in partial response to the actual parts case of Galileo's famous 1638 *Dialogues Concerning Two New Sciences*. The *Two Treatises* were widely read in mid-seventeenth-century England, and would certainly have been familiar to (for instance) Walter Charleton when he composed his caustic riposte in the 1654 *Physiologia*.

Argument P1. The Concept of Division Implies the Creation of Parts

This first argument has it that the doctrine of potential parts falls directly out of our pretheoretical concepts of division and divisibility. On this interpretation of our everyday concepts, division just *is* the creation of several entities from one. One starts out with a single entity, breaks it apart and ends up with a plurality of entities. But this implies that parts are *created* by division, not merely *unveiled*. There are more entities after division than there were before. So an undivided whole is one thing, whose undetached parts must be understood as features or properties of that one thing—properties that express ways in which the whole original *could* be broken down. Only when divided from one another are parts actualized as so many distinct entities. In the *Two Treatises*, Digby puts the argument as follows:

> *Quantity or Biggnesse*, is nothing else but divisibility;...a thing is bigge, by having a capacity to be divided, or (which is the same) to have partes made of it.
>
> [Quantity] is but one *whole* that may indeed be cutt into so many severall partes: but those partes are not really there, till by division they are parcelled out: and then, the whole (out of which they are made) ceaseth to be any longer; and the parts succeed in lieu of it; and are, every one of them, a new *whole*.
>
> This truth, is evident out of the very definition we have gathered of Quantity. For since it is *Divisibility* (that is, a bare capacity to division) it followeth that it is not yet divided: and consequently that those partes are not yet in it, which may be made of it; for division, is the making of two, or more thinges, of one.
>
> Quantity, is a possibility to be made distinct thinges by division...And yet...nothing can be more manifest, then that if Quantity be divisibility (which is a possibility, that many things may be made of it) these partes are not yet divers thinges.[93]

---

[93] Digby, *Two Treatises*, 9, 10, 13.

Given this conceptual analysis, the potential parts doctrine is not the arcane and abstruse theory it may at first appear. With its account of the 'actualization' and individuation of entities *ab extrinseco*, and of the alchemical metamorphosis of parts through division, the potential parts doctrine may seem a bizarre scholastic flight of fancy—a 'phantastick position', a 'Metaphysician's dream', as Charleton puts it.[94] But, if the current argument is correct, all this is to misunderstand the doctrine and its essential content. Rather than being a far-fetched metaphysical theory, the doctrine merely spells out the content of our pretheoretical concepts of division and divisibility. It merely explicates what is already implicit in our everyday understanding of these things: division *creates* several things from one; therefore it must actualize parts, not merely unveil pre-existing parts.

Now, if this analysis of our concepts of division and divisibility is right, then the actual parts doctrine is clearly based on a conceptual confusion. If something is a plurality of independently existing entities—as the actual parts theorist claims bodies are—then it is really *already* divided. Given our concept of division, the distinctness of the parts presupposes that the whole has already been divided into these separate parts. The actual parts theorist's bodies, then, are not really divisible at all: in fact they are already divided through and through. This is clearly a nonsense: a *reductio* which show us that the actual parts theorist has misunderstood the basic concepts of division and divisibility. We can see the charge of this sort of confusion pressed against the actual parts theorist in Anthony Collins's correspondence with Samuel Clarke. In the correspondence (a flurry of pamphlets debating rival accounts of matter, spirit, and thought, published in 1706–8) Clarke affirms the actual parts doctrine over and over again. Collins is much more sceptical about the existence of actual parts. Now, in the main, he allows this to pass for the sake of the argument: their debate turns on other issues.[95] But on occasion he cannot forbear a snide remark

---

[94] Charleton, *Physiologia*, 108.

[95] For instance, Collins writes: '[Clarke's] propositions are; (1) *Every system of matter consists of a Multitude of distinct Parts*. This Proposition, which he *thinks is granted by all*, is, I am sure, denied by a great many. However, I shall not at present enter into any Debate with him concerning it, but continue to suppose it as I have hitherto done.' Anthony Collins, *Reflections on Mr. Clarke's Second Defence of his Letter to Mr. Dodwell*, in Samuel Clarke, *Works*, iii. 800–21, 817. (The entire correspondence, including all Collins's pamphlets, is included in Clarke's *Works*.) Compare also Collins, *A Letter to the Learned Mr. Henry Dodwell*, in Clarke, *Works*, iii. 749–53, 751, where he writes that 'If several Particles of Matter can be so united as to touch one another, or closely to adhere; wherein does the Distinctness or Individuality of the several Particles consist?'

or two. And it is in these asides that I think we can see Collins join Digby in suggesting that the actual parts doctrine rests on a conceptual confusion:

[B]y Mr. *Clarke*'s own account of Matter. ... he makes [it] to consist of *actually separate and distinct Parts* (though I wonder that at the same time he should make it *divisible*, when by its consisting of separate and distinct Parts it is actually divided.)[96]

Collins seems to be obliquely endorsing our current argument: parts are only distinct once they are divided, the concepts of division and divisibility imply the potential parts view, and the actual parts doctrine accordingly rests on a conceptual confusion.[97]

It should be clear by now that this argument is excessively abrupt. The actual parts theorist will just maintain that the concept of division does not involve the literal creation of new parts, but rather the separation (or 'unveiling') of pre-existing parts. And if division is understood this way—as the separation of distinct entities that were previously contiguous and perhaps bonded in some way—then it seems perfectly consistent for the actual parts advocate to allow that bodies are *divisible*, not necessarily already *divided*. All this seems a perfectly coherent way of conceptualizing division and divisibility, so the actual parts theorist has no reason to embrace Digby's and Collins's rival account.

It is interesting to note that the current argument for potential parts (P1) is essentially a mirror image of the equally forlorn argument for actual parts (A1). Each of these arguments presents an account of what the concept of division involves. In (A1) it was claimed that—as a conceptual matter—division presupposes the prior existence of distinct parts. Now we have (P1), which claims that—as a conceptual matter—division *creates* parts, and the existence of distinct parts presupposes that they have been already been divided one from another. Each of these arguments fails because they start with a claim about the concept of division which the advocate of the rival view rejects and is given no reason to adopt. The account of division presented in (P1) seems perfectly coherent, and its availability to the potential parts theorist means that (A1) is a *petitio*.

---

[96] Collins, *Letter to Dodwell*, in Clarke, *Works*, iii 749–53, 751.

[97] For a similar insinuation in the recent literature, see Anthony Flew, 'Infinite Divisibility in Hume's *Treatise*', in D. W. Livingston and J. T. King (eds.), *Hume: A Re-Evaluation* (New York: Fordham University Press, 1976), 257–69, 259–60. (This is Flew's passage about the cake, which I quoted back in Ch. 1, s. VI.)

Likewise, the account of division presented in (A1) seems perfectly coherent, and its availability to the actual parts theorist means that (P1) is a *petitio*. For all that has been said so far (it seems to me), each of these accounts of the content of our concept of division is perfectly intelligible and internally coherent.

We are now also in a position to address the residual objection that lay behind argument (A4). I argued that (A4), which tells us that, since parts can exist separately, they must be actual parts, is a *petitio*. But I also noted that behind it lay the deeper complaint that the potential parts doctrine seems *ad hoc* in its assertion that parts metamorphose when divided from the whole. The suggestion was that the potential parts doctrine's account of actualization through division is an unnatural and tortuous one—a baroque conjuring trick avoided by the more common-sensical actual parts theory. But I hope we can now see that that this objection underlying (A4) is unfair. It assumes that the actual parts theory really does set out a more intuitive—more natural—account of division and parthood than the potential parts doctrine. But in the light of Digby's and Collins's account discussed here, I think we can see that the actual parts theorist is not entitled to this. These potential parts advocates have argued—not unconvincingly—that the notion that division creates parts captures a perfectly familiar way of thinking, and not a convoluted and alien system. The very deadlock—the symmetrical *petitio*—of arguments (A1) and (P1) lies in this fact that each appeals to a familiar and intelligible way of thinking about division, even though the two accounts are incommensurable with one another.

Argument P2. Given Actual Parts, the Sum of the Parts is Greater than the Whole; But this is Absurd

I cannot forbear to briefly review and criticize this short argument, even though (unlike all the other arguments I address) it is not drawn directly from Enlightenment sources but rather from a more recent advocate of the potential parts doctrine. But, while I have not found a version of this argument in the period literature, it nonetheless captures a pattern of reasoning that often seems tempting to potential parts sympathizers. I draw the following statement of the argument from William Charlton's recent defence of the Aristotelian–scholastic variant of the potential parts theory. (This Charlton should not be confused with Walter

Charleton, our seventeenth-century Epicurean and arch-enemy of potential parts.)

To deny that the parts of an undivided whole exist [as actual parts] is to deny that 'The beam could be sawn into five cubes' is more perspicuously rendered by 'There are five cubes and the beam could be sawn into them'. If I say 'I might have driven my car into five large cubes of wood' I do indeed claim that there were (right in the middle of your drive) five large cubes of wood, and that I might have driven my car into them. But dividing a beam into cubes is not like driving a car into cubes. The cubes do not exist in the beam in the same way in which obstacles can exist in drives. *That may be more evident if we reflect that I might have sawn the beam into twenty-five squares or three planks. If the cubes and the squares and the planks all exist, the beam is three times the size we supposed.*[98]

Charlton's first argument here can be quickly put aside. In most contexts it certainly would be unnatural to speak of 'the five cubes of wood' in the driveway, or to catalogue any of the other various parts (squares, planks, etc.) in the whole, rather than to speak of the one whole beam. But this is easily explained away in terms of the pragmatics of those everyday contexts. This fact does not really seem to engage the issue of whether the beam is an aggregate or a noncomposite simple at the ultimate metaphysical level (even if it is indeed more useful to speak of the beam, in most contexts, as if it were one thing).

But Charlton's main argument here is not this initial linguistic point, but rather the *reductio* argument I have italicized in his text. Given the actual parts doctrine, the cubes, the squares, and the planks all exist as fully fledged concrete entities. But then 'the beam is three times the size we supposed'. Since this is absurd, it follows that the cubes, the squares, and the planks—the parts into which the whole original may be divided—are only *possible* or *potential* existents, and do not really exist until actualized by division.[99]

But this main *reductio* argument is also a failure. Of course the actual parts theorist is not committed to saying that the cubes exist *in addition to* the squares existing *in addition to* the planks existing. All of these parts do indeed exist as fully fledged *concreta*, but it does not follow that they each

---

[98] William Charlton, 'Aristotle's Potential Infinities', in Lindsay Johnson (ed.), *Aristotle's Physics: A Collection of Essays* (Oxford: OUP, 1991), 129–49, 133–4. Emphasis added.

[99] See Roberto Casati and Achille C. Varzi, *Parts and Places: The Structure of Spatial Representation* (Cambridge, Mass.: MIT Press, 1999), 100–1, for a similar argument.

exist in addition to all the other existent parts. Recall the original definition of actual parts: these are parts that can be m-divided one from another and which each exist independently of all other *non-overlapping* parts. Of course they do not exist independently of their own subparts or of other parts that partially overlap them—but this does not compromise their concrete existence. Charlton's underlying thought may be that actual parts must exist in addition to all other parts (including overlapping ones) if they are to truly qualify as concrete existents. But this is simply not so, as can be brought out by a paradigmatic example of an aggregate of concretely existing parts. All parties would agree that an army is an aggregate of independently existing parts. It derives whatever reality it has from them. Now, an army typically includes several regiments and numerous battalions as parts. All the regiments and the battalions do indeed each exist, and each exists independently of the whole army. But it does not follow that the regiments exist *in addition to* the battalions, since the battalions are themselves subparts of the regiments.[100]

Argument P3. The Actual Parts Doctrine Leads to the Paradoxes of Material Structure, Establishing the Potential Parts Account by Elimination

As we saw in some detail in Chapter 1, the actual parts doctrine threatens to precipitate the paradoxes of material structure (Chapter 1, sections VI–VIII). This suggests an argument by elimination to establish the alternative potential parts doctrine.

As the reader will recall, a whole host of paradoxes was thought to threaten the actual parts doctrine. The fundamental problem is to reconcile the existence of actual parts with the fact that bodies are extended in infinitely f-divisible space. Bodies occupy regions of space and—in virtue of this fact—share certain of the topological features of space such as its infinite f-divisibility. This infinite f-divisibility of matter then seems to entail its infinite m-divisibility: the fact that it has spatially distinguishable parts within parts ad infinitum seems to entail that it has logically separable parts within parts ad infinitum. And, at least according to most Enlighten-

---

[100] One could easily make the same point about *partially* overlapping parts. 'The soldiers on horseback' may refer to a part of the army that genuinely exists, even though it partially overlaps with other genuinely existent parts of the same army, such as the first regiment.

ment thinkers, this infinite m-divisibility of matter threatens a whole host of paradoxes when set alongside the actual parts doctrine. As I showed in Chapter 1 (especially sections VII and VIII), the conflict between infinite m-divisibility and actual parts appears to generate the full range of classic antinomies of material structure.

Notice that the potential parts doctrine does not trigger these same paradoxes. With the potential parts account, one starts with the entire undivided body as one individual whole—'a *Simple entity, before division,* Indistinct'.[101] Parts are then created only as one divides and subdivides, and infinite m-divisibility implies only an ever-expandable, top–down sequence of ever-smaller parts. With the potential parts doctrine, there is no determinate number of parts *in toto*, and there need be no ultimate, m-indivisible, smallest elements. The potential parts doctrine entails neither a determinate totality of parts nor ultimate parts, and so it does not set off the classic paradoxes of material structure (see Chapter 1, section VII).

Given that the actual parts doctrine generates the classic paradoxes of material structure whereas the potential parts doctrine does not, we now have an argument by elimination for potential parts. If the paradoxes cannot be disarmed, the actual parts doctrine must be rejected as leading to absurdity. This leaves the potential parts doctrine as the only account that can accommodate the fact that bodies occupy space. So (we are told) the potential parts doctrine is established by elimination.

This sort of argument is at least implicit in Aristotle, Hobbes, Spinoza, and the later Kant. Each of these philosophers claims that various traditional problems of the structure of continua can be resolved only if the potential parts doctrine is adopted.[102] It is also Digby's favourite argument for potential parts, and it is worth quoting his particularly explicit statement of this classic scholastic pattern of reasoning. Having introduced the potential parts doctrine and briefly suggested that it is guaranteed by our concept of division (argument (P1) above), Digby then proceeds to his main argument.

---

[101] Charleton, *Physiologia*, 108. Of course Charleton is merely reporting the doctrine, not endorsing it.

[102] Each of these philosophers holds that this analysis applies to all continua—not just bodies in space. See Aristotle, *Physics*, in *Complete Works*, 315–446, esp. books 3, 4, and 8. (And see the commentary cited in n. 54 above.) Hobbes, *Works*, i. 63–4. Spinoza, 'Letter on the Infinite', in *Spinoza: The Letters*, tr. Samuel Shirley (Indianapolis: Hackett, 1995), 103–4; compare also *Ethics*, in *The Collected Works of Spinoza*, ed. and tr. Edwin Curley (Princeton: Princeton, 1995), part 1, proposition 15 scholium. Kant, *Critique of Pure Reason*, B552–5.

126 / Actual Parts and Potential Parts

[W]e must apply our selves, to bring some more particular and immediate proof of this assertion [i.e. the potential parts doctrine]. Which we will do, by showing the inconvenience, impossibility, and contradiction that the admittance of the other [i.e. the actual parts doctrine] leadeth unto: For if we allow actuall partes to be distinguished in Quantity, it will follow that it is composed of points or indivisibles, which we shall prove to be impossible.[103]

The actual parts doctrine (Digby tells us) entails that 'Quantity' is composed of 'points or indivisibles': it is built up from an ultimate constitution of metaphysical atoms. Digby thus focuses on one particular type of classic paradox: the problem that the actual parts doctrine leads to metaphysical atomism, which is absurd when set alongside the fact that bodies are extended in infinitely f-divisible space. Two pages later, after setting out his argument that the actual parts doctrine entails metaphysical atomism,[104] and his argument that this is an absurd result,[105] Digby confidently sums up the structure of his reasoning as follows: 'it [is] firmly established, *That Quantity is not composed of indivisibles* (neyther finite nor infinite ones) and consequently, *that partes are not actually in it*'.[106] Extended entities cannot be constructed from metaphysical atoms; so the actual parts doctrine must be rejected. We must endorse the potential parts analysis of matter instead.

Of course, this argument will only be successful if the potential parts doctrine does indeed present the only way of disarming the paradoxes of material structure. In terms of my survey of the four doctrinal camps that emerge in response to the problems of material structure (see Chapter 1, section IX), the argument will be successful only if faction (1) is correct. This first group maintains that infinite m-divisibility and the actual parts doctrine are indeed in inexorable conflict, and—since it regards infinite m-divisibility as undeniable—therefore rejects the actual parts doctrine for a potential parts account. But this still leaves our three other factions. There is faction (2), which accepts the existence of a conflict between infinite m-divisibility and the actual parts doctrine, but then upholds actual parts and rejects infinite m-divisibility instead. Then there is faction (3),

---

[103] Digby, *Two Treatises*, 10.

[104] His case for this is a version of the 'argument from the definiteness of parts', which I introduced under corollary (4) of the actual parts doctrine in s. II of the current chapter, and discuss in detail in Ch. 3, s. III.

[105] Here Digby sets out versions of the paradoxes I entitled classic paradox (1) and classic paradox (3A). See my Ch. 1, s. VIII above.

[106] Digby, *Two Treatises*, 12.

which rejects the idea that infinite m-divisibility and the actual parts doctrine are truly in conflict with one other. And finally there is faction (4), which maintains that there is indeed a conflict between infinite m-divisibility and the actual parts doctrine, but that—notwithstanding the sceptical consequences of such an antinomy—one must nevertheless endorse both doctrines all the same.

To flesh this out a little, recall some of the more prominent theories that fall under the rubric of the factions that resist potential parts. Berkeley and Hume's system of granular *minima* and Boscovich's force-shell atom theory stand as accounts that fall within the scope of faction (2). Or think of Galileo with his system of actual infinities of elemental *parti non quante* as a representative of faction (3). Or, again, recall Boyle and Bayle as philosophers who flirt with the radical scepticism of faction (4). The current argument for potential parts will stand up only if *none* of these responses presents a viable way of dealing with the problems of material structure. Clearly I cannot assess all these alternative responses to the problems of material structure here. But it does seem to me that at least *some* of these theories do present logically and conceptually viable accounts of material structure: accounts that offer ways of avoiding the paradoxes of material structure without adopting the system of potential parts. (I present a full survey of these 'logically viable' models of material structure in the conclusion of this book.) If this turns out to be correct—and I have not proved it here, but merely offered a promissory note—then of course the current argument by elimination for potential parts will fail.

VI. Summary and Conclusion

We are left, then, with something of an uneasy stalemate between the cases for and against our rival doctrines. The arguments on both sides of the historical debate all appear unsuccessful. Most of them simply beg the question, albeit sometimes in a fairly covert fashion. This may seem a frustrating result. But, it seems to me, the standoff is itself of philosophical interest. If neither doctrine can be demonstrated at the expense of the other, this may suggest that we are faced with two equally tenable but mutually incommensurable ways of thinking about the world. And this in turn may suggest a form of scepticism about the supposed conflict between them.

128 / Actual Parts and Potential Parts

Table 2 Arguments for actual parts and potential parts

Arguments for actual parts

(A1) How could m-division proceed unless there were distinct parts prior to division?

(A2) Whatever occupies a distinct place is a distinct thing.

(A3) Incompatible properties require distinct subjects.

(A4) The parts into which a body is m-divisible can exist separately; therefore they are actual parts.

Arguments for potential parts

(P1) The concept of division implies the creation of parts.

(P2) Given actual parts, the sum of the parts is greater than the whole; but this is absurd.

(P3) Rival doctrines lead to the paradoxes of material structure, establishing the potential parts doctrine by elimination.

Let us briefly review and take stock of the arguments for the rival doctrines of actual and potential parts (see Table 2). I have said that (P3) cannot be comprehensively assessed here (though I have suggested it will ultimately fail, asserting that there *are* other logically consistent models of material structure outside of the potential parts system). And so, noting that it is yet to be finally adjudicated, let us put it to one side. Now, I have claimed that the remaining arguments (A1), (A2), (A3), and (A4), along with (P1) and (P2) are one and all question-begging. Each of these attempts to demonstrate the one theory at the expense of the other turns out to presuppose—more or less surreptitiously—what it purports to prove. None of these arguments succeeds in embarrassing the rival theory with an internal contradiction or a conflict with some other mutually admitted principle. And in the light of this the historical success of the actual parts doctrine in replacing the potential parts theory as the orthodox account in natural philosophy now looks like a *de facto* usurpation rather than a *de jure* assumption of the throne. Neither theory seems to evince any claim to this seat that sets it above its rival. But if neither theory has been demonstrated, it is also true that neither has been refuted. Our rival doctrines have both proved remarkably robust, at least in the face of the classic criticisms presented in arguments (A1), (A2), (A3), (A4), (P1), and (P2). For all the dust that has been raised by the likes of Charleton, Bayle, and Digby, neither doctrine has been shown to be either internally inconsistent or in conflict with some mutually admitted further fact about the world. And given this resilience, one may begin to wonder if they are *both* equally viable: perhaps each theory captures an intelligible and coherent way of conceptualizing division and the status of parts. Notwithstanding their mutual incompatibility, each doctrine may on its own terms be acceptable. If one chooses to think about division as the actual parts theorist does, then one must think about the parts of bodies as so many distinct existents. And if one chooses to think about division as the potential parts theorist does, then one must think about the parts of bodies as aspects or features of the whole, requiring division in order to actualize. But each of these ways of thinking about bodies and their parts is, on its own terms, equally feasible. If this is right, then the conflict between the two doctrines starts to look bogus. There is no metaphysical fact of the matter as to which account of the parts of bodies is correct. Rather there is just the choice of how to conceptualize division and parthood. One can count entities with the actual parts theorist if one chooses, or one can choose to count entities with the more parsimonious potential parts theorist. But this is just a matter of bookkeeping. The choice of which system of entity-enumeration to use here is purely conventional or terminological: it answers to no objective, absolute way the world ultimately is prior to our decision.

How compelling is this interpretation of the standoff? It is, I think, worth sounding two notes of caution before we rush to conclude that the dispute simply boils down to a systematic terminological difference. First, while it may turn out (let us suppose) that each of our theories is consistent with all the mutually admitted data about divisibility and parthood, it need not yet follow that they have no further implications for our physical and metaphysical systems. For instance, if we follow Galileo's account, the actual parts doctrine is ultimately defensible, but only if we are prepared to follow him in admitting the physical realization of actually infinite collections. Or, to give another example, Boscovich and the younger Kant each maintain that the actual parts doctrine can evade the challenge of argument (P3) only if we accept the finite m-divisibility of matter and the system of f-divisible but m-indivisible force-shell atoms. On their view, the actual parts doctrine then entails a certain view of the ultimate architecture of matter—a view, notice, which the potential parts doctrine certainly does *not* entail. So the first point here is that, if one thinks that the actual parts doctrine can be squared with the problems of material structure and defended against argument (P3), then, depending on just *how* one thinks the problems of material structure are to be disarmed, one may end up thereby committed to substantive metaphysical theses about the nature of matter that go beyond mere terminological differences with the potential parts view.

Second, it should be clearly stated that the actual parts doctrine and the potential parts doctrine have *not* here been proved to be equally coherent. There may well be a further argument that I have not addressed here that does decisively refute the one and establish the other. For instance, to mention one line of argument found in the recent metaphysical literature (though not one I personally find convincing): some maintain that the actual parts system must be rejected, since it entails a form of mereological essentialism that they find unacceptable.[107] That said, I believe that I have shown that none of the classic arguments do succeed in refuting either doctrine. And if these classic attempts to prove the one theory at the expense of the other are all failures, then it seems at least worth entertaining the possibility that each doctrine captures an equally reasonable way of thinking about division and parthood.

Suppose that this diagnosis is correct. How radical a result would this be? To my mind, whether one takes this sort of result to constitute a deep metaphysical insight into the conventional nature of individuation, or whether one takes it simply to show that our two rival doctrines were all along 'pieces of solemn and elaborate trifling'[108] is ultimately more a matter of temperament than of serious philosophical disagreement. But here I shall express the result in the former, metaphysical tone of voice.

If there is indeed no convention-independent metaphysical fact that determines which doctrine is correct, and rather we just have a choice between different ways of conceptualizing division, this introduces a form of anti-realism that is at once reminiscent of and importantly different from the anti-realism of Hobbes outlined in section III above. Recall that Hobbes endorsed a version of the potential parts doctrine according to

[107] Van Inwagen, 'Doctrine of Arbitrary Undetached Parts', 123–37. Actual parts systems certainly entail mereological essentialism (the view that no material body can retain its identity through the loss of any of its parts) but, contra van Inwagen, there is no need to regard this as an absurd outcome. For a pre-emption of van Inwagen's objection and an embracing of mereological essentialism, see Hume, *Treatise*, 1. 4. 6. 8–12; SBN 255–7.

[108] The phrase—though not the thesis—is from H. H. Price, 'Universals and Resemblances', anthologized in Dean W. Zimmerman and Peter Van Inwagen (eds.), *Metaphysics: The Big Questions* (Oxford: Blackwell: 1998), 23–40, 39.

which i-division (division in thought) actualizes parts. This theory was radically anti-realist in that it relativized the individuation of material beings to the observer's mode of thought. Particular, determinate material beings were thus rendered mind-dependent. Now we have an interpretation of the standoff between the actual parts and potential parts doctrines that suggests that the whole question of which is correct is, in effect, a matter of (systematic) choice or convention. One can reasonably adopt either of these mutually incommensurable theories as ways of conceptualizing division and parthood, and the decision of which to adopt determines whether parts exist as *concreta*, independent of the whole, or merely as potential parts, dependent on the whole. If one elects to think about division in the one way, then each body contains as many distinct individuals as it can be m-divided into. If one elects to think division in the other way, then each body is a single individual until division creates parts out of it. So we then have—at a different, deeper level to that of Hobbes—another type of anti-realism, according to which key questions concerning individuation and determinate being are once again mind- or convention-dependent.

I return to this issue of the status of the actual parts-potential parts conflict—and speculate on the historical reasons behind the popularity of the actual parts theory—in the conclusion of this book.

3

The Actual Parts Doctrine and Shortcircuit Arguments

> We are all like the ladies of Paris; they live sumptuously without knowing what goes into the stew. Similarly we enjoy bodies without knowing of what they consist. What is the body made of? Of parts, and these parts resolve themselves into other parts. What are these last parts? Always bodies: you divide endlessly and never advance.
>
> Voltaire, *Philosophical Dictionary* (1764), entry on 'Corps'

I. Introduction

In Chapter 1 and again in Chapter 2 we saw that two corollaries were traditionally supposed to follow from the actual parts doctrine. The actual parts doctrine was thought to entail (1) that each body has a determinate number of parts, and (2) that each body has ultimate parts. These two corollaries then underpin the array of classic paradoxes of infinite m-divisibility, each of the paradoxes resting on the one or the other corollary, or perhaps on both taken together. We also saw (in preliminary overview) one traditional argument from the actual parts doctrine to corollary (1), and two traditional arguments from the actual parts doctrine to corollary (2). Each of these arguments enjoyed a wide currency during the Enlightenment.

First there was (i) *the argument from actual parts to a determinate number of parts*. This stressed that, since all the parts into which a body can be m-divided are given in advance (assuming the actual parts doctrine), the entire collection of parts must be determinate in number rather than indefinite or open-ended. This establishes corollary (1): given the actual parts doctrine, each body has a determinate number of parts.

We then saw two arguments for corollary (2). There was (ii) *the argument from the definiteness of parts to ultimate parts*, according to which, since *all* the parts are given in a determinate, complete totality, it follows that smallest, ultimate parts must be given. Then there was (iii) *the argument from composition to ultimate parts*, which maintained that, if bodies are aggregates or composites (as the actual parts doctrine has it) then they are derivative entities, depending for their existence on the ontologically prior existence of their parts. And if the parts are themselves composite, then they must in turn depend for their existence on *their* subparts. But (according to the current argument) if the whole original is to exist at all, then this regress must terminate at some point in a ground floor: there must be ultimate, non-composite elements whose existence is not so derivative.

In this chapter I examine (i) *the argument from actual parts to a determinate number of parts* and (ii) *the argument from the definiteness of parts to ultimate parts*. It is well to deal with these two arguments together, since the latter clearly piggybacks on the former, and they are often presented side by side or even interwoven in the Enlightenment texts. My aim is to present the versions of these arguments advanced in the Enlightenment literature and to reconstruct and assess their key moves. I also want to contrast these arguments with certain other subtly different forms of argument also found in the period literature. This comparison will help us to see the relative merits of arguments (i) and (ii) and bring out their dependence on the actual parts system. (I defer discussion of the rather different (iii) *argument from composition to ultimate parts* to Chapter 4 below.)

First a word about the basic strategy that underlies each of these two arguments. As the reader will recall, the actual parts doctrine asserts that the parts into which a body can be m-divided (that is, its logically separable parts, the parts God could break it into) each exist as concrete, distinct entities even prior to division. Every part into which the whole can be m-divided is present from the very start, embedded in the architecture of the whole. Even before any act of division, the entire collection of these parts is already fully given. (Contrast the potential parts doctrine, according to which parts are created by acts of division and subdivision. Here the collection of parts is given successively, enlarging as division proceeds.) And it is just this feature of the actual parts metaphysic that our current arguments (i) and (ii) seek to exploit. Since the entire collection of parts is completely given from the start, it seems that the number of these parts must be determinate. Or so argument (i)—*the argument from actual parts to a determinate number of parts*—has it. And, again, if the entire collection of all parts is fully given from the start, then it seems that smallest parts must be given. Or so argument (ii)—*the argument from the definiteness of parts to ultimate parts*—has it.

Now, the real trick that these arguments try to pull off is to mobilize this feature of the actual parts metaphysic even when a body is thought of as infinitely m-divisible. I take it that arguments (i) and (ii) are obviously successful in the case of a body that is only finitely m-divisible (if there is any such body). But the real hope behind arguments (i) and (ii) is that they will go through even in the more interesting case of a body that is m-divisible ad infinitum. The central idea is as follows. Given the actual parts doctrine, even if the whole can be divided ad infinitum, and no successive series of divisions and subdivisions will ever reach a termination in final parts, nevertheless the entire collection of parts into which it can be divided must be fully present from the start, laid up in the whole. It is as if the actual parts doctrine allows us to shortcircuit the successive series of ever-smaller parts within parts, and jump all at once to the point where the entire series is completely given. And since the entire collection is fully given from the start, it cannot be open-ended, indefinite, or merely *potentially* infinite in number. Rather, it must—according to argument (i)—be determinate in number, and it must—according to argument (ii)—include ultimate parts.

I proceed as follows. I begin by documenting and examining the two shortcircuit arguments that invoke the actual parts metaphysic. In section II I reconstruct and assess argument (i): *the argument from actual parts to a determinate number of parts*. In Section III I reconstruct and assess argument (ii): *the argument from the definiteness of parts to ultimate parts*.

In the second half of this chapter I then turn to examine two rival sorts of argument that purport to establish the same conclusions (i.e. that bodies have determinate numbers of parts, and that they have ultimate parts), but attempt to do so without employing the actual parts doctrine as a shortcircuiting device. Section IV introduces these rival forms of argument in outline. It also emphasizes that they challenge my overall argument in this essay by making the actual parts doctrine—which I place at the heart of the Enlightenment travails over infinite divisibility—altogether redundant. If these rival arguments are compelling, then—contrary to my overall interpretation—the classic paradoxes do not rest on the actual parts doctrine after all.

I then look at these rival strategies in detail. In section V I examine arguments that attempt to infer from the infinite divisibility of the whole to the possibility of a complete, through-and-through infinite division and hence the presentation of the complete collection of parts. These arguments come in versions that envisage a *successive* infinite division and versions that envisage an *all-at-once* infinite division. Then in section VI I look at another sort of argument from infinite divisibility to the existence of a complete collection of all the parts into which the whole is divisible. Whereas the previous type of argument (in section V) deployed the notion of a through-and-through infinite division, this argument attempts to show that, even without an actual infinite division, the sheer fact of infinite divisibility already presents us with the complete collection of all the parts. Section VII then wraps up with a conclusion. I argue that neither of the rival strategies (outlined in sections V and VI) is successful. The actual parts doctrine thus really is an essential premise in generating the principles that bodies have a determinate number of parts and have ultimate parts, and thus is at the core of the classic paradoxes of infinite m-divisibility after all.

II. The Argument from Actual Parts to a Determinate Number of Parts

We have already seen the essential strategy of this argument, and can now fill in the detail. (see Table 3).

The reasoning is quite straightforward. According to the actual parts doctrine, *each* of the parts into which a body can be m-divided is already present in the body prior to division. Since each of these parts is present from the very start, we can infer that *all* the parts must be there from the start: the entire collection of parts into which it can be m-divided is given in advance as a complete totality. As opposed to the potential parts doctrine (according to which 'the parts are not really there, till by division they are parcelled out'[1]), there is nothing indefinite or open-ended about the number of parts: each and every one is there from the start; no

---

[1] Kenelm Digby, *Two Treatises, in the one of which the Nature of Bodies is expounded; in the other, the Nature of Man's Soule; is looked into* (1644; facsimile edn.: Stuttgart: Friedrich Fromman Verlag, 1970), 10.

136 / Actual Parts and Shortcircuit Arguments

Table 3 The argument from actual parts to a determinate number of parts

(D1)	Each part into which a body can be m-divided is given from the start. (*From the actual parts doctrine, which is assumed.*)
(D2)	If each member of a collection is given from the start (rather than their being given sequentially or over time), then the entire collection is given from the start. (*Assumed.*)
So: (D3)	The entire collection of parts is given from the start. (*From D1 and D2.*)
(D4)	If the entire collection of parts is given from the start, then the number of parts in the collection is determinate (i.e. fixed and equal to some specific number, rather than open-ended and indefinite). (*Assumed.*)
So: (D5)	The number of parts into which a body can be m-divided is determinate. (*From D3 and D4.*)
(D6)	If a body is infinitely m-divisible, and the number of parts into which it can be m-divided is determinate, then there must be an actual infinity of these parts (i.e. a completely given greater-than-finite number). (*Assumed.*)
So: (D7)	If a body is infinitely m-divisible, then there must be an actual infinity of parts in that body. (*From D5 and D6.*)

additional parts are subsequently generated anew. It follows that the number of parts is determinate. As Kant puts the general principle underlying this move: 'as soon as something is assumed as *quantum discretum*, then the multitude of units within it is determinate and hence, by the same token, is always equal to some number'.[2] The actual parts doctrine, in asserting the distinct existence of each part of matter, takes them as *quanta discreta*, and hence as determinate in number. This gives us (D5): the general conclusion that the number of parts into which a body can be m-divided is determinate.

Now, of particular relevance to our present study is how this bears on the case of infinitely m-divisible matter. Here, again, one will have to say that all the parts into which a body can be m-divided are given as a determinate collection prior to division. So, if bodies are m-divisible ad *infinitum*, they must then have an *actual infinity* of parts (a completely given greater-than-finite number of parts). They cannot merely be said to have an Aristotelian *potential infinity* of parts (a number of parts that is always

---

[2] Immanuel Kant, *Critique of Pure Reason*, tr. Werner S. Pluhar (Indianapolis: Hackett, 1996), B555.

finite at any stage, but may be increased without limit). This indeterminate potential infinity would only make sense if parts were generated as division proceeds. But if the parts are already one and all laid up in the whole prior to division, then they must be an actual infinity of pre-existent parts if the whole is endlessly divisible. Thus we have (D7): if a body is infinitely divisible, then there must be an actual infinity of parts in that body.

There are numerous instances of this reasoning given in the Enlightenment literature. Typically the argument is presented as setting up the classic paradoxes of infinite divisibility: if bodies are infinitely m-divisible, then—by (D1)–(D7)—they must each contain an actual infinity of parts; but this proves paradoxical, so bodies cannot be infinitely m-divisible.[3] We can start with two statements of the argument in which the mobilization of the actual parts doctrine is exceptionally clear. First, here is Walter Charleton, the arch-Epicurean and foe of infinite divisibility, in his 1654 treatise *Physiologia*. Charleton is, as ever, in loquacious mood:

[I]f we intend, that a Continuum hath therefore two parts actually, because it is *capable* of division into two parts actually: then it is necessary, that we allow a Continuum to have parts actually infinite, because we presume it capable of division into infinite parts actually ... [A]s in a Continuum two parts are not denied to exist, though it never be divided into those two parts [*notice the actual parts doctrine here*]: so likewise are not infinite parts denied to exist therein, though it be never really divisible into infinite parts. ... [H]ow can a Continuum be superior to final exhaustion, unless in this respect, that it contains infinite parts, *i.e.* such whose Infinity makes it Inexhaustible. Because, as those parts, which are deduced from a Continuum, must be præexistent therein before deduction (else whence are they deduceable?) [*again: the actual parts metaphysic is clearly assumed*] so also must those, which yet remain deduceable, be actually existent therein, otherwise they are not deduceable from it. For, Parts are then Infinite, when more and more inexhaustibly, or without end, are conceded Deducible.[4]

The actual parts assumptions underpinning the argument are also quite clear in the version presented in Antoine Le Grand's 1672 *Institutio philosophicae* (here in Blome's popular 1694 translation as *An Entire Body of Philosophy*). Le Grand attributes the argument to 'the *Epicureans*'; he does not himself endorse it.

---

[3] See Ch. 1, s. VIII, on these classic paradoxes.
[4] Walter Charleton, *Physiologia Epicuro-Gassendo-Charletoniana* (London, 1654; facsimile edn.: New York: Johnson Reprint Co., 1966), 93.

[E]very thing which is *Divisible*, can only be divided into those Parts, which are *actually* in it; and consequently, that if *Bodies* be infinitely divisible, they must have *actually* infinite Parts.[5]

Other statements of the argument found in the Enlightenment literature tend to be less scrupulous in registering the crucial role played by the actual parts doctrine. The doctrine is typically implicitly assumed rather than stated overtly as a premise. Of course, this should come as no surprise. As we have seen, at least by the time the rival scholastic system is in retreat, the actual parts doctrine is simply a banal truism for most of the Enlightenment's new philosophers, hardly worth mentioning as an explicit premise, if it is even consciously noted at all. In the statements of the argument that follow, I will try to point out the steps where the actual parts doctrine is being tacitly assumed. Here, for instance, is Henry More in the 1653 *Antidote Against Atheism*:

[N]othing can be divisible into parts it has not [*note: the actual parts doctrine is assumed here; a potential parts theorist would not accept this first claim*]: therefore if *Body* be divisible into infinite parts, it has infinite extended parts.[6]

And Pierre Bayle in the 1697 *Historical and Critical Dictionary*:

[I]f matter is divisible to infinity, it actually contains an infinite number of parts. It is not then a potential infinite; it is an infinite that really and actually exists. The continuity of parts does not prevent their actual distinction [*note: the actual parts doctrine*]. Consequently their actual infinity does not at all depend upon division.[7]

I think it is clear that each of these thinkers is presenting a compressed version of the argument that I have set out in full as (D1)–(D7) above, notwithstanding their failure to be as explicit as Charleton and Le Grand about their actual parts presuppositions.

Hume's versions of the argument in the 1739–40 *Treatise* are yet more abrupt. (See Chapter 1, section VI, for my case that Hume's arguments implicitly assumes actual parts premises.)

---

[5] Antoine Le Grand, *An Entire Body of Philosophy*, tr. Richard Blome, 2 vols. (London, 1694), ii. 2.

[6] Henry More, *Antidote Against Atheism*, in *Philosophical Writings of Henry More*, ed. F. I. Mackinnon (New York and Oxford: OUP, 1935), 11.

[7] Pierre Bayle, *Historical and Critical Dictionary*, tr. and ed. Richard H. Popkin (Indianapolis: Hackett, 1991), 356.

> [W]hatever is capable of being divided *in infinitum*, must consist of an infinite number of parts.
>
> Every thing capable of being infinitely divided contains an infinite number of parts.[8]

We also find a tight summary of argument (D1)–(D7) towards the end of our period in Kant's 1786 *Metaphysical Foundations of Natural Science*. Here Kant attributes the reasoning to 'the dogmatic metaphysician', that is (in this context), someone who admits actual parts:[9]

> If... matter is infinitely divisible, then (concludes the dogmatic metaphysician) it consists of an infinite multitude of parts; for a whole must in advance contain within itself all the parts in their entirety into which it can be divided.[10]

Finally, I must mention the versions of the argument that we find in Galileo, right back at the start of our period. I mention these last since their status as versions of the current argument (D1)–(D7) is contested by no less an authority than Galileo himself. The situation is as follows. In the 1638 *Dialogues Concerning Two New Sciences*, Galileo explicitly mentions the conflict between the actual parts and potential parts doctrines, and then states categorically that his arguments are meant to stand outside this controversy and that they do *not* assume actual parts. Now, *if* this is correct, then of course these statements are not really versions of the current argument (D1)–(D7) at all. But as I indicate in my annotations to Galileo's argument, it looks to me as if he is really *is* assuming the actual parts doctrine, despite his public disavowals. (I return to this interpretative question and Galileo's claim that these arguments function without the actual parts doctrine in section VI below.) In any case, here is the text:

> [A] division and subdivision that can be carried on indefinitely presupposes that the parts are infinite in number, otherwise the subdivision would reach an end. [*note: the actual parts doctrine seems to underpin this last claim*] ... [T]hus we have a continuous quantity built up from an infinite number of indivisibles.

---

[8] David Hume, *A Treatise of Human Nature*, ed. David Fate Norton and Mary J. Norton (Oxford and New York: OUP, 2000), 1. 2. 1. 2, 1. 2. 2. 2. Additional references abbreviated 'SBN' give the corresponding page numbers in David Hume, *A Treatise of Human Nature*, ed. L. A. Selby-Bigge, 2nd edn. with text revised and notes by P. H. Nidditch (Oxford: OUP, 1978). Here the SBN references are 26, 29.

[9] According to Kant, the actual parts doctrine is unavoidable for all transcendental realists. See Ch. 1, s. IX (faction (1)), and Ch. 2. s. III.

[10] Immanuel Kant, *Metaphysical Foundations of Natural Science*, in Kant, *Philosophy of Material Nature*, tr. J. W. Ellington (Indianapolis: Hackett, 1985), 53.

The very fact that one is able to continue, without end, the division into finite parts, makes it necessary to regard the quantity as composed of an infinite number of immeasureably small elements [*again: the actual parts doctrine?*].[11]

But enough historical documentation. Our next question must be: is the argument any *good*? Well, it is transparently valid, and the assumptions (D2), (D4), and (D6) all seem (to my mind) unobjectionable truisms, so this leaves only the initial assumption (D1) as open to serious challenge. Of course (D1)—the actual parts doctrine—*is* contested by partisans of the rival potential parts metaphysic such as the Aristotelians, Hobbes, and the later Kant. But it does seem at least that, *if* the actual parts doctrine is correct, then each body does indeed have a determinate number of parts, and infinitely divisible bodies each have an actual infinity of parts. This conditional claim at least is sound, and so the actual parts doctrine does entail a determinate number of parts as a corollary.

But is there nothing to be said against (D1)–(D7) other than to challenge the actual parts assumption (D1)? What about those philosophers we saw above—such as Le Grand and Kant—who reject the argument? Don't they have anything further to say against it? In fact the only objections I have come across in the Enlightenment literature either challenge the actual parts assumption (D1) (this is Kant's move, for example), or simply take refuge in the mysteries of the infinite. We find this second sort of response in Cartesians such as Le Grand, for instance. He clearly accepts actual parts[12] and then reports the '*Epicurean*' argument (D1)–(D7) that this doctrine, alongside infinite divisibility, entails an actual infinity of parts.[13] He is unable to locate anything specifically wrong with this argument, but nonetheless clings to the claim that bodies contain only an indeterminate potential infinity of parts. As for (D1)–(D7), well, 'we need not trouble our selves too much, to explain or extricate these Difficulties, which are not proportionate to our Intellect, and which cannot perspicuously and distinctly be conceived by us. For it appertains only to *Infinite Mind*, to understand that which is *Infinite*, and to determine anything concerning it.'[14] As Le Grand himself notes, this sort of appeal to our feeble comprehension of

---

[11] Galileo Galilei, *Discourses Concerning Two New Sciences*, tr. Henry Crew and Alfonso de Salvio (New York: Macmillan, 1914), 34.

[12] '[A]n aptitude to be separated...is sufficient to make one thing not to be another'; 'whatsoever is extended must be conceiv'd to have distinct Parts'. Le Grand, *Entire Body*, i. 96, 97.

[13] Ibid. ii. 2.   [14] Ibid. ii. 8.

the infinite as a way of eluding the force of arguments like (D1)–(D7) was standard among the Cartesians.[15] But to my mind this sort of strategic humility is simply an evasion. No step of (D1)–(D7) seems to defy comprehension: the reasoning is brief and altogether limpid. If no flaw can be found in it, it should be accepted. One is tempted to reach for the *Treatise* and Hume's impatient retort (to this very objection) that 'nothing can be more absurd, than this custom of calling a *difficulty* what pretends to be a *demonstration*, and endeavouring by that means to elude its force and evidence'.[16] In any event, if we put this sort of response to the argument to one side, then the other criticisms in the literature simply challenge the actual parts doctrine (D1) and therefore do not challenge our conclusion that the actual parts doctrine does indeed entail that bodies have a determinate number of parts.

III. The Argument from the Definiteness of Parts to Ultimate Parts

We move now to the second of our two shortcircuit arguments: *the argument from the definiteness of parts to ultimate parts*. This pattern of reasoning was no less popular than its sister argument from actual parts to a determinate number of parts: in fact we often find the two arguments run together or interwoven in the period literature. However, as we shall see, it turns out to be a good deal more precarious.

The clearest statement of the argument from the definiteness of parts to ultimate parts in the Enlightenment literature is in Kenelm Digby's 1644 *Two Treatises*, and, since this will be the primary blueprint for my reconstruction of the reasoning, let us begin with a look at his text. Digby is of course a potential parts theorist. He hopes to show, first, that the actual parts doctrine entails m-indivisible ultimate parts, and, second, that since ultimate parts prove paradoxical alongside infinite divisibility, the actual parts doctrine must therefore be rejected. In fact this two-step argument was a classic scholastic pattern of reasoning against the actual parts

---

[15] Le Grand gives us the reference to the *locus classicus*, Descartes's *Principles*, part 1, s. 26.
[16] Hume, *Treatise*, 1. 2. 2. 6; SBN 31 We also know that Spinoza would have resisted this Cartesian manœuvre: see Lodewijk Meyer, preface to Spinoza, *The Principles of Cartesian Philosophy*, tr. Samuel Shirley (Indianapolis: Hackett, 1988), 6, and compare 54.

doctrine, deriving ultimately from Aristotle: the only way of avoiding the absurd conclusion that continua have ultimate parts is to reject the actual parts analysis for a potential parts approach.[17] As we shall see, many actual parts theorists also accepted the first step—actual parts entail ultimate parts, via the argument from the definiteness of parts—though of course (contra Aristotle, Digby, and the scholastics) they then reject the second step and the view that ultimate parts are absurd. Our extract from Digby gives the first stage of the traditional scholastic argument: if we assume the actual parts doctrine, then continua such as bodies must be built out of *'indivisibles'*, m-indivisible first parts.

[I]f we allow actuall partes to be distinguished in Quantity, it will follow that it is composed of points or indivisibles, which we shall proove to be impossible.

The first will appear thus: if quantity were divided into all the partes into which it is divisible, it would be divided into indivisibles (for nothing divisible, and not divided, would remain in it) but it is distinguished into the same partes, into which it would be divided, if it were divided into all the partes into which it is divisible; therefore it is distinguished into indivisibles. The major proposition is evident to any man that hath understanding. The minor, is the confession or rather the position of the adversary, when he sayeth that all its parts are actually distinguished [*i.e. when he asserts the actual parts doctrine*]. The consequence cannot be calumniated, since that indivisibles, whether they be seperated or joyned, are still but indivisibles; though that which is composed of them be divisible. It must then be granted that all the partes which are in Quantity, are indivisibles; which partes being actually in it, and the whole being composed of these partes onely, it followeth, that Quantity is composed and made of indivisibles.[18]

Digby's version of the argument is admirably clear, and can be straightforwardly lifted without serious interpolation into the reconstruction in Table 4.

According to the actual parts doctrine, *all* the parts into which a body may be m-divided are already distinct from one another, even prior to division. Now, if the parts of a body were m-divided as far as down m-division can possibly go, then the body would be m-divided into m-indivisibles. But the parts are already distinct from one another thus far down, even in advance of any act of division. As Digby puts it, given the actual parts account, the whole 'is distinguished into the same partes, into

---

[17] See 'faction (1)' in Ch. 1, s. IX.
[18] Digby, *Two Treatises*, 10.

Actual Parts and Shortcircuit Arguments / 143

Table 4 The argument from the definiteness of parts to ultimate parts

(E1)	If a body contains distinct parts as far down as far as m-division can go, then it contains m-indivisible metaphysical atoms: ultimate parts. (*Assumed.*)
(E2)	Bodies contain distinct parts as far down as m-division can go. (*From the actual parts doctrine, which is assumed.*)
So: (E3)	Bodies contain m-indivisible metaphysical atoms: ultimate parts. (*From E1 and E2.*)

which it would be divided, if it were divided into all the partes into which it is divisible'. So the parts are already distinct from one another as far down as m-division can possibly go: they are distinct from one another all they way down to the level of m-indivisibles or metaphysical atoms. The whole body is (so to speak) already 'distinguished' into independent beings through and through, even if it is not yet m-divided through and through. Here it is important to notice a claim made for this shortcircuit argument (E1)–(E3). It may well be impossible to successively m-divide the whole through and through. If the whole is *infinitely* m-divisible, then no successive process of m-division and subdivision could ever exhaustively m-divide the body through and through. (Even if any individual division in the infinite sequence of divisions and subdivisions is makeable, still the whole infinite sequence could not be completed.) But this does not matter to our current argument. It is as if the actual parts doctrine—in giving us *all* the parts prior to division—distinguishes the whole into all its parts *all at once*. If one had to advance through the series of ever-smaller parts sequentially, then one could never complete the process. But the actual parts doctrine allows us to shortcircuit this sort of successive procedure: we *begin* at the point where all parts are already distinct from one another through and through; we do not have to reach it through any sequential process.

Here is another way of intuitively thinking about the argument from the definiteness of parts to ultimate parts. (The essential point of this second version is the same: it is just that the reader is given a new way of imaginatively grasping the 'all at once' leap to ultimate parts. This second version simply replaces the intuitive notion of parts that are distinct from one another 'through and through' or 'all the way down' with the intuitive notion of a complete set of distinct parts.) Given the actual parts doctrine, the entire collection of parts in any body is pre-given in determinate totality. No part awaits subsequent generation; one and all are fully present in the body from the start. Even if the number of parts is infinite, it is (as we saw in section II above) a determinate actual infinite, not an open-ended, ever-increasable potential infinite. But, if we already have the *complete set* of parts fully given, then we must have the elements of that set that are the smallest members of that set. Since *all* the parts are given, smallest parts must be given. The idea is that one cannot avoid this reasoning by taking refuge in the claim that there is an indefinitely expandable, open-ended series of successively smaller parts, and hence no smallest part. This would be to appeal to an open-ended potential infinite. But what we have here is a completely given determinate totality of parts. So even if division can proceed endlessly, and no successive process of division and subdivision will ever *reach* the smallest parts, they must nonetheless be there, already laid up in the completely given collection.

In addition to Digby's exceptionally explicit version, we also have certain other statements of the argument in the period literature. For instance John Keill reports it in his 1733 *Introduction to Natural Philosophy* (from his Oxford lectures of 1700):

[Opponents of infinite divisibility] farther urge, If all Quantity is divisible *in infinitum*, and the Parts are actually in the Extension [*note: the actual parts doctrine*], there will be actually given a Part infinitely small, and consequently not farther divisible.[19]

Similar reasoning also appears in Galileo, where his statements of argument (D1)–(D7) (the argument from actual parts to a determinate number of parts) are woven together with versions of (E1)–(E3) (the argument from the definiteness of parts to ultimate parts). (As I noted in the previous section, we need to weigh my interpretative claim that the actual parts doctrine underpins Galileo's argument against his explicit denial. I return to this interpretive issue in section VI below.)

I say it is most true and necessary that the line is composed of points, and the continuum of indivisibles ... Open your eyes, for goodness' sake, to this light that has remained hidden until perhaps now, and recognize clearly that the continuum is divisible into parts always divisible only because it is constituted of indivisibles. For if the division and subdivision must be able to go on forever, it must necessarily be that the multitude of parts is such that one can never go beyond it, and therefore the parts are infinite, otherwise the subdivision would

---

[19] John Keill, *An Introduction to Natural Philosophy* (London: Senex, Innys, Manby, Osborn, & Longman, 1733), 40.

Actual Parts and Shortcircuit Arguments / 145

come to an end [*note: the actual parts doctrine?*], and if they are infinite then they must be without magnitude, because an infinity of parts endowed with a magnitude compose an infinite magnitude, and we are speaking of terminated magnitudes; therefore the highest and ultimate components of the continuum are indivisibles infinite in number.[20]

Finally, the current argument is at least tacitly endorsed by Walter Charleton, Pierre Bayle, George Berkeley, and David Hume. Each of these actual parts theorists employs it (albeit at an implicit level) in their claim that infinite divisibility is paradoxical, since it clashes with the ultimate parts mandated by the actual parts metaphysic.[21]

But to move to critical assessment. First, the argument reconstructed as the straightforward syllogism (E1)–(E3) is clearly valid. But is it sound? Well, premise (E2) simply asserts the actual parts doctrine: as Digby says, it 'is the confession or rather the position of our adversary, when he sayeth that all its parts are actually distinguished'. So this only leaves us with premise (E1): the claim that, if a body contains distinct parts as far down as m-division can go, then it contains m-indivisible ultimate parts. Is this true? Digby assures us that it is—indeed, it 'is evident to any man that hath understanding'. But this step may be a little too quick.

Assume the actual parts doctrine and suppose that bodies are infinitely m-divisible. Grant that, given the actual parts doctrine, the entire collection of parts into which the whole can be m-divided is pre-given as a determinate totality. And also grant that, given the actual parts doctrine, bodies indeed contain distinct parts through and through. No part into which the whole may be m-divided remains yet to become distinct from the others: each part is already distinct from the very beginning. This much does indeed follow. However, even if we allow all this, the argument (E1)–(E3) still fails, because the conditional claim (E1) is a non-sequitur. We could have a completely given actual infinity of parts—a determinate, completely given collection—and we could have distinct parts through and through, but we still need not have any m-indivisible ultimate parts.

---

[20] Quoted from an unpublished manuscript in Francois de Gandt, *Force and Geometry in Newton's Principia*, tr. Curtis Wilson (Princeton: Princeton University Press, 1995), 172–3. See also Galileo, *Two New Sciences*, 34.

[21] Charleton, *Physiologia*, 91–3. Bayle, *Dictionary*, 362. George Berkeley, *Notebook B*, s. 352 and *Principles*, s. 47, in *The Works of George Berkeley*, ed. A. A. Luce and T. E. Jessop, 9 vols. (London: Thomas Nelson & Sons, 1948–9), i. 42, ii. 283. Hume, *Treatise*, 1. 2. 2. 3; SBN 30. See also my discussion of the classic paradoxes in Ch. 1, s. VIII.

To see this, it may help to consider an analogous case. Suppose I put one plum into a first basket, two into a second basket, three into a third, and so on. I continue on and on up sequentially through the natural numbers (the baskets will have to get very big, but let this pass). Now suppose that I can somehow shortcircuit this sequential, basket-by-basket strategy and *all at once* assign plums to baskets for the entire series of natural numbers. What we then have is a completely given collection of baskets (one for each of the natural numbers) that is actually infinite in size, not merely an open-ended, indeterminate potential infinity. Moreover, each of the baskets is distinct from all the others. But now notice that, even though all the baskets are distinct from one other 'through and through' the entire series, and even though all the baskets are given in a determinate totality or complete set (no baskets are subsequently added to the collection), still there is no *largest* basket with more plums than any other. Just as the total collection of all the natural numbers is actually infinite in size and yet there is no largest natural number, so it is for our baskets. And so it is, *mutatis mutandis*, for the parts of an infinitely m-divisible body. Given the actual parts doctrine, the whole body contains distinct parts through and through ('all the way down') and yet there still need be no smallest, ultimate part. (The possibility envisaged here marries an actual parts account to what current metaphysicians call 'atomless gunk': material that is infinitely divisible and whose parts *all* have further proper parts, thereby altogether lacking a ground floor of atomic parts.[22]) Here we learn something peculiar about actually infinite collections: every member of such a collection can be given, but it does not follow that smallest members of the collection will thereby be given. We can have all the members, but not have any smallest members. So the conditional claim (E1) fails in the case of infinitely m-divisible body.

I conclude that the argument (E1)–(E3) only goes through in the case of bodies that are merely finitely m-divisible; it fails if we are dealing with an infinitely m-divisible body. So, while I claimed that the argument from actual parts to a determinate number of parts was a success, it seems that the argument from the definiteness of parts to ultimate parts is a failure. Of course this does not show that there is *no* successful argument from actual parts to ultimate parts. For instance, we have yet to look at *the*

---

[22] On atomless gunk, see David Lewis, *Parts of Classes* (Oxford: Blackwell, 1991), 20. Theodore Sider, 'Van Inwagen and the Possibility of Gunk', *Analysis*, 53 (1993), 285–9. Dean W. Zimmerman, 'Theories of Masses and Problems of Constitution', *Philosophical Review*, 104 (1995), 53–110.

argument from composition, another popular attempt to forge a link between the actual parts metaphysic and ultimate parts. (I examine this rather different argument in Chapter 4 below.)

IV. Overview of Alternative Strategies: Arguments that do Not Invoke Actual Parts

In the previous two sections I assessed a pair of arguments that enjoyed much currency in the Enlightenment literature. I suggested that the argument from actual parts to a determinate number of parts was a success, whereas the argument from the definiteness of parts to ultimate parts was a failure.

Notwithstanding the respective success and failure of these two arguments, I do want to stress that they share an essential core strategy. Each argument mobilizes the actual parts doctrine to present us 'all at once' with the complete collection of parts into which the whole can be m-divided. This neatly bypasses any series which presents the parts successively. (Given infinite m-divisibility, such a series might prove quite literally interminable.) This is why I have constantly insisted—perhaps at the risk of irritating the reader—on the way the actual parts doctrine lurks in our various statements of these two arguments in the period texts. And the obvious contrast here is with the potential parts doctrine, according to which the parts into which the whole can be m-divided are given *successively*, as the process of division and subdivision generates them. Here the parts are given in a series rather than all at once, and so, if the whole is infinitely divisible, then it seems the entire collection can never be completed: it would have to be merely *potentially* infinite.

In this section and the next I want to look at two rather different sets of arguments that eschew the actual parts strategy. According to these new arguments, *whether or not a body has actual parts*, the sheer fact that it is m-divisible is enough to guarantee that we can be presented with the complete collection of parts. And this applies even in the case of an infinitely m-divisible body. So if there is something paradoxical about the entire collection of parts into which a body can be m-divided being completely given (perhaps because this is supposed to entail a determinate number of parts, or m-indivisible ultimate parts) then we have these paradoxes with or without the actual parts doctrine. In short, the actual parts doctrine is something of a red herring: even if we ignore it altogether, any body that is m-divisible (even infinitely m-divisible) is such that we can be presented with the complete collection of its parts.

It is important to appreciate that these alternative arguments challenge my overall interpretation of the problems of material structure and my assessment of the Enlightenment debate. A central argument of this study has been that the classic paradoxes of infinite divisibility only arise in the context of the actual parts metaphysic. However, if these alternative arguments are convincing, then infinite divisibility will lead to the various classic paradoxes whether or not bodies have actual parts. The actual parts doctrine is simply beside the point. Clearly this result would compromise my view that the actual parts metaphysic underpins the classic paradoxes of infinite divisibility. (My historical claim that the early moderns were led to the paradoxes via the actual parts doctrine *might* still stand. But my philosophical claim that the paradoxes simply do not arise without actual parts assumptions would fall.)

In section V I examine the first alternative type of argument. According to this approach, whether or not a body has actual parts, we can in any case infer from the infinite m-divisibility of the whole to the possibility of performing a through-and-through infinite m-division. This would present us with the complete collection of parts into which the whole can be m-divided. This type of argument comes in two versions: one envisages a *successive* infinite m-division; the other, an *instantaneous* or *all-at-once* infinite m-division. In section VI I then look at the second alternative type of argument. According to this second approach, the sheer fact that a body is infinitely m-divisible already presents us with the complete collection of its parts, regardless of whether or not we can perform an infinite through-and-through division, and regardless of whether or not it has actual parts.

V. Arguments that Envisage the Completion of an Infinite Division

These arguments begin by reminding us that whatever is divisible can be divided: the potency of divisibility can be reduced to the act of division. This harmless verbal truism is then applied to *infinitely* m-divisible bodies with the result such bodies can be completely m-divided through and through into an infinite number of parts. The potency of infinite m-divisibility can be reduced to the act of infinite m-division. Notice that this argument does not appeal to the actual parts doctrine to give us the complete collection of parts into which the whole can be m-divided. Rather the sheer fact of m-divisibility already guarantees that we can generate the complete set of parts into which the whole can be m-divided, simply by m-dividing as far as the whole is m-divisible.

Robert Boyle gives us a crisp version of this argument in his 1662 *Examen of Mr. Hobbes*. Hobbes had argued in *De Corpore* and his *Dialogus Physicus* that if one adopts the potential parts doctrine then infinite m-divisibility implies only a potential infinity of parts, not a complete actual infinity. The potential parts theorist can thus easily avoid the paradoxes that come with admitting an actual infinity of parts: given their doctrine, 'Infinite division cannot be conceived, but infinite divisibility easily can be.'[23] And it is just here that Boyle takes issue with the Monster of Malmesbury. Boyle argues, contra Hobbes, that the potency of infinite m-divisibility already entails that the number of parts into which the whole can be m-divided is actually infinite. The actual parts–potential parts controversy is simply beside the point.

I see not, why an infinite division cannot be as well conceived as an infinite divisibility, since sure an omnipotent agent is able to do what is possible to be done; and why else should a body be called infinitely divisible?[24]

Notice that Boyle clearly indicates that this reasoning is intended to be a straightforward application of our simple verbal truth. Whatever is divisible can be thus far divided (at least by God), and so whatever is infinitely m-divisible can be m-divided into an infinity of parts—'why else should a body be called infinitely divisible?' We can also see this same reasoning in the younger Newton's 1661–5 Trinity notebook. The inference of current interest is compressed inside Newton's parentheses in the following larger argument:

I shall use one argument to show that [matter] cannot be divisible *in infinitum*... if any finite quantity were divided into infinite parts (and certainly it may if it be so

---

[23] Thomas Hobbes, *Dialogus Physicus de Natura Aeris*, in *Opera Philosophica Quœ Latine Scripsit*, ed. William Molesworth, 5 vols. (Aalen: Scientia, 1961), iv. 244. See also Ch. 2, s. III, esp. n. 55

[24] Robert Boyle, *Examen of Mr. Hobbes*, in *The Works of the Honourable Robert Boyle*, 6 vols. (London: J. and F. Rivington, 1772), i. 236.

far divisible), those infinite parts added would make the same quantity they were before²⁵

Newton explicitly credits this line of argument to 'the excellent Dr. More'—and we do indeed find similar reasoning in Henry More's slightly earlier *Immortality of the Soul* (1659), which was familiar to the younger Newton. According to More, this sort of argument establishes not merely that the complete set of parts can be given via through-and-through m-division; it also shows that there must be ultimate m-indivisible parts: that '*Matter* consists of parts *indiscerpible*'.

[T]hat there are such *indiscerpible* particles into which *Matter* is divisible, *viz.* such as have *Essential* extension, and yet have parts utterly *inseparable*, I shall plainly and compendiously here demonstrate ... by this short Syllogism.

That which is actually divisible so farre as division any way can be made, is divisible into parts *indiscerpible*.

But *Matter* (I mean that *Integral* or *Compound* Matter) is actually divisible as farre as actual division any way can be made.²⁶

(Notice the structural similarity of this argument to Digby's argument (E1)–(E3), in which matter is distinguished into 'all the partes into which it is divisible'. The crucial difference is, of course, that Digby argues from the actual parts doctrine to the claim that matter already contains distinct parts as far as it is m-divisible, whereas here More eschews actual parts premises to simply argue that matter *can be m-divided* as far as it is m-divisible.)

Later in the *Immortality of the Soul* the argument is restated and explicitly applied to the case of infinitely m-divisible matter. Even here the potentiality of infinite m-divisibility entails that there can be a complete m-division through to ultimate parts:

For though we should acknowledge that *Matter* were discerpible *in infinitum*, yet supposing a Cause of Infinite distinct perception and Infinite power (and God is such), this Cause can reduce this capacity of infinite discerpiblenes of *Matter* into act.²⁷

---

[25] Isaac Newton, 'Certain Philosophical Questions', tr. and ed. J. E. McGuire and Martin Tamny, in McGuire and Tamny, *Certain Philosophical Questions: Newton's Trinity Notebook* (Cambridge: CUP, 1983), 330–489, 341.

[26] Henry More, *Immortality of the Soul*, ed. A. Jakob (Dordecht: Kluwer, 1987), 7.

[27] Ibid. 40.

And again in More's later *Enchiridium Metaphysicum* (1671), the same argument:

By physical monads I understand particles so minute that they cannot be further divided or discerped into parts. Body, however, is composed entirely of these, and can be resolved into the same, at least by divine power. *For it is entirely contradictory that body cannot be actually divided to the degree that an actual division can be made in it.* Since, indeed, it has been arrived at that all the parts are rightly said to be physical monads, inasmuch as in them the physical division comes to a halt, they themselves cannot be discerped into further parts, that is, physically divided.[28]

Finally, we also find reports (though not endorsements) of the argument in Pierre Bayle and John Keill.[29]

There seem to be two possible ways to understand this argument, the difference between them turning on just how we are supposed to get from infinite m-divisibility through to a completed infinite m-division. On the one hand, perhaps the idea is that (1) the potency of infinite m-divisibility can be reduced to the act of infinite m-division via a *successive process*. Here the claim is that it is logically possible to completely traverse an infinite series of m-divisions successively, eventually arriving at the point at which the infinitely m-divisible whole is completely m-divided through and through. On the other hand, perhaps the idea is that (2) the potency of infinite m-divisibility can be reduced to the act of infinite m-division in a *one-step, all-at-once* operation of m-division. Here the idea is that the task of

---

[28] Henry More, *Enchiridium Metaphysicum*, tr. Alexander Jakob (Hildesheim: Georg Olms Verlag, 1995), 71–2. Emphasis added.

[29] Pierre Bayle reports a very similar argument: 'On objecte en troisieme lieu que ce qui a des parties est divisible en toutes ces parties, & que par conséquent le contenu est divisible en toutes les parties qu'il y a. Supposé donce que Dieu le divisé en toutes ces parties, on demande s'il est encore divisible ou non. S'il est divisible, donc il n'a pas été divisé en toutes ces parties. S'il n'est plus divisible, donc il peut être réduit en atome, & il n'est pas composé de parties divisibles à l'infini.' (*Système Abrégé*, in Bayle, *Œuvres Diverses* (1731; facsimile edn.: Hildesheim: Georg Olms, 1968), 303.) 'They object in the third place that that which has parts is divisible into all its parts, and, as a result, the continuum is divisible into all the parts that are there. Supposing then that God divided it into all of its parts, they ask are these parts further divisible or not? If these parts are divisible then it has not been divided into all of its parts. If they are not divisible, then it is possible to reduce it to an atom, and it is not composed of parts divisible to infinity.' Likewise, John Keill relates the following argument: 'Another Argument against infinite Divisibility of Matter, is fetched from the Divine Omnipotence. God, they say, can resolve any Quantity into its infinitesimal Parts and separate those Parts from one another: but if so, then we may be given an ultimate part and the Divisibility would be exhausted, therefore Quantity is not divisible *in infinitum*.' (*Introduction*, 39).

infinite m-division is performed instantaneously: God (at least) can splinter the infinitely m-divisible whole into a powder of *all* its parts in a single one-shot step.

Each of these approaches raises a different array of philosophical issues. Let us take them in turn.

1. Successive Infinite Division

First consider the suggestion that one could complete an infinite m-division via a successive, step-by-step process. This countenances the possibility of sequentially dividing and subdividing an infinitely m-divisible body 'through and through'—all the way down to the point at which *all* of the parts into which it is m-divisible are completely given. But this would be to completely traverse an infinite series of ever-smaller parts—to run, step by step, all the way through a sequence of parts within parts that continues ad infinitum. And this looks impossible. Any attempt to proceed, one step at a time, all the way through the series to finally arrive at a point at which every last member has been ticked off must surely fail, for an infinite series has no last member. No one (not even God) can completely traverse an infinite series one step at a time. Thus, no one (not even God) could m-divide an infinitely m-divisible body through and through in this fashion. In sum, here we can grant the simple verbal truism that whatever is divisible can be divided—or even (with More) that 'it is entirely contradictory that body cannot be actually divided to the degree that an actual division can be made in it'—without yet agreeing that this entails the possibility of successively dividing an infinitely divisible whole through and through into a complete collection of parts. Rather, an infinitely divisible whole can be divided ad infinitum only in the sense that division can continue endlessly, with no prospect of ever being completed. Infinite divisibility thus entails a potential infinity of parts. But it does not entail the possibility of ever completing the successive division of the whole. The underlying point here is that an infinite series cannot be successively run through, and therefore, if we are ever to have a completely given infinite collection (i.e. an actually infinite collection, not just an open-ended, potentially infinite one), then it must be given all at once, in a single stroke—it cannot be given successively.

Boyle's, More's, and Newton's texts offer no reply to the obvious objection that one can never complete an infinite series by successively running through its members step by step. Principles of charity might then suggest that we should interpret them as thinking in terms of an *all-at-once* m-division rather than this obviously problematic *successive* m-division. (I assess this alternative all-at-once version in the next subsection.) But it is of course also possible that these philosophers were simply unclear in their own minds about this question. The obvious weakness of the naïve successive argument certainly presents a plausible explanation of the fact that this sort of direct argument from infinite divisibility to the possibility of an infinite division fell from favour by the end of the seventeenth century and, for instance, does not appear in Newton's later work. It seems to be gradually edged out during our period by the more plausible alternative arguments (D1)–(D7) and (E1)–(E3), which attempt to establish much the same conclusion, but by way of the actual parts doctrine and the prior existence of all the parts as distinct entities.

But perhaps someone might try to resuscitate the successive version of the argument by way of the recent literature on supertasks and the logic of infinite series. I have objected that one can never complete an infinite series by successively running through its members step by step: an infinite series cannot be traversed. But it has been argued in the recent literature that an infinite series *can* in fact be traversed. It may seem at first blush that such a traversal is impossible, but this (we are told) is only because one is supposing that it would take forever to run through an infinite series, and thus that the process would never end. But suppose one *accelerated* the pace at which one proceeds through the series, taking half as much time for each successive step in the sequence. For instance, suppose that it takes me a second to divide a body into two halves. In the next half second I then split each of those halves in two, leaving me with quarters. In the next quarter second I divide again, leaving me with eighths. In the next eighth of second, I divide into sixteenths. And so I continue, dividing and subdividing at this exponentially accelerating pace. At least if time is also infinitely divisible, it seems that after two seconds I would have completely traversed the entire series successively, even if it is infinite. The lesson we are supposed to draw is that if one accelerates fast enough, then one *can* successively traverse an infinite series one step at a time: in the jargon of the literature, one can complete a 'supertask'.

A serious overview and assessment on the literature concerning supertasks would take me beyond the scope of this present study. But at least let me indicate in broad terms why I am sceptical about this sort of response.

Like most sceptics about supertasks, I certainly do not wish to challenge the mathematical coherence of the completion of an infinite series via an exponential acceleration.[30] But while this may be perfectly consistent in the mathematical realm, its translation into the real world of material objects and actual physical processes seems to rapidly generate absurdities. The main concern is that, if we are ever to divide an infinitely divisible body through and through via a successive, step-by-step approach, then we will have to *complete* the division. But of course we know that there is no last member of an infinite series. And how can we successively complete the series if it has no last member? The suggestion is that we are to advance along an infinite sequence sequentially, completing *each step in turn* before advancing on to the next. But how could we ever *finish* such an assignment, arriving at a point at which every step of the series is behind us, if we never encounter a *last* step? At least if we are thinking of this as an actual physical process in the real world, it seems clearly impossible. (It does not seem that exponential acceleration will help us get out of *this* problem.)

Some proponents of the possibility of supertasks have attempted to face down this problem, declaring that an infinite series can be successively run through and completed, even though it has no last member.[31] But if we are considering a concrete one-step-at-a-time sequential process in the real world, this seems (to my mind) too much to swallow. Certainly, if time is infinitely divisible, we can then take any finite interval of time and assign within it a separate subinterval to each and every one of the infinite number of steps that make up the supertask. (This is the key point on which defenders of supertasks insist.) But for all this admitted mathematico-logical coherence of a supertask, I do not yet see how such a process

---

[30] Current opinion about the feasibility of supertasks is divided. The following commentators are sceptical about the possibility of traversing an infinite: Pamela Huby, 'Kant or Cantor? That the Universe, if Real, must be Finite in Both Space and Time', *Philosophy*, 46 (1971), 121–32. W. L. Craig, *The Kalam Cosmological Argument* (London: Macmillan, 1979), esp. 83–7, 97–9. G. J. Whitrow, *The Natural Philosophy of Time* (Oxford: OUP, 1961), 27–33. Jean Paul Van Bendegem, 'In Defence of Discrete Space and Time', *Logique et Analyse*, 150–1 (1995), 127–50. The following commentators allow the possibility of supertasks, or the traversal of an infinite series: Paul Benacerraf, 'Tasks, Super-Tasks and the Modern Eleatics', *Journal of Philosophy*, 59 (1962), 765–84, Richard Sorabji, *Time, Creation and the Continuum* (Ithaca, NY: Cornell University Press, 1983), 327–8. See also the useful collection Andrew Morton and Stephen P. Stitch (eds.), *Benacerraf and his Critics* (Oxford: Blackwell, 1996).

[31] Sorabji, *Time, Creation and the Continuum*, 328–9. John Earman and John D. Norton, 'Infinite Pains: The Trouble with Supertasks', in Morton and Stitch, *Benacerraf*, 231–61, 233.

could be realized in the real world. *For how could one completely run through a real-world one-step-at-a-time process if that process lacked a last step?* Suppose the task was to polish a 1/2 inch piece of brass, then a 1/4 inch piece, then an 1/8 inch piece, etc., accelerating exponentially all the while in order to complete the task in a finite but infinitely divisible interval of time. Suppose the task then completed. Which is then the last piece of brass I polished? None of them. But each piece was polished in turn, one at a time. None were polished simultaneously. (I could even have taken a cigarette break between polishing each separate piece, so long as the time allowed for the break itself also diminished exponentially as the task progressed.) And now each piece in the series is fully polished. So how can there fail to be a piece that was polished last? How is this aspect of the completion supposed to work? This is what we are never told.[32]

Another traditional objection to the possibility of successively traversing an infinite series—to my mind a strong one—is the point that this would be parallel to counting to infinity.[33] Imagine that at each stage of the series—at each act of division, say—we count off another number. Now, if we can traverse the infinite series we can complete the count to infinity. But this is agreed to be impossible; and hence we cannot accomplish the parallel task of successively traversing an infinite series. This objection makes vivid the previous problem (i) that traversal would require us to successively complete a series that has no last member. It also brings out the further problem (ii) that a successive traversal would allow us to generate an infinite from a step-by-step addition of finites—violating the maxim that (in Russell's words) 'no succession of steps from one number to the next will ever reach from a finite number to an infinite one'.[34] Somehow we manage to build an infinite number, but at any successive stage we only ever add one more to our count; so how—and *when*—does the infinite appear?[35]

---

[32] Compare José Benardete's stimulating discussion of open continua in his *Infinity: An Essay in Metaphysics* (Oxford: OUP, 1964).

[33] See Aristotle, *Physics*, book 8, tr. R. P. Hardie and R. K. Gaye, in *The Complete Works of Aristotle*, ed. Jonathan Barnes, 2 vols. (Princeton: Princeton University Press, 1984), i. 315–446 (esp. s. 8).

[34] Bertrand Russell, *Our Knowledge of the External World* (London: Allen & Unwin, 1926), 202.

[35] Sorabji also resists this argument: 'For *counting* the [members of the series] is unlike *traversing* them in a crucial respect. To count is to recite numerals, aloud or to oneself, while correlating them with something else. But the time needed with the recitation does not

2. All-at-Once Infinite Division

The second way of reading the claim that an infinitely divisible body could be through and through divided involves the idea that this could be performed in an instantaneous, one-shot operation. The whole could be completely divided in a single act, presenting us with the entire collection of parts all at once. One can see straightaway that this sort of instantaneous superdivision would avoid the chief objection to the successive approach. If the entire division is performed all at once then there is no need for a last step in order for the task to be completed. The need for a last step disappears with the device of a sequential series. With one-shot infinite division we thus avoid the central difficulty of successive infinite division altogether.

Alas, neither Boyle nor More nor Newton explicitly states whether they are thinking of the envisaged task as a successive or an all-at-once operation. But they do (of course) appeal to 'an omnipotent agent' or 'divine power' as a device to get us to the possibility of a complete infinite division, and this invocation of God may suggest they have in mind an all-powerful being who could perform the operation instantaneously. Moreover, I have found one statement of the argument in the early modern literature where the all-at-once interpretation is at least strongly suggested. This occurs in Spinoza's geometrical reconstruction of Descartes's *Principles of Philosophy*, where we get a report (though not an endorsement) of the argument that

> If two quantities—say A and its double—are divisible to infinity, they can also be divided in actuality into an infinite number of parts by the power of God, who understands their infinitely many parts *with a single intuition*.[36]

---

diminish as we progress, in the same way as does the time needed for traversing... There should be no surprise, therefore, that the task of *counting* cannot be completed; it does not follow that the task of *traversing* cannot be.' (Sorabji, *Time, Creation and the Continuum*, 324.) But this move will not help. Sorabji wants to take the counting-to-infinity objection as a version of the old claim that there is not enough time to traverse an infinite series (just as there is not enough time to count to infinity). So interpreted, he can then respond by simply reminding us that supertasks employ the device of exponential acceleration in infinitely divisible time. But this misses the point of the objection: even if we allow the process of counting to be accelerated in exactly the same way that the traversal is, the idea of counting to infinity still remains objectionable as ever—as problems (i) and (ii) (which certainly survive the acceleration of counting) show.

[36] Benedict de Spinoza, *The Principles of Cartesian Philosophy*, tr. Samuel Shirley (Indianapolis: Hackett, 1988), 54. Emphasis added.

Finally, even if there is something of a silence in the early modern period over whether the infinite division is supposed to proceed successively or instantaneously, it is worth noting that there are versions of the argument in the medieval period that do quite clearly adopt the all-at-once approach—for instance, in John Buridan, Henry of Harclay, Walter Chatton, and Gerald of Odo.[37]

Is the all-at-once version of the argument any stronger than the successive version? First, I think it is worth noting that there is something quite peculiar about the position that a particular task such as infinite division *cannot* be performed over a period of time, but that that same task *can* be performed at an instant (at least if the description of the task itself makes no reference to how much time one must take to complete it—which the description of the task of infinite division does not). If there is something impossible about the performance of a specified task over time, one would usually think that that there must a fortiori be something impossible about the all-at-once performance of that same task. So there is something suspicious about the idea that one can compress an otherwise impossible task into an instant and thereby make it possible. But perhaps someone might claim that operations with infinities are an exception to this otherwise plausible general rule. In any case the point noted here is so far simply a suspicion, not an adequately articulated objection.

A second sort of worry about the all-at-once version of the argument is that the notion of an all-at-once superdivision itself seems somewhat suspect. One moment we have the undivided whole before us; the next we have an infinite powder of parts But how did we arrive at this entire collection of all the parts into which the whole can be m-divided? Certainly not through any process we can reconstruct, tracing the various movements of the parts rupturing one from another, and following each part of the body as it is separated from the whole original and in turn splintered into its own subparts—for this involves motion and any such motion would have had to have taken *time*. Which part of the body went where? How could one ever say? It almost looks as if the miraculous instant saw our undivided whole simply *replaced* with an infinite powder, rather than actually *divided* into it. (A suspicion that would grow stronger if one holds the popular view that spatiotemporal continuity is a necessary

---

[37] See John E. Murdoch's essay 'Infinity and Continuity', in A. Kenny, N. Kretzmann, and J. Pinsborg (eds.), *The Cambridge History of Medieval Philosophy* (Cambridge: CUP, 1982), 568–76.

condition of the identity of objects across time. In fact this view makes the idea of any instantaneous m-division—any instantaneous rupturing and separation of parts—look incoherent.) So the idea of an instantaneous superdivision can itself seem quite problematic. Again, I raise this second point simply as a suggestive worry—not yet as a fully fledged objection.

For the purposes of this study I can pass over these initial qualms and make a more general point about this version of the argument. The whole point of this argument was to present us with the *entire* collection of parts into which a whole can be m-divided. If this can be done without invoking the actual parts doctrine, then I will have embarrassed the overall argument of this work by showing that that doctrine is something of a false lead in the study of the problems of material structure. Since we can generate the entire collection of parts without invoking the actual parts doctrine, we will still have to deal with the various classic paradoxes even if we ignore that doctrine altogether.

But now consider the argument from the perspective of a potential parts advocate who thinks that there is something incoherent in the idea of our ever being given the entire collection of parts in an infinitely m-divisible whole. Such a potential parts advocate might reverse the present all-at-once argument, deploying it as a *modus tollens* against the logical possibility of instantaneously dividing such a whole through-and-through. The envisaged all-at-once superdivision of an infinitely m-divisible whole would present us, all-at-once, with the entire collection of parts into which the whole can be m-divided. Such a collection, completely given in its entire totality, will of course have to be greater than any finite number: it will have to be actually infinite in size. Now, as we know, for most Enlightenment thinkers, this greater-than-finite complete collection precipitates insuperable paradoxes and is thus thought impossible.[38] And if it *is* impossible, then we have the basis for a *reductio* argument against the possibility of God's one-step superdivision. If one thought that there was something paradoxical or incoherent in the concept of a completely given greater-than-finite collection, one could thence infer that the proposed task is not logically possible after all. Not even God can all-at-once divide an infinitely divisible whole through and through, since this would be, *per impossibile*, to create an actual infinity. (The same reply could, of course, equally have been made to the earlier suggestion that then be a *successive*

---

[38] See the array of classic paradoxes detailed in Ch. 1, s. VIII.

through-and-through superdivision. Through-and-through infinite divisions are simply impossible by these lights, be they successive *or* all-at-once.)

This point is made in the period literature by Pierre Bayle and John Keill. Each of these philosophers holds that, if God were to completely divide an infinitely divisible whole through and through, this would leave us with ultimate parts. But this (they say) is absurd, so such a through-and-through division is impossible, even for God. Now, while these two philosophers make this argument in terms of the impossibility of ultimate parts, here they are simply implicitly accepting the popular inference from a completely given collection of all the parts to ultimate parts: if we have all the parts, then we must (they think) have smallest parts. The deeper point here is that both of these thinkers are agreed that, in the case of an infinitely divisible whole, the notion of an entirely given collection of parts is itself impossible, and hence so is God's proposed through-and-through division. Thus Pierre Bayle in the 1675–7 *Système*:

Ainsi nous disons qu'à cause de cette contradiction, Dieu ne peut diviser en un même instant le continu autant qu'il est divisible, parce qu'il est de l'essence d'une chose divisible à l'infini, qu'elle puisse être toûjours divisible à l'infini. Or toutes les divisions du continu ne peuvent pas être faites collectivement, de même que toutes les figures prises distributivement sont séparables du corps, & ne le sont point, prises collectivement.[39]

And John Keill in his Oxford lectures of 1700:

Another Argument against the infinite Divisibility of Matter, is fetched from the Divine Omnipotence. God, they say, can resolve any Quantity into its infinitesimal Parts, and separate those Parts one from another: but if so, then may be given the ultimate part and the Divisibility would be exhausted, therefore Quantity is not divisible *in infinitum*. I answer, without doubt God is able to do whatever is possible, or what is not repugnant to his immutable Nature; but since we already have demonstrated that there cannot be given any Particle of Matter, however small, which may not be still divided into other infinite Particles; it is thence manifest, that God cannot so divide Matter.[40]

---

[39] i.e. 'We say, therefore, that God is not able, because of a manifest contradiction, to divide simultaneously a continuum, as much as it is able to be divided, because it is the essence of a thing divisible into infinity that it should always be divisible to infinity. But not all divisions are able to be placed collectively, just as each figure taken distributively is separable from a body, but not collectively.' Bayle, *Système Abrégé*, in *Œuvres Diverses*, 303.

[40] Keill, *Introduction*, 39–40. A very similar argument appears in Anne Conway, *Principles of the Most Ancient and Modern Philosophy* (Cambridge: CUP, 1996), 18–19.

So if there is something incoherent in the idea of all the parts in an infinitely m-divisible body being given as a complete totality—as nearly all early modern thinkers believed—then one might then reject the proposed superdivision as a logically impossible task. Not even God can through-and-through m-divide an infinitely m-divisible whole, since this task asks us—*per impossibile*—to realize a completely given greater-than-finite collection. The impossibility of a completely given greater-than-finite collection presents us with a principled reason for rejecting the logical possibility of a through-and-through m-division, be it successive or all-at-once. (Someone pressing this line against *instantaneous* through-and-though m-division might, perhaps, also draw additional support from the two suspicions or qualms about the notion of a one-shot superdivision outlined above.)

Here an objection will arise, and in dealing with it we shall see just why the actual parts–potential parts controversy is essential after all. Let us grant (for the moment) that there is something paradoxical in the notion of a completely given greater-than-finite collection of parts. It follows then that the proposed task of m-dividing an infinitely m-divisible whole through and through is impossible. But, one might well ask, what warrants us as taking this as a demonstration that the *act* of through-and-through m-division is impossible (as Bayle and Keill do), rather than as a demonstration that there could *be* no infinitely m-divisible whole in the first place? To get to our paradoxical completely given greater-than-finite collection of parts, we require the assumption (*a*) that the whole is infinitely m-divisible, and the assumption (*b*) that God could m-divide this whole through and through. Bayle and Keill take this as a *modus tollens* argument that rules out the possibility of (*b*). But why not take it instead to rule out the possibility of (*a*), as is clearly the intention of those whom Bayle and Keill are attacking in the above passages?

This is where the contrast between the actual parts and potential parts perspectives makes all the difference. A potential parts advocate who rejects the possibility of actual infinities (Aristotle, Hobbes, or the later Kant, for instance) has a principled and consistent reason to take the present argument as a *reductio* against (*b*), rather than against (*a*). Given the potential parts metaphysic, they can perfectly well allow that matter is infinitely m-divisible—*sequentially, successively* infinitely m-divisible—since this entails only an unparadoxical open-ended potential infinity of parts (an ever-increasable but always actually finite collection). So they can make good sense of infinite m-divisibility so long as it is understood that this m-divisibility is always sequential and open-ended. But this same potential parts theorist will then reason—in a perfectly consistent way—that the proposed task of completing an infinite m-division is logically impossible, since it effectively asks us to create a (paradoxical) greater-than-finite number of parts. According to this potential parts theorist, we can generate any given finite number of parts, and indeed can always generate more on top of that. But we cannot generate a completely given greater-than-finite number of parts, since this is an incoherent notion.

Here it is instructive to compare the position of the actual parts advocate. Someone who accepts the actual parts doctrine and infinite m-divisibility cannot avoid admitting that we must already have the entire collection of parts given from the very start. We *already have* every last part into which the whole can be m-divided—as Digby says, this is 'the confession, or rather the position of the [actual parts theorist] when he sayeth that all [the] parts are actually distinguished'.[41] If there is something paradoxical in such a completely given greater-than-finite collection, then this theorist must reject either infinite m-divisibility or the actual parts doctrine. Simply rejecting the possibility of through-and-through m-division will not get them out of the paradox, since they are already presented with the entire collection of parts even without any superdivision. And here there is a clear contrast with the potential parts theorist, who can consistently hang onto his potential parts analysis of matter, and his claim that matter is infinitely m-divisible, while at the same time rejecting the proposed superdivision as a logically impossible task.

The moral here is that the actual parts–potential parts controversy does matter after all. The current argument invoked the possibility of an all-at-once through-and-through m-division in order to get us from the sheer fact of infinite m-divisibility to a completely given greater-than-finite collection of parts. If this were successful, then my emphasis in this study on the role of the actual parts doctrine in generating the problems of material structure would be misguided. We are led to the same paradoxes even without the actual parts doctrine, simply by way of the possibility of an all-at-once superdivision. But it now turns out that the actual parts–potential parts conflict does make a clear difference after all. With the potential parts account of Aristotle, Hobbes, and the later Kant, one can

[41] Digby, *Two Treatises*, 10.

make the move (seen here in Bayle and Keill[42]) of rejecting the logical possibility of a through-and-through m-division, and thereby avoid the admission of a completely given greater-than-finite number of parts. But the actual parts theorist cannot consistently make this move, since, if they accept infinite m-divisibility, then they are *already* committed to accepting a completely given greater-than-finite collection and to dealing with the classic paradoxes that come with this.

VI. Galileo's Argument from Infinite Divisibility to a Complete Collection of Parts

I now turn to the second of the two alternative argument strategies—a second way of arguing that, *whether or not a body has actual parts*, the sheer fact of m-divisibility guarantees that we can have the complete collection of parts into which a body is m-divisible fully given.

In the previous section I looked at the first alternative argument, where the strategy was to show that the entire collection of parts can be given via a through-and-through infinite division. My second alternative argument, by contrast, attempts to reason from the infinite m-divisibility of the whole to conclude that we must *already* have a complete collection of all the parts into which the whole can be m-divided, even without any actual operation of m-division, and regardless of whether or not the body has actual parts. Instead, the sheer fact that a body is m-divisible is already enough to give us all the parts from the very start: the actual parts doctrine is superfluous.

The argument can be introduced as follows. Suppose we have a body forty cubits long. It can (of course) be m-divided into forty parts, each one cubit long. Now, according to an actual parts theorist, forty distinct beings (at least) exist here, even prior to any act of division. (Division would simply separate or unveil these pre-existing parts.) On the other hand, according to a potential parts theorist, there is only one distinct being here prior to division. Were the whole to be m-divided into forty parts, *then* we would have forty (freshly created) distinct beings—but not until then.

---

[42] Not that Bayle and Keill are themselves potential parts theorists. Bayle is in fact the most vociferous advocate of actual parts in our period (see Ch. 2, s. IV), and, while Keill does toy with the potential parts approach (see Ch. 1, s. IX), he does not ultimately endorse it as his settled answer to the problems of material structure. But the point is that their move of rejecting an instantaneous infinite division is consistently open to the potential parts theorist.

Now, an advocate of our current argument is well aware of this difference of opinion and wishes to side-step it. Their key move is to step back from the question of how many distinct beings there are in the undivided whole, but then to insist that, nevertheless, *in some sense*, there surely are (at least) forty parts in the undivided whole. This is not to say that there are forty distinct beings (which would be to take sides with the actual parts theorist). But surely there is *some sense* in which even the potential parts theorist must agree there are forty parts in an undivided forty cubit body. The sheer fact that it is m-divisible into forty parts shows that the whole contains forty parts in some sense, even if it does not yet contain forty distinct beings. (A potential parts theorist could, if they wanted, think of these forty parts as forty aspects or features of the whole—as, if you like, forty *possible* or *potential* parts—if they will not think of them as forty distinct beings or independent entities.) But then, of course, the trap is sprung. Once it is granted that each of the parts into which a body can be m-divided is given (in some sense) from the very start and prior to any act of division, then we have the complete collection of parts given from the very start (not necessarily given as distinct beings, but given in some sense). Even if we do not have the parts as so many distinct beings, we nevertheless have them given from the very start in some sense that allows their identification and enumeration. So we have the complete collection of parts into which the whole is m-divisible fully given (in some sense) from the very start, shortcircuiting any need to present them successively.

Clearly this argument will need tightening up. In particular, we will want to see the ubiquitous and evasive phrase 'in some sense' unpacked. But now we are ready to look at Galileo's statement of the argument in the 1638 *Dialogues Concerning Two New Sciences*. First, Galileo is quite careful to show that he is fully aware of the controversy between actual parts advocates and potential parts advocates, and to state that his argument is meant to stand outside that dispute. Here is Galileo's mouthpiece Salviati cross-questioning Simplicio, the representative of orthodox scholastic natural philosophy and the potential parts account:

Salviati: ... tell me whether, in your opinion, a *continuum* is made up of a finite or an infinite number of finite parts (*parti quante*).
Simplicio: My answer is that their number is both infinite and finite; potentially infinite but actually finite (*infinite, in potenza; e finite, in atto*); that is to say,

164 / Actual Parts and Shortcircuit Arguments

potentially infinite before division and actually finite after division; because parts cannot be said to be in a body which is not yet divided or at least marked out; if this is not done we say that they exist potentially.
Salviati: So that a line, which is, for instance, twenty spans long is not said to contain actually twenty lines each one span in length except after division into twenty equal parts; before division it is said to contain them only potentially.[43]

What we have here is Galileo clearly and explicitly reporting the traditional scholastic potential parts account. The parts into which continua can be divided do not exist until they are 'actualized' by some positive step of division; hence infinite divisibility implies only an indeterminate and ever-increasable potential infinity of parts, not a paradoxical completely given actual infinity. Now, Galileo does not explicitly reject this sort of potential parts analysis of the ontological status of the parts of continua. But he does think of it as a red herring when it comes to the traditional Zenonian problems of the make-up of continua and the paradoxes of their internal structure. He begins by noting that the distinction between actual and potential parts makes no difference to the *size* of the whole that ('in some sense') contains these parts:

Salviati: ... Suppose the facts are as you say [*i.e. accepting the potential parts doctrine, for the sake of the argument*]; tell me then whether, when the division is once made, the size of the original quantity is thereby increased, diminished, or unaffected.
Simplicio: It neither increases nor diminishes.
Salviati: That is my opinion also. Therefore the finite parts (*parti quante*) in a continuum, whether actually or potentially present, do not make the quantity larger or smaller; but it is perfectly clear that, if the number of finite parts actually contained in the whole is infinite in number, they will make the magnitude infinite. Hence the number of finite parts, although existing only potentially, cannot be infinite unless the magnitude containing them be infinite; and conversely if the magnitude is finite it cannot contain an infinite number of finite parts either actually or potentially.

Alarm bells should be ringing at this point, though the bumbling Simplicio allows all this to pass. To finish off Galileo's reasoning and come to our current argument:

Salviati: ... I grant ... to the philosophers, that the continuum contains as many finite parts as they please and I concede that it contains them, either actually or potentially, as they may like; but I must add that just as a line ten fathoms in

[43] Galileo, *Two New Sciences*, 34.

length contains ten lines each of one fathom and forty lines each of one cubit and eighty lines each of half a cubit, etc., so it contains an infinite number of points; call them actual or potential, as you like, for as to this detail, Simplicio, I defer to your judgement.[44]

The actual parts–potential parts controversy is beside the point, a mere 'detail'. The parts might be there 'either actually or potentially, as [the philosophers] may like'; it makes no difference. Either way, if a whole is divisible into forty parts then it contains those forty parts (not necessarily as so many distinct beings, but it certainly contains them *in some sense*). And likewise, if it is infinitely divisible then it contains a completely given actual infinity of point parts: thereby confirming Galileo's doctrine that 'a continuous quantity [is] built up of an infinite number of indivisibles'.[45]

Now, I have already indicated my suspicion that Galileo is in fact surreptitiously relying on the actual parts doctrine, despite his clear and open disavowals (see sections II and III above). But for the moment let us take him at his word and see how the current argument plays out if it is interpreted (as Galileo demands) as standing outside the actual parts—potential parts controversy. In fact the argument rapidly collapses once we cash out the sense in which the undivided whole is supposed to 'contain' each of the forty one cubit parts within it—that is, once we unpack the claim that the forty parts are given 'in some sense'. The claim that the whole 'contains' these forty parts—or that the forty parts are given 'in some sense'—is supposed to be an obvious truth that both the actual parts and potential parts theorist can agree on. So 'contains' cannot be taken to mean 'is compounded of as so many distinct beings'. Rather, if we are to keep the potential parts theorist as well as the actual parts theorist on side, 'contains' can only mean 'can be m-divided into': thus 'the whole contains forty parts' just means 'the whole can be m-divided into forty parts'. Both the actual parts and potential parts theorist can agree on this. In *this* sense, then, the whole obviously does 'contain' forty parts. But then of course the potential parts advocate will resist Galileo's inference from infinite divisibility to the conclusion that the whole 'contains' a completely given actual infinity of parts. Yes, a whole that is forty cubits in length 'contains'

---

[44] Ibid. 34–5, 35–6.
[45] Ibid. 34. For sympathetic reconstruction of Galileo's argument, see Maurice Clavelin, *The Natural Philosophy of Galileo*, tr. A. J. Pomerans (Cambridge, Mass.: MIT Press, 1974), 314–15, and Gandt, *Force and Geometry*, 172–3. Andrew Pyle presses a similar argument in *Atomism and its Critics* (Bristol: Thoemmes Press, 1995), 390–1.

166 / Actual Parts and Shortcircuit Arguments

(in our specified sense) forty parts each of one cubit length, and eighty of half a cubit length, and so on. This just means it can be m-divided into forty one cubit parts, eighty half cubit parts, and so on. But it does not 'contain' an infinity of parts, since this would mean that it could, *per impossibile*, be m-divided into an infinite number of parts.

Here is the real underlying philosophical point. The whole purpose of the Aristotelian–scholastic potential parts account was to avoid admitting a (paradoxical) actual infinity of parts. On this account, parts only exist once they are actualized through some process of division and subdivision, and so we will only ever have an (unparadoxical) potential infinity of them. Crucial to this strategy is the idea that we can only enumerate parts once they exist or (perhaps) in so far as we are considering bringing them into existence—that is, *vis-à-vis* some actual or proposed sequence of division. But what Galileo suggests, in effect, is that the potential parts theorist must allow a *full* enumeration of *all* the parts into which the whole can be divided, thus pushing them (he hopes) into admitting an actual infinity. And of course no potential parts theorist worth his salt should accept this. (This is where the unassertive Simplicio should have put his foot down.) Instead, on the potential parts view, parts can only be enumerated in so far as we are considering some actual or proposed sequence of division—and so will only ever be potentially infinite in number.

So if we are really to read Galileo's argument as neutral between the actual and potential parts theories, it is a failure. My diagnosis is that Galileo believes it valid only because (despite his open disavowals) he is in fact implicitly thinking in terms of the actual parts metaphysic. He wants us to allow a *complete* enumeration of *all* the parts into which the whole is divisible. But this simply betrays his actual parts thinking, for no such enumeration is possible in the potential parts framework. It is as if he thinks it uncontroversial that all the parts into which the whole can be divided are *really* all there prior to division—it is just that the actual parts theorist holds that they are each there as a distinct concrete entity, whereas the potential parts theorist holds that they are each there in some shadowy half-life of potentiality, awaiting a fully fledged distinct existence once division goes through and it is actualized. But, contrary to Galileo's straw-man caricature, a true potential parts theorist decidedly does *not* think of a forty cubit body as an amalgam of forty pseudo-entities called 'potential parts', each of which only becomes a fully fledged distinct being when division actualizes it. Rather, to say that a body contains forty potential parts is just to say that that whole—the only entity that is there to be counted—could be broken down into forty parts through division. Talk of potential parts is really just talk about the capacities of the whole—it is not talk about so many diverse but ethereal not-yet-distinct beings.

There is further evidence that Galileo is really thinking in terms of the actual parts metaphysic despite his avowed neutrality. While still claiming that 'the philosophers' may have their parts 'either actually or potentially, as they may like' and that he is standing back from this particular controversy, he in fact deploys reasoning that clearly depends on actual parts assumptions. We saw this reasoning back in section III, where I argued that it was a version of our argument (E1)–(E3). But to review it briefly:

I do not see how it is possible to avoid the conclusion that [continuous quantities] are built up of an infinite number of indivisible quantities because a division and a subdivision that can be carried on indefinitely presupposes that the parts are infinite in number, otherwise the subdivision would reach an end.[46]

This argument clearly assumes the actual parts doctrine. A potential parts theorist would just reject the claim that 'the subdivision would reach an end' if there is no endless supply of pre-given parts—since they hold that new parts are generated afresh as division proceeds.[47]

VII. Summary and Conclusion

First we looked at two arguments found throughout the Enlightenment literature that employ the actual parts doctrine to present us with all the parts into which the whole can be divided. I suggested that the *argument from actual parts to a determinate number of parts* is successful. The *argument from the definiteness of parts to ultimate parts*, on the other hand, is ultimately a failure. Yes, a whole with actual parts will have the complete collection of all its parts pre-given. But this need not imply that smallest parts are given. If the collection of parts is actually infinite, then it can be fully given yet lack smallest parts.

I then looked at a pair of alternative argument strategies that attempt to give us the complete collection of parts but to do so without invoking the

---

[46] Galileo, *Two New Sciences*, 34. See also the extracts quoted in section III above.
[47] Here I agree with Zev Bechler, *Newton's Physics and the Conceptual Structure of the Scientific Revolution* (Dordrecht: Kluwer 1991), 138.

actual parts doctrine. These were also drawn from the period literature, although they enjoyed less currency than the arguments that invoke actual parts. I concluded that arguments that envisage a complete through-and-through division of an infinitely divisible whole fail, whether that division is thought of as successive or as instantaneous. I also looked at an argument drawn from Galileo that attempts to present us with all the parts from the very start without invoking the actual parts doctrine. I argued that this also fails—and that in fact Galileo seemed to be surreptitiously smuggling in actual parts assumptions despite his various disavowals.

Since each of these alternative argument strategies gets nowhere, I take it that they do not show that my emphasis on the actual parts doctrine is mistaken. Had they shown a clear way to generate the classic paradoxes of infinite m-divisibility without the actual parts doctrine, then they would indeed have rendered that doctrine something of a red herring in a study of the problems of material structure. But I think I have now shown that it is with good reason that the actual parts doctrine is assumed over and over again in the Enlightenment arguments for the paradoxes. Any serious attempt to generate the classic paradoxes of infinite m-divisibility must adopt the actual parts metaphysic.

4

The Actual Parts Doctrine and the Argument from Composition

> Wise men who are asked about the soul answer that they have no idea what it is. If they are asked what matter is, they give the same reply. It is true that some professors, and above all some schoolboys, know it all perfectly; and when they have repeated that it is extended and divisible, they think they have settled everything; but if they are asked to say what this extended thing is, they find themselves in difficulty. 'It is composed of parts,' they say. And these parts, what are they composed of? Are the elements of the parts divisible? Then they are dumb or talk a lot, which is equally suspect.
>
> Voltaire, *Philosophical Dictionary* (1764), entry on 'Matière'

I. Introduction

In the previous chapter I looked at one attempt to argue from the actual parts doctrine to the supposed corollary of ultimate parts: *the argument from the definiteness of parts to ultimate parts*. I concluded that this argument fails. In the present chapter, I want to look at another sort of argument from actual parts to ultimate parts that also enjoyed much currency in the early modern period: *the argument from composition*.

The argument from composition runs as follows. Given the actual parts analysis of matter, material bodies are essentially composite or compound structures. We then add the claim that such complex, composite entities are ontologically derivative, depending for their existence on the prior existence of their parts. And if the parts are themselves also composite, then they in turn depend for their existence on the prior existence of *their* parts. But (according to the current argument) if the whole original is to exist at all, this ontological regress cannot go on forever with no ground floor. Given that the original composite whole exists, there must be an atomic base of noncomposite first parts whose own existence is not so derivative. And of course these noncomposite, simple first elements must be *m-indivisible*. (Were they m-divisible they would—given the actual parts doctrine—be composite.) So we have a ground floor of ultimate, m-indivisible parts: metaphysical atoms.

This sort of argument enjoyed wide popularity during the early modern period, finding endorsement most famously in the rational cosmology of Leibniz[1] and Wolff,[2] but also in the neo-Epicurean metaphysics of Walter Charleton[3] and the younger Isaac Newton,[4] and in the dynamical system of the pre-critical Kant.[5] Even Hume[6] and Condillac,[7] usually thought of as

---

[1] G. W. Leibniz, *The Principles of Nature and Grace, Based on Reason*, s. 1; *Monadology*, s. 2; correspondence with Arnauld, 30 Apr. 1687; all in Leibniz, *Philosophical Essays*, tr. and ed. Roger Ariew and Daniel Garber (Indianapolis: Hackett, 1989), 207, 213, 85.

[2] Christian Wolff, *Ontologia*, ss. 686, 792, and *Cosmologia Generalis*, s. 176, in *Gesammelte Werke*, ed. Joaness Ecole (Hildesheim: Georg Olms, 1962).

[3] Walter Charleton, *Physiologia Epicuro-Gassendo-Charletoniana* (London, 1654; facsimile edn.: New York: Johnson Reprint Co., 1966), 109.

[4] Isaac Newton, 'Certain Philosophical Questions', tr. and ed. J. E. McGuire and Martin Tamny, in McGuire and Tamny, *Certain Philosophical Questions: Newton's Trinity Notebook* (Cambridge: CUP, 1983), 330–489, 337.

[5] Immanuel Kant, *Physical Monadology* (1756), in *The Works of Immanuel Kant: Theoretical Philosophy 1755–1770*, tr. and ed. David Walford and Ralf Meerbone (Cambridge: CUP, 1992), 1: 473–87, 1: 477. See also n. 13 below. Sophisticated statements of the argument are also found in the Amphiboly and Second Antinomy sections of the *Critique of Pure Reason*—although in this later, critical period Kant is merely reporting the argument without endorsing it. Immanuel Kant, *Critique of Pure Reason*, tr. Werner S. Pluhar (Indianapolis: Hackett, 1996), B463–70; B321–2, B330, B338.

[6] David Hume, *A Treatise of Human Nature*, ed. David Fate Norton and Mary J. Norton (Oxford and New York: OUP, 2000), 1. 2. 2. 3. Additional references abbreviated 'SBN' give the corresponding page numbers in David Hume, *A Treatise of Human Nature*, ed. L. A. Selby-Bigge, 2nd edn. with text revised and notes by P. H. Nidditch (Oxford: OUP, 1978). Here the SBN reference is 30–1. Hume admits he has borrowed the argument from Nicolas de Malezieu, the Sun King's mathematician royal. (For Malezieu's text, see Norman Kemp Smith, *The Philosophy of David Hume* (London: Macmillan, 1941), 341.)

[7] Etienne Bonnot de Condillac, *Les Monades*, in *Studies on Voltaire and the Eighteenth Century*, 187 (Oxford: Cheney & Sons, 1980), 109–211, 168–9. (Condillac's apparent endorsement of the argument from composition in *Les Monades* should be contrasted with his attack on that same argument in his antimetaphysical *Treatise on Systems*, published just a year later in 1749. This has led some to regard *Les Monades* as simply a debater's exercise, rather than an expression of Condillac's own view. For more on this interpretative question, see Laurence L. Bongie, introduction to *Les Monades*, in *Studies on Voltaire and the Eighteenth Century*, 187 (Oxford: Cheney & Sons, 1980), 7–107, 15–20.)

antimetaphysical and positivistic philosophers, ratify versions of this a prioristic piece of ontological theorizing. The versions of the argument given by these diverse thinkers naturally differ in their precise formulation, in ways that reflect not only the various dialectical strategies of these writers, but also their deeper metaphysical agendas. For instance, these philosophers certainly disagree sharply over the precise nature of the metaphysical atoms or simples that serve as the elemental building blocks of the material realm. (Contrast the hylozoic monads of Hanover with the *minima sensibilia* of Edinburgh and the force-shells of Königsberg.) But these differences should not obscure an important unanimity among these admittedly disparate thinkers. They are each committed in their own way to the same basic argument, proceeding from the premise that material bodies are composite and hence ontologically derivative to the conclusion that the material realm must be constructed from elemental metaphysical atoms.[8]

For the purposes of this chapter I will be focusing on what is perhaps the clearest and most systematic presentation of the argument, the version given in Kant's pre-critical *Physical Monadology* of 1756. This incarnation of the argument is relatively straightforward and goes a good way in spelling out the actual parts premises that remain implicit in (say) Leibniz and Hume. It is thus a good candidate for close scrutiny and critical assessment. I hope that this close attention to one particular formulation of the argument will lend clarity to the discussion and help throw the key issues into sharper focus. (Note, by the way, that the later, critical-period Kant of the 1781/7 *Critique of Pure Reason* and the 1786 *Metaphysical Foundations of Natural Science* in fact renounces the argument from composition. The later Kant still accepts the *validity* of this argument from actual parts to ultimate parts; it is just that he now rejects its actual parts premises and accepts the potential parts system instead. For details, see Chapter 2, section III.)

In this chapter, then, my primary goal will be to assess this specific argument found in the 1756 *Physical Monadology*. However, my broader hope is that an examination of this particular token of the argument from composition will shed light on the merits of that type of argument, a type of argument to which Charleton, the younger Newton, Leibniz, Wolff, and

---

[8] The essential core of the argument can also be found—in inchoate and underdeveloped form—in the works of Renaissance natural philosophers Bruno, Gorlaeus, and Magnen. See G. B. Stones, 'The Atomic View of Matter in the XVth, XVIth, and XVIIth Centuries', *Isis*, 10 (1926), 450–1, 455, 459; also Ivor Leclerc's *The Nature of Physical Existence* (London: George Allen & Unwin, 1972), 169.

Hume are equally committed. Those who are sceptical about my claim that this text can indeed serve as representative of a shared pattern of reasoning common to all these thinkers can simply take this chapter's assessment the merits of the *Physical Monadology* argument to stand alone, without taking it as a generalizable assessment of a broader pattern of argument. However, let me serve notice here that, in discussing possible objections to this particular argument, I do draw on commentaries from the broader literature on the argument from composition where these seem relevant.

I proceed as follows. In section II I give a summary overview of Kant's *Physical Monadology* version of the argument and situate that argument in its historical context. Section III presents a detailed reconstruction of Kant's argument and defends my reading against various rival interpretations. In Section IV I address possible objections and responses. Finally in section V I offer a review and concluding assessment.

II. Overview of Kant's *Physical Monadology* Argument from Composition

The overall aim of Kant's 1756 *Physical Monadology* is to 'marry metaphysics and geometry'[9] by reconciling two prima-facie incompatible theses: on the one hand, the actual parts doctrine, which entails an atomic base of m-indivisible simples (or so Kant argues), and on the other, the doctrine that space, and bodies in space, are infinitely divisible. As we saw in Chapter 1, the tension between these doctrines had troubled natural philosophy for over a hundred years. But the more immediate background to Kant's discussion of this problem was an astonishingly bitter debate over the theory of matter that had preoccupied the academies of Germany throughout the 1740s and 1750s. All parties to this debate held that the two theses were mutually inconsistent, and the clash between their respective advocates was a key flashpoint in the struggle between what we would now identify as the rationalist and the empiricist schools of eighteenth-century German philosophical thought. Those of a rationalist bent embraced the Wolffian arguments for metaphysical atomism (or 'monadism') and hence rejected the infinite divisibility of body. Newtonians such as Euler, on the

---

[9] Immanuel Kant, *Physical Monadology*, in *Works*, 1: 475, 1: 480.

other hand, endorsed the infinite divisibility doctrine and hence rejected the Wolffians' metaphysical atomism.[10]

Kant's proposed compromise was to reject the supposition—shared by both camps—that the two theses are indeed inconsistent. In the *Physical Monadology*, this third way is developed and expounded. Kant argues that, properly understood, the two theses are compatible. He thus agrees with the Wolffians that bodies are composed of elemental f- and m-indivisible simples, but then goes on to maintain that each of these unextended point-atoms dominates an extended volume of space through a projected shell of repulsive force.[11] So finite arrays of these point-atoms can concatenate to build extended bodies without detriment to their own m-indivisibility and simplicity. Metaphysical atomism can thus be embraced *alongside* the infinite *f*- (thought not *m*-) divisibility of matter. (I discuss Kant's force-shell atoms and this proposed solution to the problems of material structure in detail in Chapter 6.)

For the purposes of the inquiry in this chapter it is not so much Kant's proposed reconciliation of metaphysical atomism with infinite f-divisibility that concerns us, as the positive case he presents for metaphysical atomism at the start of the *Physical Monadology*. This is the section of text that I want to examine in order to offer a focused reconstruction and assessment of the argument from composition. Kant would have inherited this form of argument from the Wolffians and ultimately Leibniz, but it has rarely been formulated with such clarity and succinctness, and with such an explicit emphasis on its actual parts framework.

Kant's *Physical Monadology* Argument from Composition: the Text

The text of Kant's argument runs as follows:

PROPOSITION I. DEFINITION. A simple substance, which is also called a monad, is one which does not consist of a plurality of parts, any one of which could exist separately from the others.

---

[10] On the history of the controversy and its culmination in the Berlin Academy dispute, see Ronald S. Calinger, 'The Newtonian–Wolffian Controversy', *Journal of the History of Ideas*, 30 (1969), 322–33. Irving Polonoff, *Force, Cosmos Monads and Other Themes of Kant's Early Thought* (Bonn: Grundmann, 1973), 77–89. Laurence L. Bongie, introduction to Etienne Bonnot de Condillac, *Les Monades*, in *Studies on Voltaire and the Eighteenth Century*, 187 (1980), 9–107. Also see Leonhard Euler's first-hand, partisan anecdotes in his *Letters on Different Subjects in Natural Philosophy*, tr. Henry Hunter, ed. David Brewster (New York: J. and J. Harper, 1833), 39–40, 61–2.

[11] Kant, *Physical Monadology*, in *Works*, 1: 480.

PROPOSITION II. THEOREM. Bodies consist of monads.

Bodies consist of parts, each of which has a separately enduring existence. Since, however, the composition of such parts is nothing but a relation, and hence a determination which is in itself contingent, and which can be denied without abrogating the existence of the things having this relation, it is plain that all composition of a body can be abolished, though all the parts continue to exist. When all composition is abolished, moreover, the parts which are left are not compound at all; and thus they are completely free from plurality of substances, and thus they are simple. All bodies, whatever, therefore, consist of absolutely simple fundamental parts, that is to say, monads.[12]

Bodies (Kant tells us) are simply so many distinct parts standing in a relation of 'composition' with one another. Since this relation can be 'denied' without detriment to the existence of the parts that stand in that relation, it follows that 'all composition can be abolished'. The 'abolition' of composition leaves us with parts, but these parts cannot themselves be composite (for then all composition would not truly have been abolished). They must therefore be simple. So all bodies must ultimately be concatenations of these 'absolutely simple fundamental parts': metaphysical atoms.

In the *Physical Monadology*, Kant offers no further defence of this reasoning. He clearly thinks that propositions I and II provide a compelling self-contained case for metaphysical atomism with no need for further argumentation. A similar confidence permeates the brief references to the argument from composition found in Kant's other pre-critical works. He tells us that the argument establishes atomism 'easily', 'without difficulty', and 'with the greatest certainty', even going so far as to commend the proof as 'self-evident'.[13]

---

[12] Kant, *Physical Monadology*, in *Works*, 1: 477.

[13] '[W]hen a substantial compound has been given, we arrive without difficulty at the idea of things which are simple by taking away the concept of *composition*... For the things which remain when every element of conjunction has been removed are *simple* things.' (Kant, *Inaugural Dissertation*, in *Works*, 2: 385–419, 387, 389; see also 2: 378–9.) '[A]ll bodies must consist of simple substances. Without determining what a body is, I nonetheless know for certain that it consists of parts which would exist even if they were not combined together. And if the concept of a substance is an abstracted concept, it is without doubt one which has been arrived at by a process of abstraction from corporeal things which exist in the world. But it is not even necessary to call them substances. It is enough that one can, with the greatest certainty, infer from them that bodies consist of simple parts. The self-evident analysis of this proposition could easily be offered, but it would be too lengthy to present here.' (Kant, *Inquiry*, in *Works*, 2: 273–301, 286–7.)

Preliminary Qualms about Kant's *Physical Monadology* Argument from Composition

Notwithstanding Kant's confidence in the force of this 'self-evident' reasoning, the reader may be less than persuaded. Before presenting a more detailed reconstruction of Kant's reasoning, it may be helpful to indicate some of the traditional objections facing this sort of argument. A number of objections to the argument have gained currency and some of them have no doubt already occurred to the reader. At this point I summarize only the most classic objections.

First, some hold that in assuming that bodies are composite entities, the argument immediately begs the question, for the predicate 'composite' here, when unpacked, simply means 'built up from simples'. The argument straightaway commits a grotesque *petitio principii*, assuming matter is composite—that is, built up from simples—in order to demonstrate that it is built up from simples. This is the objection of Kemp Smith, and is also found in Schopenhauer and the writings of eighteenth-century opponents of metaphysical atomism such as Euler and Justi.

A second popular *petitio* charge runs as follows. Kant seems to argue directly from the claim that composition is a 'contingent' relation to the claim that 'all composition of a body can be abolished'. The objection has it that this follows only if it is implicitly presupposed that the process of successively decomposing a composite entity into ever smaller parts must terminate at some point at which all composition is indeed removed. But this is just to assume that decomposition must be a finite procedure, terminating with fundamental noncomposite elements. And this implicit presupposition begs the question against the rival view that bodies are just so much atomless gunk, containing distinct actual parts, perhaps, but then parts within parts ad infinitum, without ever coming to a ground floor of atomic first constituents.

A third classic objection. The whole argument trades on the concept of a first part that is *absolutely* simple, rather than the concept of an entity that is merely 'simple' relative to some conventional standard of simplicity and complexity. But (according to this objection) the concept of an absolute simple is incoherent. Thus the argument fails. The essential core of this objection can be retrieved from Hobbes, or so I will argue. It is developed more explicitly in the twentieth century by Wittgenstein.

176 / The Argument from Composition

III. Kant's *Physical Monadology* Argument from Composition: Reconstruction and Exegesis

As with the rest of the *Physical Monadology*, Kant presents his reasoning in a style clearly meant to recall the 'geometrical method' pioneered in Euclid's *Elements* and revived by Descartes and Spinoza. The argument comes in two 'propositions': one stipulative 'definition', and one derived 'theorem', defended in a compact subjoined argument. I have said that this text presents one of the clearest versions of the argument from composition I am aware of in the Enlightenment literature. But however explicit we think Kant's reasoning, he certainly has not laid bare all of his inferences in full keeping with the rigorous norms of the geometrical method. Certain key definitions are given implicitly at best, and various postulates or unargued axioms are made without being explicitly identified as such. Thus there is room for a reconstruction that sets out all the steps of Kant's argument a little more systematically and explicitly than his text does.

For the purposes of the critical assessment of the argument to be developed in this chapter, I will be reconstructing Kant's *Physical Monadology* argument as in the table. I develop and defend this interpretation of the argument further below.

Kant's overall strategy at least should be clear. Material bodies are assumed to be composite entities, where such entities *consist of*—that is, just *are*—so many (actual) parts standing in a relation of composition with one another. Each of these actual parts can exist independently of the present composition relation: they could continue to exist even if broken asunder from the whole. Even if we cannot physically break them apart, there certainly is no logical reason why they could not exist in the absence of the relation of composition in which they presently stand. But if composition can be removed without compromising the existence of the parts, then (Kant tells us) we should be able to remove *all* composition. And this could only leave us with simple parts: the elemental, partless *minima* from which bodies are ultimately composed. Metaphysical atomism is thereby established.

K1 and K2. Kant's Actual Parts Account of Body and Composition

The first two premises make clear Kant's commitment to the actual parts analysis of the structure of matter. Kant tells us that *bodies are composite*

Table 5 Reconstruction of Kant's *Physical Monadology* argument from composition

---

(K1) Bodies are composite entities (i.e. bodies consist of a plurality of parts, where these parts stand in a relation of composition with one another). (*Assumed.*)

(K2) Composition is a contingent relation (i.e. the relata—the parts which jointly comprise the whole composite entity—could each exist in the absence of that relation). (*Assumed.*)

(K3) All composition can be removed from a composite entity. (*From K2.*)

(K4) What is left would have to be simple. (*Assumed.*)

(K5) So all composite entities are ultimately collections of simple elements. (*From K3, K4.*)

(K6) So bodies are ultimately collections of simple elements. (*From K1, K5.*)

---

entities—that they consist of a plurality of parts, where (K1) those parts all stand in a relation of composition with one another, and (K2) each of those parts could exist independently of that relation. This establishes the actual parts framework of the argument. Undivided bodies are not simple units whose parts ('potential parts') are merely dependent aspects or features of the whole until actualized by division. Rather, bodies (even while undivided) are structured aggregations of so many distinct parts, each of which exists prior to division and independently of the current composition relation. The one imperfection in Kant's statement of the actual parts doctrine is his failure to assert explicitly that composition and simplicity are understood to track m-divisibility and m-indivisibility. But I think it is quite obvious that this is what is intended: bodies are composite because m-divisible, and, sure enough, when Kant comes to describe his simple first elements, their defining feature is their m-indivisibility. Kant's argument then presupposes the actual parts account of matter, and K1 and K2 simply set out this framework.

The idea behind K1 is quite familiar then. It tells us that bodies are structured concatenations of diverse parts: they are composite or compound entities. Premise K2 then spells out a second crucial aspect of the actual parts metaphysic: the parts of which a body consists could each continue to exist even if the composition relation between them ceased to obtain. They are *actual parts*. 'Bodies consist of parts,' as Kant puts it, 'each of which has a separately enduring existence' (proposition II). This immediately recalls the antithetical definition of a 'simple substance' given in proposition I: 'A simple substance...is one which does not consist of a plurality of parts, each of which can exist separately from the others'. Bodies then have *actual parts*, parts that can exist independently; simple substances precisely do not.

One way Kant puts this point about the parts of a body being *independent* or *self-subsistent*—that is, able to exist even if the composition relation between them failed to obtain—is by stating that the relation of composition is 'contingent'. For the relata which stand in a relation of composition (that is, the parts which compose a certain body) their present participation in this particular composition relation is not necessary for their continued existence. Rather, so far as each part is concerned, the composition relation is 'contingent': it could either obtain or fail to obtain without this compromising the existence of any of the parts which may—or may not—stand in it at different times. As Kant puts it in proposition II, composition is a relation 'which can be denied without abrogating the existence of the things having this relation'. This is what Kant means when he claims that composition is a 'contingent relation' or 'a determination which is in itself contingent'. Again the basic idea is familiar from the framework of the actual part metaphysic. It is also the key idea in play when other Enlightenment writers variously describe composition as an 'external relation', an 'accidental relation', or an 'extrinsic denomination'.

A final point about K2. The close reader of Kant's argument will notice that he doesn't merely assert that the relation of composition is contingent in this sense, but rather asserts the much stronger claim that *all* relations are contingent. He writes that 'the composition of such parts [i.e. parts each of which has a separately enduring existence] is nothing but a relation, and hence a determination which is in itself contingent' (proposition II). The implicit major premise of this syllogism is clearly that *all* relations are contingent: that no relation is necessary for the continued existence of the things between which it holds. This major premise is extremely strong and may well be contested. But for the purpose of evaluating Kant's argument from composition we need not enter this highly involved debate about the metaphysical status of relations *tout court*. All that Kant needs to assume for his argument to proceed is the much weaker premise (K2) that the specific relation of composition is contingent. And this is just to say (with other actual parts theorists) that any part of a material body could continue to exist even if the present composition relation it stands in failed to obtain—say if that part were separated from the rest of the body or if the remainder of the body were annihilated.

K3 and K4: Kant on the Decomposition of Composite Entities

In the next stage of the argument Kant presents an account of what happens when a compound entity is decomposed into its constituent parts. We are told that (K3) all composition can be removed from a composite entity, and that (K4), when this happens, what remains has to be simple.

Kant thinks of K3 as a derived theorem rather than an unargued axiom of his argument. In particular, he thinks that it follows (at least in part) from K2, the claim that composition is a contingent relation. This becomes clear when he makes the following inference: 'Since ... the composition of such [i.e. self-subsistent] parts is nothing but a relation, and hence a determination which is in itself contingent, and which can be denied without abrogating the existence of the things having this relation, it is plain that all composition of a body can be abolished, though all the parts continue to exist' (proposition II). But this crucial step of the argument may seem puzzling. Why does K3 follow from K2?

The first thing which must be said about K3 is that when Kant talks about 'abolishing', 'removing', or 'denying' the composition of a composite entity, he must be understood to be talking about decomposition *in thought*. Kant has in mind a mental operation by which we 'think away' the composition relation, rather than the actual physical process of disassembling the whole. That Kant is considering decomposition as a mental operation is quite explicit in the version of the argument given in the *Inaugural Dissertation* and again in the Second Antinomy of the *Critique of Pure Reason*.[14] But even here in the *Physical Monadology*, I think Kant's talk of 'denying' composition makes it clear that we are talking about theoretical decomposition, decomposition 'in thought', rather than a physical procedure.

Beyond the bare assertion that K3 can be derived from K2, Kant fails to give us any explicit help in seeing *why* this is supposed to follow. Nor do the other versions of the argument he gives help to make this any clearer. So here I can only offer a somewhat speculative reconstruction of his thinking. I begin by considering three preliminary attempts to derive K3 from K2, each of which reflects an interpretation found in the critical

---

[14] Immanual Kant, *Inaugural Dissertation* (1770) in *Works*, 2: 837. The statement in the *Critique of Pure Reason* is particularly clear: '[I]f all composition were annulled in thought, then there would remain no composite part' (B462). As I've already indicated, in the later *Critique* Kant is merely reporting the argument from composition rather than endorsing it, as he does in the pre-critical *Physical Monadology*, *Inquiry*, and *Inaugural Dissertation*.)

literature on the argument from composition. These three strategies each run into their own difficulties. So, naturally enough, those commentators who read the argument from composition in one of these first three ways typically reject this stage of the argument as a failure. Once these initial attempts have been cleared away, I then present a fourth strategy which I think is both the correct interpretation of Kant's implicit reasoning, and a much more compelling argument in its own right.

(i) *First attempt to derive K3 from K2: assume there are ultimate parts*

One initial temptation is to see Kant as implicitly reasoning in the following way. (L1) Assume there are ultimate parts that jointly compose a composite entity. (L2) Given K2, these parts could each individually exist in the absence of their present relation of composition. So (L3) there is no reason why this entire set of ultimate parts could not exist in the absence of the present composition relation. So (L4), there is no reason why composition cannot be entirely removed (in thought) from a composite entity. *QED*. Now, as we will see, some critics *have* interpreted Kant to be reasoning in this way. But, if this is his strategy, then his overall argument is hopelessly question-begging. The assumption L1 presupposes that there *are* ultimate parts from which all composites are constructed—but this is precisely what the overall argument K1–K6 purports to establish.

(ii) *Second attempt to derive K3 from K2: successive decomposition*

Fortunately, there are other possible ways in which we can reconstruct an inference from K2 to K3 which do not leave Kant charged with such an embarrassingly obvious *petitio*. A second way to see Kant's reasoning runs as follows. Imagine decomposition as a *process*, at each stage of which a successive layer of complex structure is peeled away ('in thought', of course). For instance, as a first stage of decomposition we might (in thought) divide a given material body into, say, halves. Next, those halves are each subdivided, leaving us with quarters of the original body. Next, we split those quarters and are left with an array of eighths, each of which was part of some quarter or other. And so on, through sixteenths, thirty-secondths, and so on. With this model of decomposition as a process, then, each successive step breaks down the original composite entity into ever-smaller parts. Now, according to K2, every part of a composite entity can exist independently of its participation in the current composition relation. So there will be no reason why we cannot (in thought) decompose any parts that are themselves composite into smaller constituent parts. Whenever we proceed with a stage of decomposition, K2 guarantees that we will be left with parts at the end of that stage. And moreover, so long as we are left with parts that are themselves composite, K2 also guarantees that we can break those parts down further and that the process can continue.

Kant's implicit reasoning may then be that, with this picture of decomposition, according to which each successive stage leaves us with parts, and which is only stopped short if we encounter noncomposite simple parts, it follows from K2 that (K3) all composition can be removed from a composite entity. Given K2, there is no layer of composite structure that cannot be peeled away. So (Kant may be reasoning) we can imagine this process of decomposition—no stage of which is problematic—to be completed 'in thought'. *Each* layer can be removed without difficulty, so *all* layers can be removed without difficulty. (Of course, read this way, the inference from K2 to K3 closely resembles an argument form we have seen before. See my discussion of sequential through-and-through m-division in Chapter 3, section V(1).)

However, this second sort of inference also has its own problems. Indeed, those commentators who have interpreted Kant to be arguing in this second way from K2 to K3 have typically pressed another sort of *petitio* charge. Here the concern is that, while K2 may indeed show that *any individual stage* of the process of successive decomposition can proceed so long as composite parts remain, it does not follow that *all* the stages can be completed. This would only follow if it is tacitly assumed that there are only a finite number of stages in the process. After all, if the process continued to infinity, then merely showing that each individual layer of compositional structure can be removed does not suffice to show that that every last layer can be removed—for this would be to reach the end of an infinite sequence. So, if Kant thinks that K3 follows from K2 in this second fashion, then he is tacitly assuming that there are only a finite number of layers of composition to be removed from a body. And this assumption clearly begs the question against Kant's dialectical foes—the anti-atomists, advocates of atomless gunk—whose position is precisely that material bodies can be broken down into ever-smaller parts ad infinitum, with no prospect of ever coming to an atomic base. (For more on this objection, see my related discussion of the argument from infinite m-divisibility to the possibility of a successive through-and-through infinite m-division in Chapter 3, section V(1).)

(iii) *Third attempt to derive K3 from K2: all-at-once decomposition*

I think it must be granted that Kant is begging the question against anti-atomists if he is indeed arguing that, since *each* layer of structure can be removed without difficulty, then *all* layers can be removed in a successive process of decomposition. This does indeed assume that the successive process will terminate eventually, which is just what the anti-atomist denies. But perhaps it will be thought that Kant can avoid this objection if he is thinking of decomposition not as a step-by-step process of peeling away successive layers of structure, but rather as an all-at-once, one-step matter. With this approach, every composition relation in the body is denied *simultaneously*, rather than in a successive process of step-by-step decomposition.

On the current reading Kant is taken to be arguing as follows. Since composition is a contingent relation, *each individual* composition relation in a body can be denied (including each of those which obtain between parts which jointly compose the whole body, *and* each of those which obtain between subparts which jointly compose proper parts of the whole). Therefore *all* composition relations within a body can be denied *simultaneously*. So 'all composition of a body can be abolished', and K3 follows from K2. (Just as attempt (ii) gave us an imaginative decomposition that mirrored the successive m-division discussed in Chapter 3, section V(1), so the current imaginative decomposition mirrors the *all-at-once* m-division discussed in Chapter 3, section V(2).)

But this third sort of argument can avoid the *petitio* of attempt (ii) only by falling back into the *petitio* committed by attempt (i). Although it may sound tempting to move from the deniability of each individual layer of structure to argue the *simultaneous* deniability of *all* layers of structure, in fact this inference once again tacitly assumes that there are ultimate parts. For suppose that there are no ultimate parts and the whole exhibits nothing but endless composition with no ground floor. What would it be to remove all structure from such an entity? However much structure is simultaneously removed, if (as Kant himself claims) something remains, then given that there are no ultimate parts, that remainder must be itself composite. It is only by tacitly assuming that any remainder is noncomposite that the argument could establish (K3) the removability of all layers of composition. But this assumption once again begs the question against the anti-atomist hypothesis. In fact the assumption is just that of L1 in our attempt (i) argument. So this third attempt, when stripped of its specious plausibility, is really just a more elaborate version of the (question-begging) attempt (i). So this third attempt to prove (K3) that all composition can be removed from a composite entity also founders. Only if it is tacitly assumed that one can isolate the last members of the series of ever-smaller parts does K3 follow. But this is just to assume that there *are* ultimate parts—and this assumption simply begs the question against someone who holds that the whole lacks such an ultimate constitution.

(iv) *Fourth attempt to derive K3 from K2: argue from the inadmissibility of relations without ultimate relata*

Given the difficulties with our first three attempts to derive K3 from K2, it is natural to look for a more compelling way of making this inference viable. I think that there is a fourth such strategy, which is plausible both as a reading of Kant's *Physical Monadology* and as a reading of the other versions of the argument from composition found in Kant's corpus. Moreover, this interpretation finds strong textual support in the broader tradition of the argument from composition in the Enlightenment, for instance in the work of writers such as Leibniz and Hume. Finally, as I will argue, this derivation of K3 from K2 is much more powerful than those previously considered, and is of significant philosophical interest in its own right.

I will elaborate on what this reasoning is in a moment. But first let me be clear that I see this inference as the crucial manœuvre in the argument from composition, as found in both the wider Enlightenment literature and here in Kant's *Physical Monadology*. This is the cornerstone of the argument from composition, and it is the endorsement of this key step that ties the likes of Leibniz, Hume, and the pre-critical Kant together as advocates of that argument type. Moreover, it is symptomatic of many misplaced criticisms of the argument from composition to fail to properly identify this reasoning as the argument's fundamental move. Many traditional criticisms of the argument from composition (I will argue) turn out to be based on interpretations which, in essence, see the move from K2 to K3 along the lines of the attempt (i), attempt (ii), or attempt (iii) considered above. Once it is appreciated that the move from K2 to K3 is instead motivated by the sort of considerations to be elucidated here under attempt (iv), many of these classic objections simply evaporate.

My task is to show that there must be a ground floor or atomic base to any compound entity by way of 'thinking away' all of its composite structure. The key to showing this is to remind us that composition is a relation (cf. K2) and then to stress that the instantiation of a relation depends—*ontologically depends*—upon the existence of the relata which enter into that relation. A relation—I mean the *particular instantiation* or *token* of a certain relation type—depends upon there being various relata which stand to each other in a certain fashion. Were the relata not there, there would be no relation. The instantiation of a relation just consists in certain things standing to one another in certain ways. It is in this sense ontologically dependent upon the prior existence of the relata. Now, a relation can certainly obtain between things that themselves consist of items in such-and-such further relations, and these lower level relations can certainly obtain between further things that themselves again consist of yet further items in such-and-such yet further relations.... But it is unsettling to think of this sort of ontological dependence running on forever with no *ultimate* relata, relata that do not themselves depend on further lower-level relations obtaining. This looks like a structure with no ground floor. Each level is parasitic on the lower level, which is in turn parasitic on a yet lower level. And it is at least difficult to see how there could exist such an ontological parasite on parasites on parasites ad infinitum, with no ultimate host or hosts.

Let me now apply this general qualm about relations with no ultimate relata to the case of material bodies conceived along the actual parts lines set out in K1 and K2. Recall that, according to K1, a body is a composite entity—that is, just so many parts standing in a relation of composition with one another. Now, suppose (for *reductio*) that a given body has no ultimate parts, as the anti-atomist has it. The existence of the composite whole would then consist in the relatedness of an array of parts,[15] whose

---

[15] Jonathan Bennett resists this analysis: 'Granted that all the facts about the army are facts about how the soldiers are interrelated, that still does not make it a relation among (a "state of being of") the soldiers. A general cannot lead a state of being; a quartermaster cannot feed a polyadic relation. To treat the army as a relation among the soldiers is to disregard their status as parts of it.' Jonathan Bennett, *Learning from Six Philosophers: Descartes, Spinoza, Leibniz, Locke, Berkeley, Hume*, 2 vols. (Oxford: OUP, 2001), i. 226. But I do not see the problem here. To my mind, if the reductive analysis is carried through thoroughly, then a composite entity *is* fundamentally a state of being of its various parts—and a general *can* lead a state of being if that state of being is, say, the Army of the Potomac. But even if we stress (with Bennett) that the existence of the whole requires the existence of the parts in addition to a structural relation obtaining between them, the overall argument, emphasizing the inadmissibility of relations without ultimate relata, can still proceed.

existence in turn consists in the relatedness of their own subparts, whose existence in turn consists in the relatedness of their own (sub-)subparts, and so on, with no ultimate or final constitution. Its existence would then have to derive from an endless chain: it depends on a certain relations obtaining between relata, which relata in turn depend on further relations obtaining between yet further relata, and so on in an infinite ontological regress with no ground floor. But nothing could exist in this way. So any composite entity that actually exists must bottom out in ultimate, noncomposite parts. The buck must stop somewhere. Now, I have already shown that (given K2) any individual layer of composite structure can be peeled away in thought. And now we know that there must be an atomic base of noncomposite first elements. So we can focus on these ultimate parts and deny their current relation of composition, thereby removing (in thought) *all* composition from any composite entity. '[A]ll composition of a body can be abolished' and K3 is established.[16]

More formally, this reasoning runs as shown in Table 6.

M1–M5 get us to where we really wanted to be all along: at an array of simple first parts or metaphysical atoms. But we need M6–M8 to tie off the argument (albeit somewhat redundantly) and explicitly spell out the way in which it gives us a bridge from K2 to K3 in the master argument K1–K6.

Before I defend this interpretation, it is important that I pre-empt a common misunderstanding. This argument is supposed to establish the necessity of an atomic base of ultimate parts (and hence the removability of all layers of composition). It purports to show us that there must be noncomposite first elements. But, properly understood, it does *not* also purport to establish that bodies are only finitely complex. This remains a further question. It is often thought that this sort of argument—emphasizing the impossibility of relations without ultimate relata—also proves the impossibility of infinite regresses of relations, and hence the impossibility of infinite complexity. Likewise it is often thought that the existence of an atomic base of ultimate parts *ipso facto* entails a merely finite complexity. But in fact the issue of finite versus infinite complexity may not track

[16] Compare James Van Cleve's similar reading of the Amphiboly of the Concepts of Perception in the *Critique of Pure Reason* (B321, B330, and B339), and Michael Radner's reading of the thesis argument of the Second Antinomy (B462–5). James Van Cleve, 'Inner States and Outer Relations: Kant and the Case for Monadism', in Peter H. Hare (ed.), *Doing Philosophy Historically* (Buffalo, NY: Prometheus Books, 1988), 231–47. Michael Radner, 'Unlocking the Second Antinomy: Kant and Wolff', *Journal of the History of Philosophy*, 36 (1998), 413–41.

186 / The Argument from Composition

Table 6 A demonstration of K3 from the inadmissibility of relations without ultimate relata

---

(M1) Any relation that exists must be traceable back to ultimate relata that are not themselves dependent on further relations obtaining. (*Assumed.*)

(M2) The existence of a composite entity consists in the relatedness of its parts. (*From the elaboration of K1 given in Table 5.*)

(M3) If those parts are also themselves composite entities, then their existence likewise consists in the relatedness of their parts. (*From M2.*)

(M4) If the original composite entity is to exist at all, it must be traceable back to ultimate parts which are not themselves composite. (*From M1, M2, M3.*)

(M5) So if any composite entity is to exist at all, there must be a ground floor of noncomposite ultimate parts that stand in a composition relation with one another. (*From M4.*)

(M6) Each individual layer of composite structure can be removed in thought. (*From K2.*)

(M7) So we can remove (in thought) this composition relation between the ultimate parts, leaving us with the atomic base of noncomposite simples. (*From M5, M6.*)

(M8) This step at once removes all layers of composite structure, since higher level layers are parasitic on lower level layers, and we have at once removed the lowest of all layers. So all layers of composite structure can be removed in thought. (*From M2, M7.*)

Which establishes principle K3 of the master argument. *QED.*

---

this current issue of ultimate parts. Consider Galileo's account of material structure, for instance. His bodies are built out of an *actual infinities* of ultimate atomic parts (the extensionless *parti non quante*) and for precisely this reason are themselves infinitely complex. A sequential process of dividing and subdividing one of Galileo's bodies could continue forever without ever exhausting its layers of structure.[17] Notice also that one could correctly describe such a body as a composite of composites of composites ad infinitum. So such a body presents an example of an entity that depends on the relatedness of parts, which parts in turn depend on the relatedness of their parts, and so on ad infinitum. But at the same time it does *not* present us with an example of a relation without ultimate relata, since its

---

[17] On Galileo's account of material structure, see Ch. 1, s. IX, faction (3).

existence does ultimately depend on the atomic base. It is a composite of composites ad infinitum, while at the same time it is also a composite of ultimate parts. This shows us something that may come as a surprise: at least if we allow actually infinite collections, a relation can depend upon further relations obtaining that depend upon further relations obtaining ad infinitum, and at the same time also depend upon ultimate relata.

Of course, not everyone would accept Galileo's account as a possible picture of material structure. Kant in fact rejects actual infinities and so *does* think that a given atomic base entails a merely finite complexity. But this would be a further step in the argument, involving contentious further claims. My point here is simply that the present argument purports only to establish the existence of ultimate parts. It does not by itself also purport to establish that bodies are only finitely complex.

Can this ontological dependence argument from composite entities to ultimate noncomposite parts be read into Kant's *Physical Monadology*? It must be immediately confessed that such reasoning is not explicitly given in proposition II. Nonetheless, I think that it can be plausibly seen as the sort of argument Kant is implicitly drawing on when he claims that K3 can be derived from K2. I have four reasons for this.

First, in the *Physical Monadology* Kant simply states *that* K3 follows from K2 while providing no explanation of *how* this is supposed to follow. Any account of why Kant thinks this follows is therefore necessarily somewhat speculative. But interpretive charity mandates that we credit him with the most plausible reasoning available—and this (to my mind) is the fourth strategy, appealing to the inadmissibility of relations without ultimate relata, rather than the obviously question-begging attempts (i), (ii), and (iii).

Second, this reading of Kant's implicit reasoning makes good sense of the key importance he attaches to actual parts analysis of matter and the premise that composition is a relation between independently existing parts when arguing that all composition can be removed. It is in crucial part *because* 'the composition of such [i.e. self-subsistent] parts is a relation' that 'all composition can be abolished' (proposition II). Here it is instructive to contrast attempt (ii) and attempt (iii) above. As was noted, (ii) and (iii) are simply dressed-up versions of the old argument from m-divisibility to ultimate parts via the possibility of a through-and-through m-division (successive m-division in the one case, all-at-once in the other). (Compare my discussion of this argument in Chapter 3, section V). And this old argument does not depend on the actual parts doctrine at all, making Kant's deployment of his actual parts premises K1 and K2 altogether redundant.

Third, this interpretation gains further credibility as it chimes with other statements of the argument from composition Kant gives in the later *Critique of Pure Reason*. Both in the case for the thesis of the Second Antinomy and especially in the Amphiboly, Kant sets out versions of the argument from composition which fit this reading well (in this later period he is merely reporting the argument, not endorsing it).[18] (Unfortunately, none of Kant's other brief *pre-critical* statements of the argument helps to spell out the inference from K2 to K3.[19])

And fourth, if we look outside Kant's corpus to the broader tradition of the argument from composition in the Enlightenment, we find statements of the argument which insist quite explicitly that it is *because* composites are ontologically dependent on their parts that they must resolve to noncomposite first elements. Here Charleton,[20] Leibniz,[21] and

---

[18] For instance: '[A]ccording to mere concepts the intrinsic is the substratum of all relational or extrinsic determinations. If, therefore, I abstract from all conditions of intuition and keep solely to the intuition of a thing as such, then I can abstract from all extrinsic relation, and there must yet remain what signifies no relation at all but merely intrinsic determinations. Now from this it then seems to follow that in every thing (substance) there is something that is absolutely intrinsic and that precedes all extrinsic determinations inasmuch as it makes them possible in the first place, and that hence this substratum is something that no longer contains any extrinsic relations and consequently is *simple*. (For corporeal things still are never more than relations, at least relations of the parts outside one another.)' *Critique of Pure Reason*, B338–9; see also B321–2 and B330 in the Amphiboly and B463–70 in the Second Antinomy, also in the first *Critique*.

[19] See n. 13 above.

[20] '[W]e are to recognize, that a *Modal Ens* cannot subsist without conjunction to an *Absolute* ... so likewise cannot *Union* be conceived without *Parts*, though on the contrary, Parts may be without Union. And hence we argue: That only which is made *independeter a subjecto*, or holds its *essence ex proprio*, is the Term of Creation; but Union is not independent *a subjecto*: therefore is not Union the Term of Creation. Since therefore the Term of Creation in the First Matter is devoid of Union; it must consist of Individuals, for Division proceeds from the solution of Union. This derives Confirmation from hence; that the subject from whence another is deduced must be praecedent in nature to that which is derived; *Ergo*, are the Parts of the First Matter in nature praecedent to all Union; and consequently they are Individuals, i.e. *Atoms*.' (Charleton, *Physiologia*, 109.)

[21] '[E]very being by aggregation presupposes beings endowed with real unity, because every being derives its reality only from the reality of those beings of which it is composed, so that it will not have any reality at all if each being of which it is composed is itself also a being by aggregation, a being for which we must seek still further grounds for its reality, grounds

Hume[22] are particularly gratifying. Leibniz's texts in particular provide us with strong supporting evidence for a reading of the *Physical Monadology*'s case for atomism. Recall that Kant wrote this early, pre-critical work in order to reconcile the metaphysical atomism of the Leibnizian–Wolffian school with the doctrine of infinite divisibility of space. Kant's argument for metaphysical atomism in propositions I and II closely follows the classic Leibnizian–Wolffian line wherever he spells it out explicitly. It is only reasonable to suggest that, where Kant leaves gaps in the argument, his implicit reasoning still likewise follows the Leibnizian model.

But is this argument successful? As the reader may have already observed, it is structurally quite similar to the cosmological argument familiar from rational theology. Both arguments maintain that, if something exists as a result of other things existing, and these latter things likewise exist as a result of yet further things existing, then the chain of parasitic dependences here invoked must ultimately have a non-parasitic starting place. Of course, the *sort* of dependence invoked in the respective arguments is very different. In our argument M1–M8, the dependence is that of a relation upon its relata; in the standard incarnation of the cosmological argument, the dependence is that of effect upon cause. Nonetheless, the structure of the two arguments is closely analogous. Given that something exists, and that its existence depends on something else's existence, the regress of dependences here set off must end somewhere in things that are not so dependent.

In his study of the cosmological argument, J. L. Mackie characterizes the key principle in play here as follows: '[w]here items are ordered by a relation of dependence, the regress must end somewhere; it cannot be either infinite or circular'.[23] This is almost but not quite right. As I noted with my example of Galileo's material bodies, there might in fact be some infinite ontological

which can never be found in this way... if we must always continue to seek for them.... if there are aggregates of substances, there must also be true substances from which all the aggregates result.' (Leibniz, correspondence with Arnauld, 30 Apr. 1687, in *Philosophical Essays*, tr. and ed. Roger Ariew and Daniel Garber (Indianapolis: Hackett: 1989), 85; see also *Philosophical Essays*, 103, 207, 213, 262.)

[22] "Tis evident, that existence in itself belongs only to unity, and is never applicable to number, but on account of the unities of which the number is compos'd. Twenty men may be said to exist, but 'tis only because one, two, three, four, &c., are existent; and if you deny the existence of the latter, that of the former falls of course. 'Tis therefore utterly absurd to suppose any number to exist, and yet to deny the existence of unites'. (Hume, *Treatise* 1. 2. 2. 3; SBN 30.)

[23] J. L. Mackie, *The Miracle of Theism* (Oxford: OUP, 1982), 90.

regresses which at the same time also have last terms. (For Galileo's bodies are composites of composites of composites ad infinitum—though at the same time also composites with ultimate, noncomposite parts.) But this leaves Mackie's key point intact: there must be an ultimate ground floor—an original source of all the parasitic entities—even if there can at the same time be an infinite regress 'above' that ground floor.

Many have found this basic principle convincing: regresses of ontologically derivative entities must ultimately rest on a ground floor of entities that are not so derivative. Mackie—who finds the principle 'at least highly plausible'—goes on to defend it by invoking various intuitive examples. We feel uncomfortable at the thought of an accelerating train whose last carriage is pulled along by the second-to-last, which is in turn pulled by the third-to-last, and so on, when we are told that the series of carriages lacks a starting place. 'Where is the engine?' we want to ask. 'Where is the accelerating force coming *from*?' Similarly, the thought of a chain that dangles vertically but, running upwards to infinity, is not hanging *from* anything, makes us uneasy.[24] Of course, such discomfort and unease do not in themselves provide us with logical proof. Infinite regresses do not amount to logical contradictions. But Mackie's examples do induce a sort of metaphysical vertigo that helps to underline the principle's intuitive appeal. I have nothing new to add in defence of this principle. It is invoked in M1 simply as an axiom, and I think it is clear that someone who simply rejects M1 cannot be faulted on logical grounds. Still, many have found this principle intuitively compelling. It was certainly overwhelmingly popular with metaphysicians of the early modern period.

Another traditional objection to the cosmological argument is the charge that it requires that we make sense of a *necessarily* existent being to halt the regress and function as the uncaused first cause—and this, it is said, is impossible: 'the words... "necessary existence" have no meaning; or, which is the same thing, none that is consistent'.[25] Perhaps this is so, but I do not see that an analogous problem applies to the regress argument from composites to simples. Here the charge would have to be that the argument M1–M8 requires that we make sense of a partless first part to halt the regress and function as the noncomposite first element. Some of course have found the notion of such simple first parts problematic (see the para-

---

[24] Mackie, *The Miracle of Theism* 90.
[25] David Hume, *Dialogues Concerning Natural Religion*, ed. Richard H. Popkin (Indianapolis: Hackett, 1980), 55–6.

doxes of ultimate parts in Chapter 1, section VIII). But I see no contradiction in the bare idea of such an entity, and have argued that there are in fact consistent models of such simples: for instance, consider Boscovich's and the younger Kant's force-shell atoms. (For a full review of all the consistent models of partless first parts, see the conclusion of this book.)

Under attempt (iv) I presented what I see as both the most plausible interpretation of why Kant thinks K3 can be derived from K2, and what seems to me the best bet for making such an inference viable. We are now ready to complete the exegesis of the overall argument K1–K6.

Recall that Kant thinks that bodies are composite entities. This is to say that they consist of a plurality of parts where (K1) those parts all stand in a relation of composition with one another, and (K2) these parts can each continue to exist in the absence of that composition relation. We are then told that (K3) all composition can be removed from a composite entity, and that (K4) when this happens, what remain are simples.

Kant offers us a brief defence of the claim K4. 'When all composition is abolished... the parts which are left are not compound at all; and thus they are free from plurality of substances, and thus they are simple' (proposition II). The reasoning here is self-interpreting. All composition has been removed from the original entity, therefore what remains cannot be composite. However, as a defence of K4 this remains incomplete. Before Kant can conclude that the total removal of composition from a composite entity leaves us with simples, he must rule out the possibility that the decomposition of a composite entity leaves us with *nothing* whatsoever.

Fortunately it is not difficult to fill in this gap. (And in fact Kant himself does so in the version of the argument he presents in the Second Antinomy in the *Critique of Pure Reason*.) It is most likely that here in the *Physical Monadology* Kant is merely taking this step in the argument to be obvious enough not to bother spelling out explicitly. One simply needs to insist that there was an entity—the composite whole—which existed prior to the removal of composition, and that such a composite entity consists of—just *is*—so many parts standing in a relation of composition with one another. And since (given K2) the removal of composition does not prejudice the existence of those things which stood in that relation, then there must be something left when composition is removed. Since something existed in the first place, there must be something left when the 'contingent relation' of composition is removed. Putting this in terms of

192 / The Argument from Composition

the insistence that there can be no relations without ultimate relata, it is obvious that if all relations of composition are removed, then ultimate relata which ground those relations will remain. So: something remains when all composition is removed from a composite entity. This something cannot be composite. Therefore what remain are simples. This establishes K5: all composite entities are ultimately collections of simple elements. And, given that (K1) bodies are composite entities, it follows that (K6) bodies are ultimately collections of simple elements. 'All bodies, whatever, therefore, consist of absolutely simple fundamental parts, that is to say, monads' (proposition II). Finally, given the actual parts framework, these simples cannot be m-divisible (for m-divisibility would compromise their simplicity). So we have an atomic base of elemental m-indivisible simples: metaphysical atomism is established.

IV. Objections and Replies

In this section I assess the leading objections to the argument from composition. Here I draw on objections articulated in the broader literature on the argument, and not simply from those few critics who focus explicitly on the *Physical Monadology* version. My aim here is to draw on objections from all relevant sources, yet to preserve the clarity and focus of our assessment by seeing how these various criticisms play out against our particular, systematically reconstructed version of the argument.

Clearly I have already outlined certain objections and responses in my exegesis of the argument in section III above. This was necessary because the development of a charitable interpretation of Kant's argument had to be sensitive to certain possible objections that might be pressed against that reasoning. In what follows I both summarize those objections and responses we have already seen, and examine objections that have yet to be considered.

Objection 1. Schopenhauer, Kemp Smith, Justi: The Assumption that Matter is Composite Begs the Question

One sort of objection we find in the literature asserts that the argument from composition commits a flagrant *petitio* in defining matter as a composite substance. This criticism underpins Schopenhauer's dismissive treatment of the argument and is also echoed in Kemp Smith's commentary.

Here is Schopenhauer in his 'Criticism of the Kantian Philosophy' in *The World as Will and Representation* (Schopenhauer is commenting on the version of the argument as it appears in the Second Antinomy in Kant's *Critique of Pure Reason*):

> In the second antinomy, the thesis at once commits a *petitio principi* that is not in the least subtle, since it begins: 'Every *compound* substance consists of simple parts.' From the compoundness here arbitrarily assumed, it of course very easily demonstrates afterwards the simple parts. But the proposition, 'All matter is compound,' remains unproved, because it is just a groundless assumption.... [I]t is really tacitly assumed that the parts existed before the whole, and were gathered together, and that in this way the whole came into existence; for this is what the word 'compound' means.[26]

And similarly in Kemp Smith (again, in a commentary on the Second Antinomy version of the argument):

> Kant here assumes by the definition of his terms, the point which he professes to establish by argument.... Kant identifies [extended matter] with 'composite substance.' Substance, he further dogmatically decides, is that which is capable of independent existence, and to which all relations of composition are accidental. If these assumptions be granted, it follows at once that composition cannot be essential to matter, and that when all composition is thought away, its reality will be disclosed as consisting in simple parts. Kant, however, makes no attempt to prove that extended matter can be defined in any such terms.[27]

I think that we can discern two nested *petitio* charges here. First, there is the charge that, in assuming the actual parts analysis of matter, and hence that material bodies are composite structures, Kant simply ignores other possible views of material structure (such as the potential parts account). Second, there seems to be the charge that the term 'composite' employed in the argument is not neutral on the question of ultimate parts but rather surreptitiously smuggles in the conclusion that there are simple atomic parts. This is because (on this reading) 'composite' doesn't merely mean 'is built up from distinct parts' but also 'is built up from distinct *simple* parts'.

---

[26] Arthur Schopenhauer, *The World as Will and Representation*, tr. E. F. J. Payne, 2 vols. (New York: Dover, 1969), i. 496.

[27] Norman Kemp Smith, *A Commentary on Kant's Critique of Pure Reason* (London: Macmillan, 1918), 489.

Let us start with the former charge. Although neither Schopenhauer nor Kemp Smith employs the language of actual and potential parts, and neither of them clearly sets out the two rival views in any other terms, I think it is clear that at least one of their concerns is that the argument from composition presupposes the actual parts metaphysic and thereby begs the question against those anti-atomists who endorse potential parts. This comes out clearest in their diagnosis of where the argument from composition goes wrong. Their suggestion is that the *metaphysical* concept of the composite has been confused with the *geometrical* concept of the extended and divisible. Bodies certainly are extended and divisible, but advocates of the argument then go on to confuse this obvious truth with the (actual parts) thesis that bodies are composite. Here Kemp Smith writes that 'On any [anti-atomist] view the extended and the composite are not equivalent terms. The opposite of the composite is the simple; the opposite of the extended is the non-extended.'[28] Similarly Schopenhauer maintains that divisibility should not be confused with composition: 'Divisibility implies merely the possibility of splitting the whole into parts; it by no means implies that the whole was compounded out of parts, and thus came into existence. Divisibility merely asserts the parts *a parte post*; compoundness asserts them *a parte ante*.'[29] In assuming the actual parts doctrine, advocates of the argument from composition assert the existence of parts *a parte ante*: they assume that the parts are distinct entities prior to division and that the whole is a composite built out of these parts. This begs the question against the anti-atomist who endorses the potential parts metaphysic instead.

A version of this first *petitio* objection was raised in the eighteenth century during the notorious Berlin Academy tumult over the structure of matter, and so should have been familiar to Kant. In 1748, the Royal Academy of Berlin assigned the ultimate parts controversy as the topic of its annual essay competition and, in awarding the prize to Justi's anti-atomist polemic, set off a political crisis in German letters. Euler at least was still crowing about this symbolic victory over Wolff and the metaphysical atomists years later.[30] In his prize essay Justi levels the same *petitio* charge against the argument from composition and lays the blame

---

[28] Kemp Smith, *A Commentary on Kant's Critique of Pure Reason*, 489.
[29] Schopenhauer, *World as Will*, i. 496.
[30] Leonhard Euler, *Letters of Euler on Different Subjects in Natural Philosophy*, letter X, 39 (see also letters XV, XVI, and XVII). See also n. 10 for the history of this controversy.

squarely on a confusion of metaphysical and geometrical concepts: '[A]s soon as one combines a geometrical notion with a metaphysical one, a false conclusion always results.... [I]t cannot be denied that the idea of the composite could not be a metaphysical one here; the conclusion is therefore necessarily false wherever such a notion is combined with the geometrical notion of a "simple." '[31]

Obviously I agree that the argument from composition does presuppose an actual parts framework that ties (in-)divisibility to composition. Bringing this out has been one of the central expository goals of this chapter, and indeed of this study more generally. So someone who accepts the potential parts system will have no reason to accept this argument for ultimate parts: as my reconstruction makes clear, they will immediately reject the opening premise K1. Still, I think we can take Kant to be offering a substantive argument to those who *do* accept the actual parts doctrine. This includes the overwhelming number of natural philosophers of the period, and certainly includes many who are hostile to ultimate parts.[32] A proof of ultimate parts from actual parts assumptions would thus be interesting. So, while the argument does indeed assume actual parts and will not appeal to potential parts theorists, this does not yet warrant our joining Schopenhauer in writing it off as stipulative and trivial.[33] At the very least we are presented with an interesting argument for the (contested) conditional claim: *if* actual parts are assumed, ultimate parts must follow.

This brings us to the second *petitio* charge nested in Schopenhauer and Kemp Smith's complaint. They each seem to imply that, once it is conceded that material bodies are composite entities, it immediately follows as a mere tautology that they are constructs from simples. This, Schopenhauer tells us, simply falls out of the meaning of the word 'compound'; likewise Kemp Smith claims it is true just in virtue of the definition of terms. The charge here is that, while Kant acts as if the predicate 'composite' (or 'compound') is neutral on the question of ultimate parts, in fact the way it is used in the argument shows that it really means 'built

---

[31] Justi, quoted in Poloroff, *Force*, 37. Notice that this 1748 prize essay of Justi's predates Kant's 1756 *Physical Monadology*. Kant could not but have known of Justi's infamous essay and its inflammatory central charges against the Wolffians.

[32] See Ch. 2, s. II, on the overwhelming popularity of the actual parts doctrine, and Ch. 1, s. IX, 'faction (3)', on those who accept actual parts but reject ultimate parts.

[33] Schopenhauer, *World as Will*, i. 497; also ii. 303.

out of simple elements'. If this is indeed what 'composite' means in the argument, then it is simply an analytic truth that whatever is composite must resolve to simples. But then the assumption that material bodies are thus composite is transparently question-begging.

But this second charge is unfair. Certainly, if 'composite' as employed in the argument just means 'has ultimate simple parts', then the argument would be a grotesque *petitio*. But if we read 'composite' as I have been employing it throughout my various statements of the actual parts doctrine and in my reconstruction K1–K4 as neutral on the question of ultimate parts, then the argument can proceed without danger. In setting out Kant's account of composition above, I said that a composite entity can be defined as an entity that consists of parts, where those parts are each distinct beings and stand in a (contingent) relation of composition with one another. Now, this does imply that anything that is composite has independent parts. But one can say this while remaining neutral on the question of whether its parts have subparts that have subparts ad infinitum with no ground floor, as the anti-atomist or gunkist has it, or whether it must ultimately resolve to simple first parts, as atomists claim.

In fact (and as we have seen) the argument from composition does not get to ultimate parts simply from analytically unpacking the meaning of 'composite', but rather also has to invoke further crucial and substantive metaphysical premises—such as the impossibility of ontological regresses without a ground floor, and the claim that a complete decomposition of a complex existent must leave some residue behind. In short, Kant in the *Physical Monadology* and actual parts theorists in general employ a concept of a composite that does not analytically entail ultimate parts. So Kant does not beg the question in his definition of terms when he asserts in K1 that bodies are composite entities.

Presumably those who press this second *petitio* charge will agree that the argument does not explicitly beg the question and will concur that the use of the concept of a composite is presented as neutral on the question of ultimate parts. But they will likely claim that when one actually follows the argument closely through all of it twists and turns, explicit and implicit, it emerges that the term is surreptitiously being read as equivalent to 'has ultimate, simple parts'. For instance, they may point to the crucial but suppressed reasoning by which K3 is derived from K2 and maintain that this key step implicitly presupposes that composites have ultimate parts. But my response to this should, by now, be obvious. This reading of the argument, in which it is implicitly presupposed that composites have ultimate parts, is just that which I reconstructed above as attempt (i) to prove K3. This is the sort of reading which Schopenhauer and Kemp Smith seem to have in mind when they claim that Kant assumes his conclusion by sleight of hand. But above I argued that the first attempt is really neither the correct interpretation of the reasoning which Kant is actually relying on, nor the best bet for actually establishing a sound argument overall. It is only the temptation to read Kant's argument as implicitly relying on the strategy of attempt (i) which makes the second charge of Schopenhauer and Kemp Smith look plausible. But when we see there is a better way to derive K3—the fourth attempt which invokes the regress of relations argument—there is no longer any temptation to think that Kant needs to implicitly presuppose that composites resolve to simples, and so no temptation to say he is (surreptitiously) reading 'composite' as 'has ultimate simple parts'.

Objection 2. The Assumption that Decomposition is a Finite Procedure Begs the Question

A second sort of objection has gained wide currency in the more recent literature on the argument from composition. This objection is once again motivated by what seems to me a misinterpretation of the key derivation of K3. I have already examined this objection in section III of this chapter, so here I will limit myself to a brief résumé of the criticism and my response to it.

Recall that both here in the *Physical Monadology* and in his other presentations of the argument, such as that in the Second Antinomy, Kant fails to explicitly spell out how the crucial derivation of K3 is supposed to run. This omission led us to consider various possible speculative reconstructions of Kant's implicit reasoning at this key point. Of these interpretations, one sort of reading—that of 'attempt (ii)' and 'attempt (iii)' set out above ((iii) being essentially a variant of (ii))—has proved very popular in recent commentaries. Indeed, the commentators to be examined here each take it as straightforwardly obvious that the second attempt strategy, or perhaps the similar third attempt, does in fact capture Kant's implicit reasoning.

The second attempt to derive K3 ran as follows. Given K2, *each individual layer of composite structure can be peeled away in thought. Therefore all layers can be peeled away in thought successively.* (The third attempt ran similarly. Given K2, *each individual* layer of composite structure can be peeled away in thought. Therefore *all* layers can be peeled away in thought *simultaneously*.) The problem with this sort of strategy was identified in section III above. The argument is valid only if one implicitly presupposes there is an ultimate ground floor of first parts, rather than an infinite number of layers of atomless composite structure. But this is just what the anti-atomist denies. So, if one interprets Kant's key move in the argument to be that of the second attempt or third attempt, then clearly the overall argument for metaphysical atomism is a *petitio*.[34]

If this is how one interprets Kant's implicit reasoning, then he is indeed begging the question in presupposing that the structure of bodies is only finitely complex. However, I have argued that this interpretation of Kant's tacit derivation of K3 is mistaken, since there is an alternative strategy (which certainly does not commit this sort of *petitio*) that can be more plausibly credited to him.

Finally, notice that the basic interpretive error that I attribute to those commentators sponsoring the current objection is a version of the more general mistake that runs through the current secondary literature on the problems of material structure (or so I argued in Chapter 1). These commentators seem to take Kant's argument to be a variant of the old argument that runs from the property of m-divisibility through to metaphysical atomism via the possibility of performing a through-and-through m-division. (See Chapter 3, section V, for analysis of this form of argument.) To be sure, the argument as they present it is explicitly framed in terms of a through-and-through 'decomposition' rather than a through and through 'division'. But here this seems simply a verbal difference: either way, as they read the argument, Kant is asking us to envisage the final completion of a process by which the whole original is fully disassembled or broken into its subparts, through and through. And either way, this sort of argument makes no appeal to considerations of the ontological dependence of a composite whole on its component parts, or on the way in which relations presuppose relata. (Notice, for instance, that this sort of argument could apply equally well to an entity with potential parts: Kant

---

[34] For versions of this reading (and this criticism) see Polonoff, *Force*, 149. James Wm. Forrester, 'If, in Thought, All Composition be Removed...', *Kantstudien*, 71 (1981), 406–17, 412–14. James Van Cleve, 'Reflections on Kant's Second Antinomy', *Synthese*, 47 (1981), 481–94, 492–3. The relevant section of Van Cleve's article is also reprinted in his *Problems from Kant* (Oxford: OUP, 1999), 64–5.

could just as easily have asked us to envisage the final completion of a process by which a whole with merely potential parts is broken into parts and subparts, through and through, all they way down.) This reading then misses the argument's essential metaphysical context: its crucial actual parts assumptions, its implicit appeal to the parasitic nature of composite structures, and its rejection of ontological regresses without ground floors. So (it seems to me) these commentators fall into the common error of interpretation that dominates the current literature on the early modern debate. They take the problems of material structure to be simply about the structure or form of bodies, and thereby miss their crucial dependence on metaphysical premises concerning the stuffing or filling of actual physical concreta.[35]

Objection 3. Hobbes, Wittgenstein: the Concept of an Absolute Simple is Incoherent

A third and final objection runs as follows. The principle that composite entities are ontologically derivative and depend ultimately on simple first parts is crucial to the argument from composition. And the notion of simplicity (and correlatively, of composition) here invoked would seem to be an *absolute* or *intrinsic* simplicity—a *per se* simplicity that is an objective rather than observer-relative or conventional matter. But this notion of absolute simplicity (or absolute complexity) is incoherent. Simplicity and complexity are necessarily relative to the context and purposes of the observer—so talk of an 'absolute', 'true', 'objective', or '*per se*' simple is so much nonsense. Now, the argument from composition employs this incoherent notion of an absolute simple explicitly in premise K4 and implicitly

---

[35] In fairness to Van Cleve (n. 34), I should add that in a subsequent paper he offers a reconstruction of an argument in the Amphiboly of the *Critique of Pure Reason* (found at B321, B330, and B339) that he takes to crucially turn on the principles that (*a*) composite structures like bodies ontologically depend on their parts, and that (*b*) relations must be traceable back to ultimate relata. So here Van Cleve shows full sensitivity to the metaphysical, actual parts context of Kant's reasoning concerning the structure of matter. See James Van Cleve, 'Inner States and Outer Relations: Kant and the Case for Monadism', in Peter H. Hare (ed.), *Doing Philosophy Historically* (Buffalo, NY: Prometheus Books, 1988). I take this argument in the Amphiboly to be a further statement of the same fundamental argument from composition also found in the *Physical Monadology* and in the Second Antinomy of the *Critique of Pure Reason*. Van Cleve presumably does not, since he reconstructs the Second Antinomy argument so differently from the Amphiboly arguments. (He does not address the *Physical Monadology* argument in either paper.)

in K1 and K2. So the argument is at best unsound, and may even be unintelligible.

Is the notion of an absolute simple incoherent? The classic statement of this view comes in Wittgenstein's *Philosophical Investigations*. While the younger Wittgenstein of the *Tractatus* seems to have endorsed a version of the argument from composition and the notion of an absolute simple,[36] in this later period he came to reject its invocation of absolute simplicity:

> But what are the simple constituent parts of which reality is composed?—What are the simple constituent parts of a chair?—The bits of wood of which it is made? Or the molecules, or the atoms?—'Simple' means: not composite. And here the point is: in what sense 'composite'? It makes no sense at all to speak absolutely of the simple parts of a chair.
>
> We use the word 'composite' (and therefore the word 'simple') in an enormous number of different and differently related ways.
>
> To the *philosophical* question: 'Is the visual image of this tree composite, and what are its component parts?' the correct answer is: 'That depends on what you understand by "composite".' (And that is of course not an answer but a rejection of the question.)[37]

Wittgenstein's claim here seems to be that, if we study our usage of the term 'simple' and the correlative term 'composite', we find that they lack an absolute sense. They are always employed within a certain framework, in which the standards of complexity and simplicity are laid down by context and convention. One and the same thing can count as simple or complex, depending on what those conventions are. So there is no absolute standard of simplicity. The point is echoed by D. W. Hamlyn, who goes on to weave this objection together with a rejection of the actual parts metaphysic along the lines of objection (1) above:

> [T]he notion of absolute simplicity is itself an incoherent one. Simplicity is always a relative matter because the notion of simplicity itself is a relative one. Hence the proper response to the question 'Is such and such simple?' is 'A simple what?' ... [A] complex F is obviously secondary to a simple F. But a simple F need not, and indeed cannot, be absolutely simple.
>
> This point covers even spatial complexity and simplicity. ... The fact that a substance is spatial, and therefore potentially divisible spatially, does not entail

---

[36] Wittgenstein, *Tractatus Logico-Philosophicus*, propositions 2.02, 2.021.

[37] Wittgenstein, *Philosophical Investigations*, tr. G. E. M. Anscombe (Oxford: Basil Blackwell, 1953), s. 47.

that it is spatially complex, let alone dependent for its existence on separable parts. Substances can therefore, be both extended and simple of their kind.[38]

This account of simplicity as necessarily relative to context and convention seems to be foreshadowed in Hobbes's *De Corpore*. Hobbes does not make the point in terms of linguistic frameworks. But he does insist, in his remarkable account 'Of Place and Time', that the concepts of one, compound, part, and whole must be given an observer-relative, *per mentem* gloss, rather than an absolute, *per se* one. If something is said to be 'one' it is said to be '*one of them*'—that is, the observer is considering it among other like entities. This is contrasted with the 'common notion of *one*, namely, that *one is that which is undivided*'—a (scholastic, potential parts) version of absolute or objective oneness that Hobbes rejects as leading to absurdity. What we seem to have here is an account of oneness or simplicity which is inescapably *per mentem*, and which mirrors his observer-dependent accounts of division, parthood, and composition.[39] So, it seems to me, the core of the later Wittgenstein's objection can be found at least as far back as Hobbes.

I think it must be granted that, *if* the concept of an absolute simple is incoherent, then the argument from composition is doomed. But is the notion of an absolute simple really incoherent? This simple-looking question rapidly unravels into a whole host of involved and far-reaching metaphysical problems. Issues of realism and anti-realism bear on this question, as does the relationship between the 'transcendental' properties of existence and unity. I cannot hope to settle all these issues here. My comments will be limited to a few central points that bear on this question.

Wittgenstein points out—rightly I think—that our typical usage of the correlative terms 'simple' and 'composite' is annexed to conventional standards of simplicity and complexity which shift from context to context, from language game to language game. His evidence for this amounts to an appeal to our everyday use of the terms. But to establish that any notion of an absolute simplicity is altogether incoherent he must do more than point to our usual or standard practices. He needs to show that nothing could *possibly* count as a simple unless it were simple *vis-à-vis* some relative, non-absolute standard of simplicity.

---

[38] D. W. Hamlyn, *Metaphysics* (Cambridge: CUP, 1984), 107–8.

[39] Hobbes, *De Corpore*, ch 7, ss. 4–9, n *The Collected Works of Thomas Hobbes*, ed. William Molesworth, 11 vols. (London 1839–40; facsimile edn.: London: Routledge Thoemmes Press, 1992), i. 95–7. See also my discussion of Hobbes in Ch. 2, s. III.

But couldn't there be some privileged standard of simplicity that is absolute or objective, and not merely a function of local human attitudes? The early moderns' metaphysical framework at least purports to present us with a recognizable and apparently objective standard of simplicity and complexity. Consider the actual parts doctrine. This asserts that the parts into which a whole can be m-divided each exist prior to that division. The whole is then considered a composite of these independently existing parts: it is complex, since it is an amalgam or aggregate of multiple independently existing entities. So tied up in the actual parts framework is what looks like an absolute, non-conventional account of complexity and simplicity. Something counts as an absolute simple if and only if it is an entity that lacks subparts that exist independently of the whole. Conversely, something counts as an absolute composite if and only if it does have such independently existing subparts. (Notice that an early modern potential parts theorist will just as much accept this definition of simplicity and complexity—it is just that they think that some m-divisible but undivided bodies do in fact satisfy the criterion for simplicity: some bodies do lack independently existing subparts, their m-divisibility notwithstanding.) This account of simplicity and complexity certainly seems to capture the sort of thing our Enlightenment thinkers intended by their absolute or *per se* simples. Galileo's *parti non quante* fit the definition well, as do Berkeley's and Hume's *minima sensibilia*, Leibniz's monads, and the force-shell atoms of Boscovich and Kant. And of course this early modern account of simplicity set out here is endorsed quite clearly in Kant's explicit definition of a 'simple substance' in the *Physical Monadology*: 'A simple substance is one which does not consist of a plurality of parts, any one of which could exist separately from the others' (proposition I).

While there certainly are other uses of 'simple' and 'complex' that are context-dependent and conventional, it does seem to me that the early modern definition of simplicity articulates a recognizable use of the term. And it does seem right to say of composites and simples, defined in this way, that the former are dependent beings (dependent on the ontologically prior existence of their distinct parts) in a way that the latter are not. Moreover, even if Wittgenstein were to insist (oddly, I would say) that such an employment of the term 'simple' is utterly alien to our actual linguistic practices, there is still no reason why we could not just stipulate that this is how we mean to use the term in the argument, regardless of how it is usually employed. We could simply legislate that this is how we are to use the term (also stipulating a correlative definition of 'composite', of course) and the argument from composition can then proceed accordingly. Only if there is some internal incoherence or self-contradiction in this definition irrespective of the common language use of the term is there a problem in employing it in the argument.

But (someone might ask), granted that the early modern definition sets out a recognizable use of 'simple' (and of 'composite'), does it really present us with an absolute or objective sense of the term? According to the proposed definition, something is simple if and only if it lacks independently existing subparts, and composite if and only if it has independently existing subparts. These criteria might seem at first glance to look non-conventional and objective. But in fact they will only really prove absolute if we can give satisfactory definitions of parthood and independent existence that are themselves absolute, lacking any reference to observers' conventional attitudes. Philosophers of a realist bent will be prepared to grant that, when it comes to the fundamental ontology of the world, such basic issues of distinct existence and individuation are indeed absolute or objective. But there will be some anti-realists who maintain that questions such as these ultimately turn on the local attitudes or conceptual scheme of the observer. Hobbes is certainly happy to relativize questions of individuation and parthood to the observer's conventional attitudes,[40] and the later Wittgenstein would presumably be similarly prepared to annex these ontological questions to whatever language game is in play. If one is prepared to join Hobbes and Wittgenstein in their relativization of individuation and fundamental ontology to conceptual scheme, then, it will of course follow that the argument from composition (which presupposes absolute standards of individuation and hence of simplicity and complexity) founders. On the other hand, if one rejects this sort of relativization of fundamental ontology to the parochial and shifting preoccupations of human attitudes, then the early moderns will have an intelligible and recognizable account of absolute simplicity.[41]

---

[40] See my discussion of Hobbes's anti-realist account of individuation and parthood in Ch. 2, s. III.

[41] Hume stands as a good example of this latter approach. He admits (of course) that there can be *per mentem*, observer-relative standards of simplicity or 'unity', but insists that this sort of simplicity is merely a 'fictitious denomination' which cannot serve to ontologically ground genuinely existing things. He writes that, while one can elect to call any given quantity 'an unite,' 'by the same rule these twenty men *may be consider'd as an unite*. The whole globe, nay the

V. Summary and Conclusion

I have presented an interpretation of an argument found in Kant's *Physical Monadology* which I take as representative of a broader pattern of reasoning common to many Enlightenment metaphysical systems. (If the reader is sceptical of my claim that Kant's argument can serve as a model or example of a type of argument also found in Leibniz, Wolff, and Hume, he can simply take this chapter's assessment of this particular argument to stand *in vacuo*.) My reconstruction may be contrasted with other readings that take Kant's argument to function without any necessary reliance on actual parts premises. I stress the role of the actual parts framework and argue for a reconstruction of Kant's implicit but crucial move from K2 to K3 that turns on the inadmissibility of relations without ultimate relata. I argued that my 'fourth attempt' reconstruction of this inference to K3 is preferable to the rival readings, both on textual-historical grounds and on grounds of interpretive charity.

I can summarize the argument briefly here. It is assumed that all composites consist in the relatedness of their parts. This means that composites are ontologically dependent on their parts. If these parts are themselves composite the regress continues. But (it is assumed) the regress cannot run on forever if the whole composite entity actually exists. So it must ultimately terminate in atomic parts which are noncomposite. This establishes the major premise that all composite entities are built up from simples. The actual parts doctrine then provides the minor premise that material bodies are composite entities, and we have our conclusion: all material bodies are built up from simples. Metaphysical atomism is established.

None of the various objections considered have (it seems to me) embarrassed the validity of Kant's reasoning. So, given Kant's background framework of metaphysical assumptions, his argument seems to be a success. Any challenge to the argument must then take the form of a critique of

whole universe *may be consider'd as an unite*. That term of unity is merely a fictitious denomination, which the mind can apply to any quantity of objects it collects together; nor can any such unity any more exist alone than number can, as being in reality a true number. But the unity, which can exist alone, and whose existence is necessary to that of all number, is of another kind, and must be perfectly indivisible, and incapable of being resolved into any lesser unity'. (Hume, *Treatise*, 1. 2. 2. 3; SBN 30–1.) There must, Hume thinks, be an absolute sense of unity or simplicity that serves as the basis for all composites. As a good actual parts theorist, Hume identifies this absolute simplicity with (m-) indivisibility. Compare also Leibniz in the correspondence with Arnauld, 30 Apr. 1687, in *Philosophical Essays*, 89.

this background framework of assumptions—assumptions which most new philosophers of the seventeenth and eighteenth centuries would have found compelling. We can now enumerate these metaphysical assumptions and hence the possible challenges to Kant's background framework. First, Kant's argument assumes the actual parts doctrine and its immediate corollary that bodies are composite structures. Most of the new philosophers would have agreed with this, though some will, of course, reject it. (See objection (1) above, and potential parts advocates in general.) Second, Kant's argument assumes that composites are ontologically derivative entities (depending on their parts), and also assumes that no ontologically derivative entity can exist unless it rests ultimately on a ground floor of entities that are not so derivative. Most would, I think, accept this former assumption, but some will reject the latter, claiming that infinite series of parasites without ultimate hosts *are* admissible. (As I have said, this is a logically respectable position, although it will induce an unsettling metaphysical vertigo in many.) Third and finally, Kant does assume that there are respectable concepts of absolute simplicity and absolute complexity. Most early moderns would certainly have accepted this: it falls out of the realist claim that the individuation of distinct entities is an objective matter rather than a matter of convention and contingent conceptual scheme. But some anti-realists will resist this claim about individuation and hence the idea that there is an absolute or objective standard of simplicity and complexity. (See objection (3) above.)

5

The Case for Infinite Divisibility

> I shall not believe that you are entirely cured of mathematics so long as you maintain that these tiny bodies, about which we were disputing the other day, can be divided in infinitum.
>
> Chevalier de Meré, in a letter to Pascal

I. Introduction

If the interpretation advanced in Chapter 1 is correct, the fundamental challenge facing Enlightenment matter theory was the apparent conflict between the geometrization of nature on the one hand, and the actual parts metaphysic on the other. Geometry seemed to mandate matter's infinite divisibility; the actual parts doctrine to preclude it. Chapters 2 to 4 dealt with the one side of this conflict: the actual parts doctrine and its various corollaries (genuine and otherwise). In the current chapter I look at the other side of the conflict: the geometrization of nature and the case for infinite divisibility.

This chapter focuses on the two central arguments for infinite divisibility. Of the various arguments for infinite divisibility presented in early modern period, these two are the most philosophically interesting and also the most popular: they can justly be considered the classic arguments on this side of the debate. First there is (i) *the argument from geometry*, which purports to establish infinite divisibility by way of certain constructions in classical geometry. Although this was by far the most popular argument for infinite divisibility during the Enlightenment, it is given short shrift nowadays: it clearly presupposes that physical space and matter conform to the axioms of classical geometry, and thus begs the question. (The objection is certainly lethal. However, as we shall see, the more sophisticated early modern advocates of the argument from geometry were well aware of this possible *petitio* charge and attempted to answer it in various philosophically intriguing ways.) The second classic argument for infinite divisibility was slightly less popular in the Enlightenment literature, but is more direct and can seem more compelling. This is (ii) *the conceptual argument*, which asserts that, as a conceptual truth, whatever is extended has parts that are themselves extended; that the concept of a minimal extension is self-contradictory; and thus briskly concludes that whatever is extended is divisible ad infinitum. Once I have covered these two classic arguments in the body of the chapter, I then briefly survey the other less popular and less interesting arguments for infinite divisibility in a short appendix.

First a word about the status of these arguments. This book addresses the question of whether or not matter is infinitely m-divisible—that is, the question of whether or not it has logically separable parts within parts to infinity. But the arguments in this chapter do not deal *ab initio* with the m-divisibility of matter. Rather, they deal with f-divisibility of space—that is, with the question of whether or not any region of space contains subregions ad infinitum. So even if these arguments prove successful, they will not immediately prove that matter is infinitely m-divisible.

To reach that further claim, two additional steps are required. First, one would have to infer from the infinite f-divisibility of a region of space to the infinite f-divisibility of any material body that occupies that space. This step looks uncontroversial: if a given region of space has spatially distinct parts ad infinitum, then surely a body that occupies that space likewise has spatially distinct parts ad infinitum, each subregion of space playing host to one such spatially differentiable subpart of the body. The second step would then be to infer from the infinite f-divisibility of the body to its infinite m-divisibility: from the claim that it has *spatially differentiable* parts within parts ad infinitum to the conclusion that it has *logically separable* parts within parts ad infinitum. Now, I think that we can grant this second step at least some initial plausibility. In the case of matter, doesn't f-divisibility entail m-divisibility? If a body has spatially distinct parts, won't those same parts each be logically separable from one another? Couldn't God rupture and distance any two such parts? So (it seems to me) the burden of proof is at least initially on those who would resist this second step, to show how they can block the inference from the f-divisibility of a body to its m-divisibility. A few philosophers have, however, attempted to meet this challenge. (This, in essence, is the strategy of Boscovich and the early Kant with their metaphysic of f-divisible but m-indivisible force-shell atoms.) So—if these thinkers are correct—this second step may prove more problematic.

To sum up the dialectical situation and set the scope of the current chapter. This chapter examines the classic arguments presented on this side of the early modern debate: (i) *the argument from geometry* and (ii) *the conceptual argument*. Each of these arguments purports to establish the infinite f-divisibility of any region of space and *ipso facto* the infinite f-divisibility of any body (since each body occupies a region of space and, if a body occupies an infinitely f-divisible region of space, it is itself infinitely f-divisible). So these arguments, if they are successful, prove the infinite f-divisibility of any body. It was also widely thought, and it is certainly initially plausible, that the infinite f-divisibility of matter would entail its infinite m-divisibility. But this is a further step that I will examine in the next chapter. In this chapter, the focus will simply be the classic arguments for the infinite f-divisibility of space and, *ipso facto*, the infinite f-divisibility of body.

The chapter proceeds as follows. Sections II–IV are devoted to the argument from geometry. Section II reconstructs this argument and presents examples of the sort of geometrical proof that were thought to establish infinite f-divisibility. In section III I present what I take to be the fatal objection to this argument, giving due credit to the period's so-called 'ungeometrical philosophers' responsible for this objection. Then in section IV I examine possible attempts to salvage the argument from geometry and show why they each must fail. This last section will take us into a brief survey of the rival Enlightenment accounts of the epistemological status of geometry.

Sections V and VI are then given over to the conceptual argument. In section V I reconstruct this argument and document its popularity with both early modern thinkers and various present-day philosophers. In section VII I then attack the argument: its initial plausibility, I think, proves specious. Neither the argument from geometry nor the conceptual argument, the two classic a priori arguments for the infinite f-divisibility of space, then, is a success. This appears to leave the question an empirical matter: for all logic mandates, space may or may not be infinitely f-divisible. Finally, an appendix documents two other (less interesting and less popular) types of argument for infinite f-divisibility also found in the period literature.

II. The Argument from Geometry

The doctrine of the infinite f-divisibility of space was orthodox in all the main traditions of Enlightenment natural philosophy. Notwithstanding their famous differences over the ontological status of space, the central factions—Cartesian, classical Newtonian, and Aristotelian–scholastic—were all in agreement on this question. Along with these traditions we could add individual luminaries such as Galileo, Hobbes, Barrow, Spinoza, Locke, Leibniz, Euler, and Kant. Even Gassendi, the great champion of p-indivisible material atoms, renounces Epicurus on this question: contra the Greek atomist, space and matter are both endlessly f-divisible.[1] (This leaves only a few dissenting figures whose rejection of infinite f-divisibility, however just, can only be described as idiosyncratic when set against the main traditions of natural philosophy. These dissenters include the hardcore Epicureans Charleton and the younger Newton, the ever-sceptical Bayle, and the champions of *minima sensibilia* Berkeley and Hume.)

By far the most popular form of argument for the doctrine of infinite f-divisibility involved the invocation of geometrical constructions. These constructions underwrote clear mathematical proofs, advancing from elementary geometrical axioms through lock-step theorems to demonstrate the doctrine. And no doubt because of geometry's celebrated status in the Enlightenment as a paradigm of *scientia*, offering self-evident step-by-step deductions, these elegant, clear, and a priori demonstrations seemed to most period philosophers to present incontrovertible proof of infinite f-divisibility. Thus Nicolas Malebranche feels comfortable assuring us that 'we have clear mathematical demonstrations of the infinite divisibility of matter', and Antoine Arnauld and Pierre Nicole that 'the divisibility of matter to infinity has been demonstrated ... [g]eometry provides proofs of it as clear as proofs of any of the truths which it reveals to us'.[2] Likewise Kant writes that 'employing infallible proofs of geometry, I can demonstrate that space does not consist of simple parts', and again that 'mathematics ... may rest in the certain possession of its evident assertions of the infinite divisibility of space'.[3]

---

[1] See Gassendi in the *Syntagma*, in *Opera Omnia* (Stuttgart-Bad: Canstatt, 1964), liber III, sectio I, caput V, 256–6. Compare Epicurus' rejection of space's infinite f-divisibility in the *Letter to Herodotus*—for discussion, see Richard Sorabji, 'Atoms and Time Atoms', in Norman Kretzmann (ed.), *Infinity and Continuity in Ancient and Medieval Thought* (Ithaca, NY: Cornell University Press, 1982), 37–82.

[2] Nicolas Malebranche, *The Search After Truth*, tr. Thomas M. Lennon and Paul J. Olscamp (Columbus, Ohio: Ohio State University Press, 1980), book I, ch. 6, s. 1; Antoine Arnauld and Nicole, *Logic or the Art of Thinking*, tr. and ed. Jill Vance Buroker (Cambridge: CUP, 1996), 231.

[3] Immanuel Kant, *Inquiry Concerning the Distinctness of the Principles of Natural Theology and Morality*, in *The Cambridge Edition of the Works of Immanuel Kant: Theoretical Philosophy 1755–1770*, tr. and ed. David

A wealth of such geometrical proofs is tendered in the period literature. I will briefly review just two, since this will be sufficient to bring out the essential strategy behind them all. Here I focus on the two proofs that turn up the most often in the literature and that (it seems to me) are also the most immediately intuitive.

Example (1). Infinitely Increasable Extension Proves Infinite F-Divisibility

This first proof crops up again and again in works of Enlightenment philosophy. It appears, for instance, in Antoine Arnauld's and Pierre Nicole's famous 1662 Port-Royal *Logique*, and again in Jacques Rohault's 1671 *Traité de Physique*, which, accompanied by Samuel Clarke's Newtonian annotations, becomes the standard textbook in mathematical physics at Oxford and Cambridge in the earlier part of the eighteenth century. It also turns up in Antoine Le Grand's popularization of Cartesian philosophy, the 1672 *Institutio philosophiae secundum principia D. Renati Descartes* (translated in 1694 as *An Entire Body of Philosophy*) and in Pierre Bayle's 1675–7 *Système Abrégé de Philosophie*. It occurs in the mathematician John Keill's Oxford lectures of 1700, and again in Kant's *Physical Monadology* of 1756. It even appears in a children's introduction to 'Experimented Physiology, or Natural Philosophy', Benjamin Martin's 1735 *Philosophical Grammar*.[4]

Let CA and DB be two parallel lines, each of which extends onwards to infinity. Let AB then be perpendicular to CA and DB. If we then draw a line from C (to the left of A) to a point E (to the right of B on the extended line DB), this line CE will cut the line AB at a point. But if we then draw a second line from C to a further point F, further out along DB than E, this new line CF will cut AB at a higher point. And so we can continue, through the points G, H, etc., each of which is still further out

---

Walford (Cambridge: CUP, 1992), 2: 273–301, 287. Kant, *Metaphysical Foundations of Natural Science*, in Kant, *Philosophy of Material Nature*, tr. and ed. J. W. Ellington (Indianapolis: Hackett, 1985), 52.

[4] Arnauld and Nicole, *Logic*, 232. Jacques Rohault, *System of Natural Philosophy*, tr. John Clarke and Samuel Clarke (New York: Johnson Reprint Corporation, 1969), 32. Pierre Bayle, *Système*, in *œuvres Diverses*, 5 vols. (1731; facsimile edn: Hildesheim: Georg Olms Verlag, 1968), iv. 300–1. Antoine Le Grand, *An Entire Body of Philosophy*, tr. Richard Blome, 2 vols. (London, 1694), ii. 8. John Keill, *An Introduction to Natural Philosophy* (London, 1733), 26. Immanuel Kant, *Physical Monadology*, in *Works*, 1: 473–87, 478. Benjamin Martin, *The Philosophical Grammar, Being a View of the Present State of Experimented Physiology, or Natural Philosophy* (London: J. Noon, 1735), 41.

The Case for Infinite Divisibility / 211

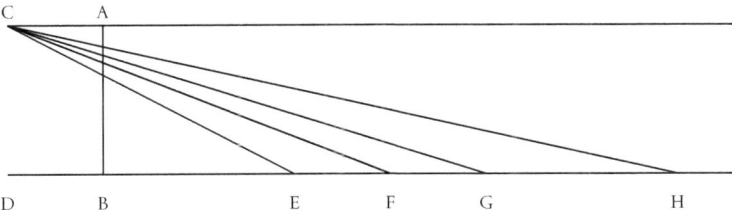

Figure 1 Geometrical proof of infinite f-divisibility from infinitely increasable extension

along the infinitely extended DB. and with each new line CG, CH, etc. cutting AB at a still higher point. Now, since DB extends to infinity, we can continue this process without end, forever drawing new lines that cut AB at points that are ever closer to A, but that never actually reach it. So AB is divisible ad infinitum. And analogous constructions could clearly show that any given extension in any spatial dimension is divisible ad infinitum. So any region of space, and anything extended in space, is infinitely f-divisible. *QED*.

Example (2). Incommensurable Magnitudes Prove Infinite F-Divisibility

Perhaps the simplest geometrical proof in the period literature appeals to the conflict between finite f-divisibility and incommensurable ('surd') magnitudes. This proof is found in the above-mentioned texts of Arnauld and Nicole, Rohault, Bayle, Le Grand, and Keill—though not, this time, in Kant. The proof is also found in Nicholas de Malezieu's 1705 *Elemens de Geometrie* and Francois Cartaud de la Vilate's 1733 *Pensées Critiques sur les Mathematiques*. It is also interesting to note that Samuel Clarke explicitly concurs with the demonstration in his running Newtonian annotations to the English version of Rohault's *Traité de Physique*.[5] In fact the proof has an ancient lineage, and can be found at least as far back as the thirteenth- and fourteenth-century work of philosophers such as Roger Bacon, John Duns

---

[5] See the section of Malezieu's *Elemens* quoted in Norman Kemp Smith's *The Philosophy of David Hume* (London: Macmillan, 1941), 341–2. Francois Cartaud de la Vilate, *Pensées Critiques sur les Mathematiques* (Paris, 1733; facsimile edn.: Geneva: Slatkine Reprints, 1971), 321–5. Samuel Clarke's endorsement is in his annotations to Rohault, *System*, 33.

Scotus, Gregory of Rimini, and William of Ockham.[6] The argument runs as follows.

We already know by a famous proof that the diagonal AC and the side AB of a square ABCD are incommensurable with one another: they share no common measure. No matter how many *aliquot* (i.e. same-sized) parts we divide the side AB into, we will never come to an *aliquot* part that can in some determinate number add up to the precise length of AC. As Rohault says, 'we may go on in the Division [of AB into ever-smaller aliquot parts] thus, for an Age together, without ever being able to come at Parts so small, as to say, that the Line AC contains a certain determinate Number of them'.[7] Now assume (for *reductio*) that space is not infinitely f-divisible but rather built up from so many f-indivisible atomic *minima*. But in this case the side AB and the diagonal AC would each have to be made up of a certain number of these *minima*. Then AB and AC would share a common measure: the fact that they each resolve to a determinate number of *minima* establishes that they are commensurable after all. But since we know that AB and AC cannot be commensurable, this constitutes a *reductio* on the assumption that space is merely finitely f-divisible: space must be infinitely f-divisible after all. *QED*.

The commonality between our two examples of the argument from geometry in action—and thus the argument's underlying point—should now be apparent. In each example the doctrine of finite f-divisibility is in conflict with one of the theorems of classical geometry. But such theorems are paradigms of demonstrative certainty and cannot be rejected without absurdity. So space cannot be merely finitely f-divisible; it must be f-divisible ad infinitum. And it should now be obvious why such a ready supply of constructions and proofs is available to the advocate of the argument from geometry. In fact *every* theorem in book I and II of Euclid's *Elements*, starting from the very first proposition, depends upon the supposition of infinite divisibility.[8] Proof after proof of the infinite f-divisibility of space can then clearly be generated, each on the back of a demonstrable theorem of classical geometry. The full authority of this most certain of sciences thus seems to preclude the granular model of space and to guar-

---

[6] See Pierre Duhem, *Medieval Cosmology: Theories of Infinity, Place, Time, Void and the Plurality of Worlds*, tr. and ed. Roger Ariew (Chicago: University of Chicago Press, 1985), 18–19.

[7] Rohault, *System*, 33.

[8] See Douglas Jesseph, *Berkeley's Philosophy of Mathematics* (Chicago: University of Chicago Press, 1993), 48–53.

antee its infinite f-divisibility. Here, for instance, is the great mathematician Isaac Barrow, Newton's predecessor in the Lucasian Chair at Trinity, clearly noting the wealth of geometrical proofs that are available:

> I could devise and produce even an Infinity of such Instances, whereby it may be proved from [the] Assertion of Composition out of Indivisibles [i.e. the denial of infinite f-divisibility], that the whole of Geometry is altogether subverted and destroyed, having nothing left in it sound and solid: and thereby a huge and deplorable Ruin, Confusion, and Inconsistency is brought upon this most Divine Science; whose Principles notwithstanding, besides their Evidence and the Rigour of their Reasonings, are so firmly established, as well by their wonderful Consonance one with another which could not happen if they were false, as by their perpetual and exquisite Agreement with Experience; so that the Poles of the World will be sooner removed out of their Places, and the Fabrick of Nature destroyed, than the Foundations of Geometry fail or its Conclusions be convinced of Falsity.[9]

And similarly in John Keill's Oxford lectures of 1700:

> There might be innumerable other Demonstrations produced, to shew the infinite Divisibility of Quantity, and entirely to overthrow the Hypothesis of Indivisibles. But what occasion is there for more? Since the arguments hitherto alleged, have not less force to compel the Assent, than any Demonstration in *Euclid's Elements*; insomuch that it is as impossible to weaken them, as to destroy the Fundamentals of Geometry, which no Age or Sect of Philosophers has ever been able to effect.[10]

III. Against the Argument from Geometry: the 'Ungeometrical Philosophers'

As the reader may already have noticed, the argument from geometry faces a devastating objection, notwithstanding its popularity with a constellation of Enlightenment thinkers. (In fact, for all Barrow's ebullience in the above quotation, we can already see a hint of the problem amidst the pomp of his own words.) The argument purports to establish that the infinite f-divisibility of space is mandated by various demonstrations in

---

[9] Isaac Barrow, *The Usefulness of Mathematical Learning Explained and Demonstrated*, tr. John Kirkby (1734; facsimile repr. London: Frank Cass, 1970), 156.

[10] Keill, *Introduction*, 30. See also Cartaud de la Vilate, *Pensées*, 296–349.

classical geometry. But of course this will only go through if classical geometry does indeed correctly describe space (I mean the physical space of the natural world, the realm of material objects)—that is, if space does indeed conform to the definitions and axioms of the classical geometry of Euclid. And this implicit assumption that classical geometry perfectly maps the structure of actual physical space—a doctrine we might call 'Pythagoreanism'—is controversial. In fact the assumption begs the very question that is supposed to be at issue in the argument. After all, no one would deny that certain figures of classical geometry—such as a square with incommensurable diagonal—can only be constructed in an infinitely f-divisible space. But one cannot very well assume that they *are* constructable in physical space (thereby presupposing that physical space is infinitely f-divisible) in order to prove that space is infinitely f-divisible. This is clearly to beg the question.

It is sometimes thought, quite mistakenly, that the first serious questioning of Pythagoreanism came only with Bolyai's and Lobachevsky's formal systematization of non-Euclidean geometries in the 1820s–1830s and Einstein's hypothesis of curved physical space in the 1910s—and thus that early modern philosophers like Arnauld, Barrow, and (most famously) Kant can, perhaps, be excused their uncritical faith in the doctrine. But this is too easy. In fact the insight that to presuppose Pythagoreanism is to make a significant assumption was quite well established in our period. Gassendi had been quite clear in his 1649 *Animadversiones* and 1658 *Syntagma Philosophicum* that the first principles of geometry are stipulative and thus that geometers are creators and legislators of abstract new thought-realms, realms at once generated, defined, and governed by the definitions and axioms laid down by fiat. The physical realm and the geometers' realms are thus quite distinct epistemological domains, and whether any geometrical demonstration proves anything at all about the physical realm depends on the prior question of whether the features of the physical realm do in fact satisfy the geometrical definitions to the requisite degree of accuracy.[11] Walter Charleton, the English popularizer of Gassendi and even more hard-core Epicurean, is as verbose and excoriating as ever in his prosecution of this point in his 1654 *Physiologia*. Finagling physical conclusions out of geometrical proofs, the advocates of the argument from geometry are guilty of 'an ambitious Affectation of extraordinary subtilty in the

---

[11] See Lynn Joy, *Gassendi the Atomist* (Cambridge: CUP, 1987), 102, 159–63.

invention of Sophisms (wherein Fallacy is so neatly disguised in the amiable light of right Reason, as to be charming enough to impose on the circumspection of common Credulity, and cast disparagement upon the most noble and evident Fundamentals)'.

For certainly, they could not be ignorant that they corrupted the state of the Quæstion; the *Minimum*, or *Insectile* of *Atomists*, being not *Mathematicum*, but *Physicum*, and of a far different nature from the Least of Quantity, which Geometricians, imagining only, denominate a *Point*. . . . [I]t is the *Naturalist*, whose enquiries are confined to sensible objects, and such as are really Existent in Nature: nor is He at all concerned, to use those *Abstractions* (as they are termed) from *Matter*; the Mathematician being the only He, who cannot, with safety to his Principles, admit the Tenet of Insectility, or Term of Divisibility. For to Him only is it requisite, to suppose and speculate Quantity abstract from Corporeity; it being evident, that if He did allow any Magnitude divisible only into Individuals, or that the number of possible parts in a Continuum, were definite: then could he not erect Geometrical, or exquisite Demonstrations.[12]

The same point is brought out in Pierre Bayle's 1675–7 *Système Abrégé* and in the article on Zeno the Epicurean in his widely read *Dictionnaire Historique et Critique* of 1697 (not to be confused with the article on Zeno of Elea, which also famously discusses infinite divisibility).[13] The question-begging nature of the argument from geometry is also condemned in Berkeley's notebooks:

Mem. To Enquire most diligently Concerning the Incommensurability of Diagonal and side. whether it Does not go on the supposition of unit being divisible ad infinitum, i.e. of the extended thing spoken of being divisible ad infinitum ... & so the infinite indivisibility [*divisibility.* surely?] deduc'd therefrom is a petitio principii.[14]

In fact the objection is almost as venerable as the argument from geometry itself. Versions of it can certainly be found in the works of Epicurus and his school, in the texts of the medieval Islamic Mutakallimum theologians, and in the fourteenth-century writings of John Buridan.[15]

---

[12] Walter Charleton, *Physiologia Epicuro-Gassendo-Charletoniana* (London, 1654; facsimile edn.: New York: Johnson Reprint Co., 1966), 95.
[13] Bayle, *Système Abrégé*, iv. 295; *Dictionary*, tr. and ed. Richard H. Popkin (Indianapolis: Hackett, 1991), 390–1.
[14] Berkeley, *Philosophical Commentaries*, notebook B, s. 263, in *The Works of George Berkeley*, ed. A. A. Luce and T. E. Jessop, 9 vols. (London: Thomas Nelson & Sons, 1948), i. 33.
[15] See Duhem, *Medieval Cosmology*, 19–20. Andrew Pyle, *Atomism and its Critics* (Bristol: Thoemmes Press, 1995), 12–13.

Finally, lest it be thought that this objection was little-known during the Enlightenment and shared only among a cabal of Epicurean apologists such as Gassendi, Charleton, and Berkeley, the problem is also clearly documented in the popular introductions to Newtonian physics of John Keill (the 1733 *Introduction to Natural Philosophy*, drawn from his Oxford lectures of 1700) and Leonhard Euler (the 1761 *Letters of Euler on Different Subjects in Natural Philosophy*). Now, Keill and Euler both explicitly *endorse* the argument from geometry: we already saw Keill's claim that such demonstrations of infinite f-divisibility 'have not less force to compel the assent, than any proof in *Euclid's Elements*'; likewise Euler insists that 'the geometrician...is warranted in asserting that every magnitude is divisible to infinity'.[16] But before advancing their geometrical arguments, each of these mathematicians expressly warns the reader about the complaints of those Keill scornfully christens 'our ungeometrical philosophers'. Thus Keill:

there are now-a-days some Philosophers who attempt to banish Geometry out of Physicks, by reason they are ignorant of that Divine Science; and as these Gentlemen would be reckoned amongst the most Learned, they leave no means untried, whereby, tho in vain, they may overturn the Force of the Demonstrations... [A]mongst the Philosophers of this Class, the famous *John Baptist du Hamel*... says then, that Geometrical Hypotheses are neither true nor possible, since neither Points, nor Lines, nor Surfaces, as Geometers conceive them, do truly exist in the Nature of Things; and therefore that the Demonstrations that are produced from these, cannot be applied to things actually existing, when none of these exist any where but in our Ideas. He desires therefore that the Geometers keep their Demonstrations to themselves, and not to make use of them in Philosophy.[17]

And similarly in Euler:

There are however philosophers, particularly among our contemporaries, who boldly deny that the properties applicable to extension in general, that is, according as we consider them in geometry, take place in bodies really existing. They allege that geometrical extension is an abstract being, from the properties of which it is impossible to draw any conclusion with respect to real objects.[18]

But notwithstanding Keill's and Euler's dismissive caricatures, the 'ungeometrical philosophers' are surely in the right. The modern reader will readily join them in agreeing that the axioms of any pure geometrical

---

[16] Keill, *Introduction*, 30. Leonhard Euler, *Letters of Euler on Different Subjects in Natural Philosophy*, tr. Henry Hunter, ed. David Brewster (New York: J. and J. Harper, 1833), 35.

[17] Keill, *Introduction*, 22   [18] Euler, *Letters*, 31.

system are simply stipulations or postulates with no necessary connection to physical space. The point is captured nicely in C. G. Hempel's witticism: 'to characterize the import of pure geometry we might use the standard form of a movie-disclaimer: No portrayal of the characteristics of geometrical figures or of the spatial properties or relationships of actual physical bodies is intended, and any similarities between the primitive concepts and their customary geometrical connotations are purely unintentional'.[19] Only when the primitive terms of the system are given a particular physical interpretation, linking them to actual physical properties, will that system make any assertions about actual physical space. But then the system in its physical interpretation may or may not present a factually correct account of real space: it all depends on whether or not real space does in fact conform to the axioms and definitions, so interpreted. As Einstein famously puts it, 'as far as the propositions of mathematics refer to reality, they are not certain; and as far as they are certain, they do not refer to reality'.[20] All this is but to stress that Pythagoreanism *is* a significant assumption, and that, for all logic mandates, classical Euclidean geometry may or may not describe physical space. Thus the argument from geometry clearly begs the question in its presupposition that classical geometry (with its stipulative assertion of infinite divisibility implicit in its basic definitions and axioms) accurately describes the physical realm.

This being the case, why then were advocates of the argument from geometry so convinced of its strength? The question is obviously particularly sharp in the case of those such as Keill and Euler who showed themselves to be well aware of the objection that Pythagoreanism is a significant assumption. An explanation will take us into the classical theories of geometry's epistemological status and into the traditional reasons for asserting the necessary concurrence of Euclidean geometry and physical space. Such attempts to ground Pythagoreanism are of course doomed to failure. But any account of the natural philosophy of the Enlightenment and its attempt to deal with the structure of space and matter must address them, if only in outline.

---

[19] C. G. Hempel, 'Geometry and Empirical Science', *American Mathematical Monthly*, 52 (1945), 12.

[20] Albert Einstein, 'Geometry and Experience', in *Ideas and Opinions* (New York: Crown Publishers, 1954), 254. Compare also Ernst Mach, *Space and Geometry*, tr. Thomas J. McCormack (Chicago: Open Court, 1906), 124. Bertrand Russell, 'Mathematics and Metaphysicians', in *Mysticism and Logic, and Other Essays* (London George Allen & Unwin, 1917), 92.

IV. Pythagoreanism and Early Modern Epistemologies of Geometry

> [I]f someone asks how truth, or the near semblance of truth, flows from metaphysics into geometry, there is to be sure no other road save the perilous gateway of the point.
>
> Giambatista Vico, *On the Most Ancient Wisdom of the Italians*, 1710

We can begin by setting aside the period's most forlorn attempt to salvage the argument from geometry. It is almost painful to report that Kant (like Keill and Euler) was aware of the question raised over Pythagoreanism and yet so embarrassingly insensitive to it. In the 1756 *Physical Monadology* Kant advances a variant of the argument from geometry, 'proving' that space is infinitely f-divisible by way of the first geometrical construction I presented above in section II. Now, throughout his version of this proof, Kant refers to the lines in his construction as 'physical lines': lines that are 'composed of the fundamental parts of matter'. A scholium following the proof then declares 'I have adapted [this demonstration], as clearly as I could, to physical space, so that those who employ a general distinction between geometrical and natural space, should not escape the force of my argument by means of an exception.'[21] But this is merely to report the words of the objection without grasping their true import. After all, 'those who employ a distinction between geometrical and natural space' will simply point out that it is question begging to proceed as if physical space shares all the properties of classical Euclidean space. Kant's 'adaptation'—which simply inserts the term 'physical' throughout the geometrical proof as if it were an honorific—misses this altogether.

If we put aside Kant's distressing pseudo-argument in the *Physical Monadology*, are there any more serious responses to the challenge that the argument from geometry begs the question by assuming Pythagoreanism? In fact the two main epistemological traditions of the Enlightenment—empiricism and rationalism—each offer reasons, consistent within their own conception of geometrical knowledge, for endorsing the doctrine of Pythagoreanism. And, of course, if independent grounds can be given for this doctrine, then the argument from geometry can go through without *petitio*: if it is already established that classical geometry does indeed accur-

---

[21] Kant, *Physical Monadology*, in *Works*, 1: 478, 479.

ately describe physical space, then the various geometrical proofs will in fact show that physical space is infinitely f-divisible after all.

Nearly all philosophers of the early modern period endorsed Pythagoreanism: the ungeometrical philosophers were definitely a minority group. Now, the defining Pythagorean view is that the physical world must conform to Euclidean geometry. There are then interesting disputes between different Pythagorean factions over the precise ontological nature of the Euclidean structures in the physical realm—for instance, over whether or not geometrical entities with less than three dimensions (i.e surfaces, lines, and points) exist in the physical realm along with three-dimensional Euclidean structures, and over whether Euclidean structures are instantiated in matter only or also in empty space.[22] But setting these aside, the main division between Pythagoreans that concerns us here is over the grounds offered for faith in the doctrine. This brings us to the two great rival epistemological traditions of the Enlightenment and the respective arguments in each tradition for Pythagoreanism: empiricist on the one hand, and rationalist on the other. Each of these traditions offered a different account of geometrical knowledge and hence different grounds for endorsing Pythagoreanism. Both traditions were agreed that the movement in geometrical demonstrations from basic axioms and definitions to derived theorems was paradigmatically deductive and a priori. But rationalists and empiricists would then differ on the question of the provenance of the axioms and definitions themselves, the starting points of any geometrical proof.

The Empiricist Case for Pythagoreanism

For those of an empiricist bent, the foundational definitions and axioms of classical geometry are *abstracted* from our experience of the world. So the basic propositions of the system are thus also truths about the physical realm, and a demonstrative system of theorems built upon those

---

[22] On the first dispute, see the following useful commentaries: Duhem, *Medieval Cosmology*, 20–32. Dean W. Zimmerman, 'Indivisible Parts and Extended Objects: Some Philosophical Episodes from Topology's Prehistory', *The Monist*, 79 (1996), 148–80, 151. Douglas M. Jesseph, *Squaring the Circle: The War between Hobbes and Wallis* (Chicago: University of Chicago Press, 1999), 76–9. On the second dispute, contrast Barrow and Newton: Barrow, *Usefulness*, 76, 176, 178; Isaac Newton, *De Gravitatione*, in *Unpublished Scientific Papers of Isaac Newton*, tr. and ed. A. R. Hall and M. B. Hall (Cambridge: CUP, 1962), 133.

propositions will establish further truths about physical reality. On this view, a material body is just a certain type of geometrical body: a geometrical body that happens to be also 'filled in' with a certain 'stuffing' of material substance, a certain solidity or impenetrability that blocks in its dimensions and shape. We experience these material bodies and then, ignoring their material stuffing and considering only their size and figure, come to the abstract idea of a geometrical body. Geometrical body is thus the genus, the broader abstracted category; material body, the species. This is most famously the account of Aristotle's *Physics* and Newton's *Principia*.[23] It is also deployed by Barrow, Keill, and Euler (and subsequently by William Drummond in his 1805 *Academical Questions*) as an explicit defence of the argument from geometry, in answer to the complaints of the ungeometrical philosophers.[24] We already saw Barrow's claim that the axioms of this 'Divine Science' are 'firmly established...by their perpetual and exquisite Agreement with Experience'.[25] But Euler is perhaps clearest on this issue:

You know that extension is the proper object of geometry, which considers bodies only insofar as they are extended, abstractedly from impenetrability and inertia; the object of geometry, therefore, is a notion much more general than that of body, as it comprehends, not only bodies, but all things simply extended, without impenetrability, if any such there be. Hence it follows that all the properties deduced in geometry from the notion of extension must likewise take place in bodies, inasmuch as they are extended; for whatever is applicable to a more general notion, to that of a tree, for example, must likewise be applicable to the notion of an oak, an ash, an elm, &c.; and this principle is even the foundation of all the reasonings in virtue of which we always affirm and deny of the species, and of individuals, every thing that we affirm and deny of the genus.[26]

But there is a compelling objection to this empiricist case for Pythagoreanism and the argument from geometry. The problem is that the process of abstraction clearly ignores certain limitations on the kind of thing met with experience. As Gassendi puts it, the geometer, in freely abstracting

---

[23] Aristotle, *Physics*, tr. R. P. Hardie and R. K. Gaye, in *The Complete Works of Aristotle*, ed. Jonathan Barnes, 2 vols. (Princeton: Princeton University Press, 1984), i. 315–446, esp. book 2, ch. 2. Isaac Newton, *Principia Mathematica*, tr. Andrew Motte (Amherst, NY: Prometheus Books, 1995), preface to the 1st edn.

[24] Barrow, *Usefulness*, 156. Keill, *Introduction*, 30–1. Euler, *Letters*, 31–2, 37–8. William Drummond, *Academical Questions* (1805; facsimile edn.: Delmar, NY: Scholars Facsimiles and Reprints, 1984), 69.

[25] Barrow, *Usefulness*, 156.   [26] Euler, *Letters*, 31.

quantity from extended material objects, with no hindrance imposed by their ultimate physical structure, fashions himself a new 'maximally free' *regnum*, a new realm or domain.[27] There is an inescapable idealization or simplification in the process of forming the basic concepts of Euclidean geometry from our observations of the physical realm. Whether one is thinking of this process as an abstraction from the private visual space of our own subjective experience, or from the objective physical space of external reality, in neither case are we infinitely discriminating. Our experience of the physical realm simply does not present us with infinite divisibility as a datum. It is thus an idealization to incorporate infinite divisibility into the axioms of our geometrical system—as Euclid does[28]—if that system is supposed to mirror physical reality. Whether the physical realm actually conforms in all its fine detail to that idealization is thus a further question. Once again we return to the modern account of geometry: even if the basic axioms of the system are suggested to us by experience, as soon as we interpret the system to present literal assertions about the structure of physical space, those axioms become empirical conjectures that may or may not be true of the physical realm in all its detail.

The Rationalist Case for Pythagoreanism

We turn now to the second great tradition in Enlightenment epistemology: rationalism. According to the rationalist, the basic concepts and axioms of classical geometry are innate 'common notions': principles that the mind can generate out of itself through pure ratiocination, independently of experience. Starting with these innate axioms as the basic building blocks, the geometer can then proceed to derive the various theorems of the system—all completely independently of experience. So this time the entire system of classical geometry is knowable a priori. Descartes is the most famous champion of this account in our period, though (of course) it ultimately derives from Plato and his doctrine that geometrical knowledge is 'recollected'—drawn out of the soul—without recourse to observation of the physical world.

---

[27] Joy, *Gassendi the Atomist*, 162.
[28] Infinite divisibility is implicit in the structure of Euclid's express axioms. For more on this, see Jesseph, *Berkeley's Philosophy*, 49.

Given such an account of the epistemological basis of geometry, the challenge to Pythagoreanism is obviously particularly sharp. Why should we think that classical geometry, a system that (we are told) simply draws out the various interrelationships among innately given concepts and propositions, has any bearing at all on the nature of the physical realm? As Barrow puts it, isn't it possible that geometrical objects are simply 'Ideas or Types formed in the Mind, which will be no more than mere Dreams or the Idols of Things no where existing'?[29] Descartes himself admits that this is 'the objection of objections' to his rationalist programme of grounding mathematical physics on innate ideas.[30]

The classic rationalist response is that we know that the innate 'common notions' do correspond to the structure of the physical realm because a non-deceiving creator deity has laid them down in our soul to concur with that external reality. It is as if God, in investing our soul with innate ideas that form the basis of classical geometry, thereby gives us a map or guidebook to the realm of external physical space. And, since (in Descartes's famous phrase) 'God is no deceiver', we know that that guidebook will accurately describe the external realm: physical space will correspond to the innately given system of classical geometry. But few have found this convincing. Even in an updated Darwinian version in which our innate ideas correspond to external reality as a result of natural selection rather than divine fiat, the argument seems insufficient to guarantee that physical space conforms to our (allegedly) innate ideas to the degree of precision required to establish infinite f-divisibility. Mightn't God (or natural selection) have given us a simplified 'guidebook' that is sufficient to allow us to navigate through and live our lives in physical space, without that guidebook being completely accurate in every last detail? In any case, we now know that, if God (or natural selection) did install the basic axioms of Euclidean geometry in our soul as a map of external reality, then he (or it) *is* a deceiver, since Pythagoreanism is now empirically refuted. So even if we grant the assumption (which is itself highly dubious) that the basic axioms of Euclidean geometry are given as innate ideas, the argument for Pythagoreanism will still fail to go through.

---

[29] Barrow, *Usefulness*, 77

[30] Descartes, letter to Clerselier, quoted in *The Selected Works of Pierre Gassendi*, tr. and ed. Craig B. Bush (London: Johnson Reprint Co., 1972), 263.

It is sometimes thought that Kant presents a more compelling rationalist argument for a necessary correspondence between physical space and our innately given geometrical ideas, at least if we grant his system of transcendental idealism. Given transcendental idealism, space is itself a product of our own innate cognitive apparatus, a form that our mind imposes on all experience. And if the nature of space is determined by our own cognitive apparatus, rather than it being a thing-in-itself wholly external to us, it might then seem that we *could* have innate knowledge of its structure after all—since this is really only knowledge of our own form of sensibility. Kant at least thinks that this is so: it explains how we can have what he regards as synthetic a priori knowledge of the properties of space. But of course this does not really help. All our previous problems about the transcendental realist's external physical space arise in analogous form for the transcendental idealist's physical-space-as-a-form-of-sensibility. For instance: just how do we know that the (allegedly) innate ideas of classical geometry correctly describe the space that our cognitive apparatus imposes on all experience as a form of sensibility?

The best efforts of both the empiricist and rationalist traditions to establish Pythagoreanism are, then, unsuccessful. So this doctrine remains questionable. And this being the case, the argument from geometry is indeed a failure. As the ungeometrical philosophers insisted, it begs the question in its assumption that classical Euclidean geometry describes physical space.

V. The Conceptual Argument

The previous three sections dealt with the argument from geometry and its problematic implicit commitment to Pythagoreanism. In this section and the next I look at the second—and simpler—of our two classic arguments for the infinite f-divisibility of space: *the conceptual argument*.

In what follows I will employ both the term 'space' and the term 'extension'. In my idiom, the term 'space' refers to three (or more) dimensional regions: volumes extended in three dimensions. The term 'extension' simply refers to distance in a given dimension. Thus 1-D lines provide examples of extension, but they do not provide examples of space; on the other hand, all spaces are extended. This distinction cuts across the further distinction between *mathematical* space and extension (space and extension defined by the mathematician's postulates) on the one hand, and *physical* space and extension (the real space and extension of the natural world of material objects) on the other.

We can introduce the conceptual argument by way of a possible objection to my discussion of the argument from geometry. The advocate of infinite f-divisibility must concede, I think, that Pythagoreanism *is* a significant assumption and that no *a priori* proof can be given to establish the complete correspondence of Euclidean space and physical space. But they may then insist that, while Euclidean space and physical space may differ in *some* respects (the parallel postulate, Euclid's fifth axiom, being the most famous) there are other respects in which they must correspond. And infinite f-divisibility, precisely the feature of concern to us, is often thought to be just such a property. Granted, certain other features of the structural geometry of physical space may not concur with Euclid's axioms. But isn't physical space necessarily infinitely f-divisible, whatever other structural properties it might have?

What stands behind this conviction that physical space must be infinitely f-divisible, notwithstanding the possibility that it differs from Euclidean space in its other properties? The argument can, I think, be put as follows. Whatever else the structural properties of physical space, any region of space is certainly extended. No recognizable description of physical space can omit this elementary fact, whatever its other details. And no recognizable description of this physical extension can avoid including the notion of parts that are themselves extended: the concept of extension includes the notion of extended subparts. But if these subparts are themselves extended, then (by the same reasoning) they must in turn contain further extended subparts. And so on, ad infinitum. The idea here is that the infinite f-divisibility of space is an analytic truth, demonstrable through conceptual analysis: the very concept of extension itself includes the notion of subparts that are themselves extended, and so anything extended is f-divisible to infinity.

Although it is much more direct than the argument from geometry, the conceptual argument is somewhat less common in the Enlightenment texts. (My suspicion is that some version of the conceptual argument does in fact underpin the faith of most early modern philosophers in the infinite f-divisibility of space, even if these philosophers tend to invoke only the more roundabout argument from geometry in their explicit defences of the doctrine. The argument from geometry, with its endless wealth of constructions and the celebrated certitude of geometry behind it, may have seemed more likely to persuade waverers than the somewhat abrupt conceptual argument. But this is speculation.) The conceptual argument was popular in the scholastic tradition, where the property of being extended was standardly defined as the having of *partes extra partes*: parts outside of parts.[31] Moreover, according to the scholastic analysis, those parts are each extended after the manner as the whole. For instance, Kenelm Digby, following the orthodox scholastic line on the question of physical continua, writes in the 1644 *Two Treatises* that 'Quantity is nothing else, but the extension of a thing, and...this extension, is expressed by a determinate number of lesser extensions of the same nature'.[32] The conceptual argument is also endorsed by Antoine Arnauld and Pierre Nicole in the 1662 Port-Royal *Logique*, where we find the claim that 'two things having zero extension cannot form an extension, and...every extension has parts'.[33] The implication would seem to be that every extension has parts that are themselves extended. Similarly, Antoine Le Grand writes in the 1672 *Institutio Philosophiae* that 'in things extended we can never come to the very least Part, because as long as it is extended it cannot but be conceiv'd Divisible. Whence proceeds that *Axiom*, that *no Quantity or Magnitude can be made of that which is not Quantitative.*'[34] And in Pierre Bayle's 1697 *Dictionnaire*: 'every extension, no matter how small it may be, has a right and a left side, an upper and a lower side.... If there is any extension then, it must be the case that its parts are divisible to infinity.'[35] Sometimes the argument is expressed in terms of the conceptual incoherence of the notion of a least part of extension and hence the impossibility of the rival model of finite f-divisibility. Thus Descartes writes in the Second Replies (1641) that 'we cannot conceive of any extended thing that is so small that we

---

[31] Edmund Whittaker, *From Euclid to Eddington: A Study of Conceptions of the External World* (New York: Dover Publications, 1958), 11 n. The scholastic definition was still current into the 18th cent.: witness the anonymous reply to Hume on infinite divisibility in the 5 July 1740 issue of *Common Sense: or the Englishman's Journal*, collected in *Early Responses to Hume's Metaphysical and Epistemological Writings*, ed. James Fieser (Bristol: Thoemmes Press, 2000), 85–91, 90.

[32] Kenelm Digby, *Two Treatises, in the one of which the Nature of Bodies is expounded; in the other, the Nature of Man's Soule; is looked into* (1644; facsimile edn. Stuttgart: Friedrich Fromman Verlag, 1970), 9.

[33] Arnauld and Nicole, *Logic*, 231.

[34] Le Grand, *Entire Body*, i. 96.

[35] Bayle, *Dictionary*, 360. See also his *Systéme Abrégé*, 300.

cannot divide it, at least in our thought';[36] similar reasoning also occurs in Hobbes's 1655 *De Corpore*.[37]

The conceptual argument can seem quite powerful: the concept of extension surely does include the notion of parts that are themselves extended, and so the concept of a least part of extension—an extended but atomic minimum of space—does look self-contradictory. To press the point: how could there be an extended region of space that lacks a right-hand subregion and a left-hand subregion, each of which is again extended? This is quite obviously impossible in the case of large-scale regions of space. But mustn't the same also apply to small-scale regions of space like the Epicureans' supposed *minima*? Epicureans sometimes attempt to take refuge in the exceptional smallness of their proposed *minima*: their miniscule scale explains why they are f-indivisible, or (more plausibly) at least explains why we have trouble conceiving of them and their atomic nature. But an advocate of the conceptual argument will insist that this is an evasion. There seems no reason to think that small-scale regions of space are any different from large-scale regions in requiring subparts that are extended. Although we might not be able to visualize a miniscule region of space, it still seems conceptually guaranteed that, so long as it is extended, it must have extended subparts. The conceptual argument can thus seem quite compelling, and is certainly endorsed by various present-day philosophers.[38]

VI. Against the Conceptual Argument

Although many philosophers have thought the conceptual argument an improvement on the argument from geometry, I do not think that it is

---

[36] René Descartes, Second Replies, in *The Philosophical Writings of Descartes*, tr. and ed. John Cottingham, Anthony Kenny, Dugald Murdoch, and Robert Stoothoff, 3 vols. (Cambridge: CUP, 1991), ii. 115. See also *Œuvres de Descartes*, ed. Charles Adam and Paul Tannery, 12 vols. (Paris: Vrin 1964–76), iii. 213–14.

[37] Thomas Hobbes, *De Corpore*, ch. 7.13, in *Collected Works*, i. 100–1.

[38] J. E. McGuire and Martin Tamny, *Certain Philosophical Questions: Newton's Trinity Notebook* (Cambridge: CUP, 1983), 74–5. Joshua Hoffman and Gary S. Rosenkrantz, *Substance Among Other Categories* (Cambridge: CUP, 1994), 101, 115. Zimmerman, 'Indivisible Parts', 152. (Zimmerman also adds a related argument: not only does the *extension* or *size* of any region of space guarantee its infinite f-divisibility, so does its *shape* or *figure*. Any region of space must have some particular shape or other. But if it has a particular shape, then surely it must have extended and figured subparts. For Zimmerman, then, '[a] partless square region is no more possible than a round square region' (152): each is equally conceptually incoherent.)

ultimately a success. In fact its fundamental flaw, though less immediately obvious, is similar in nature to the *petitio* problem we saw pressed against the argument from geometry.

First, I think it must be conceded that our concept of extension does indeed include the notion of extended subparts. Given this concept, the notion of an atomic, least extension is thus a conceptual incoherence. This much is analytic. But does this a priori, analytic argument show us anything about the nature of actual physical space? Certainly, *if* physical space answers to the concept of extension we have been discussing, then it must be f-divisible ad infinitum. But need it answer to that concept?

We must be careful here. There are (at least) two distinct concepts at issue when we talk loosely about 'the' concept of extension. The confusion of these two can lend the conceptual argument an initial—but specious—plausibility. First, there is the mathematician's or geometer's version of the concept: as should now be clear, this is simply the creature of our own stipulation, devoid of factual import about the physical realm. Second, there is the concept of extension as it is in the physical realm: this concept is at least in aim supposed to accurately represent the structure of external reality. (Most likely this second concept is simply the first, mathematical concept given a physical interpretation.) Now, if we want to show anything about physical space, then we must of course operate with this second concept of physical extension. But—herein lies the rub—the second is an *empirical* concept: it is supposed to represent something in the empirical world, in this case the nature of extension in physical space. And *all empirical concepts are revisable*. They are supposed to represent the nature of the various types of object of experience. But they may do this more or less well; and where we discover that a certain empirical concept in fact gets something wrong about whatever it is supposed to represent, then we will revise it. In sum, whether a particular empirical concept correctly maps its object is always an empirical matter. This is most obviously the case with the more mundane, low-level concepts. We had to revise our concept of a whale, for instance, when we discovered that whales are not fish but mammals. Or, as a higher level example, consider the empirical concept of space, which has undergone several dramatic revisions over the centuries. One could certainly draw out various implications from the medieval concept of space by a priori, analytic argument. But whether that concept, with its finitely bounded universe and denial of the vacuum, in fact answers to real physical space is a further, empirical question. Likewise one could also show by analysis that the Newtonian concept of space has such-and-such an implication. But, once again, the concept itself may have to be revised in the light of empirical discoveries such as Einstein's. My suggestion is that precisely the same is true of our current concept of physical extension: analysis can certainly unpack the content of that concept (perhaps, indeed, showing that it entails infinite f-divisibility)—but whether or not that concept accurately maps real physical extension is a further, empirical question.

Objection (1): But isn't it just obvious that for any extension E, no matter how small, there must always be a smaller extension $E/2$?

I certainly grant that for any number there is always a smaller number, and that for any mathematical extension (that is, extension in the first version of the concept) there is always a smaller extension. I even grant that, with the standard way of conceptualizing physical extension, there should always be a smaller physical extension. But this immediately implicates an empirical concept, and whether that empirical concept accurately represents the real structure of physical extension is always an empirical matter. (Imagine someone who argued: 'Isn't it just obvious that for any possible velocity V, no matter how great, there could always be a greater velocity $2V$? Hence there is no greatest possible velocity.' Such an armchair physicist would quickly become embarrassed if they attempted to apply this conclusion to the empirical world, where the speed of light establishes a ceiling to possible velocities in the material realm.)

Objection (2): But even though our concept of physical extension is indeed empirical, surely it is so simple and so basic to our understanding of the physical world that we could not possibly be wrong in thinking that it does accurately describe real physical extension?

I think that this is overoptimistic. Empirical investigation has already thrown up surprise after surprise about the nature of space in its fine structure: think of quantum foam, the rips and tears in the space–time fabric, and the numerous curled-up dimensions that physicists now entertain in the miniscule world below the Planck length. In a recent popularization, the physicist Brian Greene reports that in this ultramicroscopic realm space takes on a 'frothing, turbulent, twisted form', giving rise to 'an unfamiliar region of the universe in which the conventional notions of left and right, up and down (and even before and after) lose their meaning'.[39]

---

[39] Brian Greene, *The Elegant Universe* (New York: Random House, 2000), 127–9.

All this is quite alien to the everyday concept of physical extension that finds confirmation in the immediate realm of mid-sized objects, and casts doubt on the conviction that that concept is beyond revision, perfectly describing the nature of real physical extension all the way down. Once we realize that our current working concept of physical extension is drawn from interaction with mid-sized phenomena (or, if this empiricist account is rejected for a nativist one, the realization that our current concept of physical extension is pre-programmed into us to allow us to navigate mid-sized phenomena), we have good grounds to ask whether that concept applies in infallible and infinite detail to the entire physical realm.[40]

Objection (3): But there is no other *possible* concept of physical extension than our current one (which includes the notion of extended subparts and hence infinite f-divisibility). The notion of an extension without extended subparts is simply incoherent—and so the impossibility of physical extension being like this is established as an a priori, non-empirical truth.

I think that this objection makes clear the real underlying force and appeal of the conceptual argument. An advocate of this argument will insist that it is not just that our current concept of physical extension *happens* to include the idea of extended subparts. Rather, this is the only possible, self-consistent concept of extension—for how could one ever build extended regions without extended subparts?

But this claim that every extension necessarily has extended subparts is actually quite controversial. In fact the current literature boasts a variety of mathematical models of discrete space, constructing space from f-indivisible *minima*. These models build extended space from finite arrays of atomic first parts that altogether lack extended subparts—systems, for instance, where 'the number of points in any sensibly shaped (say convex) region of space–time of finite volume is finite'. For example, Peter Forrest deploys finite sets of point elements along with the dyadic, symmetric, irreflexive relation of adjacency as the basic ingredients of a system that gives rise to recognizable and well-defined notions of distance and

---

[40] Compare Ernst Mach: 'the moment we begin to operate with mere things of thought like atoms and molecules, which from their nature *can never be made the objects of sensuous contemplation*, we are under no obligation whatever to think of them as standing in spatial relationships which are peculiar to the Euclidean three-dimensional space of our sensuous experience'. Mach, *Space and Geometry*, 138. See also Edmund Whittaker, *From Euclid to Eddington: A Study of Conceptions of the External World* (New York: Dover Publications, 1958), 8.

volume.⁴¹ Certainly such discrete models lack the mathematical simplicity of continuous space. But nonetheless they do seem to show that the discrete approach remains a conceptual and mathematical possibility.

One could tie this back to the early modern period by thinking of Berkeley's system as another type of model that builds extension from finite numbers of f-indivisible first parts. Berkeley constructs extended entities from finite arrays of *minima sensibilia* that—in an important sense—really do lack extended subparts. Since the mind can perceive no smaller subparts within the *minima*, and 'esse est percipi', there *are* no extended subparts. This allows Berkeley to pull off the trick of allowing minimal parts of extension which are altogether f-indivisible: at the ontological level there just are no spatially distinct parts answering to (say) the left or right side of a given *minimum*. (Berkeley will of course allow that the mind can represent the length of any extension by an arbitrary number, and by dividing that number we can consider smaller parts of the original extension in a procedure that could (in principle) be iterated endlessly. The *minima* are in this sense i-divisible. But he will immediately insist that it is a mistake to think of this endlessly iterable mental operation as implying that there *actually exist* spatially distinct parts below the level of *minima sensibilia*.⁴²)

It is of course difficult—probably impossible—to *visualize* the atomic architecture of Berkelian space or the fine grain of discrete space in the more recent mathematical models. But, of course (and as Berkeley insists), when one tries to visualize this fine structure one inevitably ends up imagining a space characterized by the features of larger scale spatial regions. Here again we need to appreciate that our everyday picture of space is derived from the level of mid-sized objects and may well cease to apply in microscopic regions (as, in fact, quantum physics tells us it does). Finally, for all the strangeness of the discrete model of space, it is worth noting with Jean Paul Van Bendegem that 'the continuum view is about as good as the discrete view in terms of strangeness'.⁴³ We certainly cannot *visualize* the fine structure of extension as fully dense (never mind continuous) any more readily than we can visualize it as quantized or discrete.

---

⁴¹ Peter Forrest, 'Is Space–Time Discrete or Continuous? An Empirical Question', *Synthese*, 103 (1995), 327–54, 327. See also Jean-Paul Van Bendegem, 'In Defence of Discrete Space and Time', *Logique et Analyse*, 150–1 (1995), 127–50.

⁴² Berkeley, *Principles of Human Understanding*, ss. 123–7, in *Works*, ii. 97–100.

⁴³ Van Bendegem, 'In Defence of Discrete Space and Time', 127–45.

Objection (4): But the bare conceptual possibility of discrete physical space seems a rather meagre point to set against the present argument. First, no positive reason has been given to think that our current concept of physical extension does in fact fail to match the real nature of space. And, second, *all* our reasoning about the empirical realm must employ empirical concepts, so—if one were pedantic enough—one could always question any such reasoning by asking whether our current concepts accurately match reality. This looks like a singularly unproductive type of nitpicking and will usher in a really quite boring form of scepticism.

But I should stress that I am not arguing that it is *likely* that space is only finitely f-divisible. My point is simply that no purely analytic argument can deliver us concrete facts about the physical structure of space, as the conceptual argument purports to. Suppose someone were to argue that whales are warm-blooded by analysing the current concept of a whale (which includes the notion that whales are mammals and hence warm-blooded). It would certainly be boring to challenge this inference by asking whether we are certain that our concept of a whale matches the nature of whales in reality. But all the same, it would be a necessary reprimand to anyone who thought that we could establish that whales are warm-blooded simply through conceptual analysis, and without recourse to experience!

I conclude that the conceptual argument fails to deliver an a priori demonstration that space is infinitely f-divisible. Just as the argument from geometry begged the question by presupposing that Euclidean geometry accurately maps physical space, so too the conceptual argument illegitimately presupposes that the mathematical concept of extension (which includes infinite f-divisibility by stipulation) applies to physical space. Once we appreciate that the relevant concept this argument must invoke is an *empirical* concept of physical extension—extension as it is in the real world—we can then see it is a further *a posteriori* question whether or not that empirical concept matches the actual physical structure of space. I therefore conclude that neither the argument from geometry nor the conceptual argument establishes that space is infinitely f-divisible. For all that these two classic arguments have shown, the doctrine of the infinite f-divisibility of space could still be empirically refuted.

APPENDIX
TWO MINOR ARGUMENTS FOR THE INFINITE F-DIVISIBILITY OF SPACE

> They who knew not Glasses had not so fair a pretence for the Divisibility ad infinitum
>
> George Berkeley, *Philosophical Commentaries*

The central task of Chapter 5 was to examine the two arguments for the infinite f-divisibility of space that I have been categorizing as the 'classic' or 'traditional' arguments. These two arguments are by far the most popular in the Enlightenment literature and are also the most philosophically rewarding to study. However, for the sake of completeness in this appendix I briskly outline two other types of argument for infinite f-divisibility that are also present in the period literature. As we shall see, they can each be dealt with fairly quickly.

1. Smooth Motion and Elastic Impacts Require Infinitely F-Divisible Space

This first argument points out that, if space were only finitely f-divisible, then motion would have to be a series of discrete leaps as each f-indivisible part of the moving body jumps successively from one space atom to the next. This contradicts the idea that the motion of bodies in space is a continuous, smooth matter. One might also push this point and claim that 'motion' through staccato atomic leaps would really be no motion at all. Rather than actually moving, the body would simply be repeatedly disappearing and reappearing in an adjacent location. Moreover, if we then add the assumption that time is only finitely divisible (and this assumption is often made by those who argue for *minima* of space—since time seems relevantly analogous to space in terms of the arguments for its ultimate structure), it might then seem that all motion would have to proceed at an equal velocity: a constant advance of one space atom for each time atom. The conclusion we are supposed to draw is that these results are absurd, and thus that space must be infinitely f-divisible after all. Versions of this argument are reported by Charleton, and endorsed by Barrow and Keill.[44]

---

[44] Charleton, '*Empirucus*, with great Virulency of language inveighing against the Patrons of Atoms, accuseth them of subverting all Local Motion, by supposing that not only Space and

But I cannot see that this argument presents us with any genuinely new reason to accept infinite f-divisibility. It seems that the opponent of infinite f-divisibility will simply accept the view that motion, like space, is discrete in its fine structure, notwithstanding its apparent continuity at the level of mid-sized objects. The motion of objects will thus be conceptualized like the animated 'motion' of cartoon characters on a movie screen: a successive series of very small atomic leaps that looks like smooth, continuous motion at the larger scale. (The animation analogy also shows why velocity need not be constant for all moving objects: some objects will proceed at one space atom per time atom, some at several space atoms per time atom, and some at one space atom every few time atoms.) Now, I say that this argument presents no *new* reason to endorse infinite f-divisibility because the advocate at atomic *minima* of space is already committed to the view that apparently continuous quantities turn out to have a quantized base. After all, they already maintain that space is quantized, notwithstanding the fact that it appears to be continuous at the level of mid-sized objects. Similarly, they maintain that space is non-Euclidean in its fine structure, though of course it approximates to the Euclidean description at the level of everyday experience. And given that they are prepared to bite this sort of bullet in the case of space, it does not seem that the case of motion offers any additional, new problem. (It is perhaps also worth noting that most of us are ready to concur that at least some natural quantities that initially appear continuous—such as gravitational force and electric charge—turn out to be quantized on investigation. These results underwrite the particle side of wave–particle duality: individual gravitons and electrons represent the smallest quanta of their respective forces.)

Another argument for infinite f-divisibility can also be brought under the current rubric. Certain natural philosophers, including Leibniz and Huygens, maintain that impacts between objects require an elasticity that is only possible with infinitely f-divisible space and matter.[45] Without this infinite divisibility, impacts would involve *instantaneous* changes of momentum, since at the atomic level the f-indivisible parts of matter would be completely rigid, lacking any deformable internal structure that might provide elasticity and shock absorption.

Time, but also Natural Quantity indivisible beyond Insectile Parts' (*Physiologia*, 94). Barrow: 'How lastly is not all the Difference also of Motions, as to Velocity, taken away by this means. *Ex. gr.* If a moveable Point runs through five Points in one Time, how can another perform the subduple, subtriple, or subquadruple of that Space in the same Time, when the whole is not capable of being divided into these Parts?' (*Usefulness*, 156). Keill: 'If Quantity consisted of Indivisibles, it would follow, that all Motion would be equally swift, nor would a slow Snail pass over a less Space in the same Time than the swift-footed *Achilles*' (*Introduction*, 31).

[45] A. J. Snow, *Matter and Gravity in Newton's Physical Philosophy* (Oxford: OUP, 1926), 44–5, 62–3. D. A. Anapolitanos, 'The Continuous and the Discrete: Leibniz versus Berkeley', *Philosophical Inquiry*, 13 (1991), 22.

But this sort of inelastic shock impact and instantaneous change in momentum would require infinite force (given force = mass × acceleration)—and this is absurd. However, we can give the same reply to this argument. The advocate of atomic *minima* of space will simply say that the various laws of dynamics, such as F = MA, are just like the laws of Euclidean geometry: they capture approximations that are roughly true at the level of mid-sized objects, but that need not hold at the quantized base. After all, anyone who conceptualizes motion as a series of discrete leaps between adjacent space atoms has already parted from such notions as continuous velocity and acceleration. All acceleration, at the atomic level, involves these abrupt shifts in velocity, and so would require infinite force if F = MA held at this scale. But F = MA (on this view) is simply an approximation of how large numbers of atoms appear to behave when viewed as mid-sized objects at the everyday level of apparent continuity.

2. *Arguments from the Observed Subtlety of Matter*

Another argument for the infinite f-divisibility of space (and matter) that enjoyed some currency in the early modern period involved the appeal to experimental results drawn from both investigations with the microscope and tests on the ductility of various metals. Each of these experimental inquiries unveiled new miniscule realms beneath the ordinary level of observation, suggesting a complexity and fine structure in nature that had previously been undreamt of. The impact of the developments in the precision of microscope construction and in lens-grinding technologies on Enlightenment thinking is now widely appreciated.[46] Pioneering work by the likes of Swammerdam, Leeuwenhoek, and Hooke excited an intellectual craze of interest in the enthralling new worlds within worlds that the new lenses were revealing. (Parodies also abounded: think of Cyrano de Bergerac's *Voyages dans la Lune*, Swift's *Gulliver's Travels*, and Voltaire's *Micromegas*.) For many thinkers, this new ornate and exquisite microscopical realm was a monument to God's grandeur that rivalled the enormity of the heavens above, and as microscopes showed ever greater complexity the more powerful they became, it was certainly natural to suggest that God's creation might be inexhaustibly—*infinitely*—intricate in its fine structure just as it was infinite in size. The view that these observations through the microscope provided a case for the doctrine of infinite divisibility was of course encouraged by the theory of preformation:

---

[46] Catherine Wilson, *The Invisible World: Early Modern Philosophy and the Invention of the Microscope* (Princeton: Princeton University Press, 1995).

spermatazoa were thought to contain preformed miniature homonculi, which themselves would each have to contain their own preformed spermatazoa and homunculi, and so on ad infinitum. Thus we find philosophers such as Nicolas Malebranche arguing that experimental data from microscopic observations corroborates the thesis of infinite divisibility.[47] Similar arguments also appealed to the mind-boggling division of matter that had actually been accomplished in the case of certain metals. Thus we find Jacques Rohault, Pierre Bayle, and John Keill each appealing to the extreme ductility of gold as a supporting argument in their overall case for infinite f-divisibility.[48]

Now of course none of these arguments can actually prove the doctrine of infinite f-divisibility. The rival Epicurean model of granular, quantized space could equally well allow for all these experimental results, assuming that the space atoms were small enough. In fact many of those who invoke these sorts of observations while arguing for infinite f-divisibility saw this clearly: really they only intended the appeal to experimental results to break down initial resistance to the possibility that there might well be tiny uncharted realms beneath the scale of everyday experience. These experimental observations may then have some role in refuting the lazy view that there are no microworlds beneath the level of those things we view with the unaided eye. So they are best seen as a response to this sort of lazy objection to the thesis of infinite f-divisibility.[49] But they hardly amount to a positive argument for that doctrine.

---

[47] Malebranche, *Search After Truth*, book I, ch. 6.

[48] Rohault, *System*, 34–6; Bayle, *Système*, 301; Keill, *Introduction*, 46.

[49] This 'lazy objection' may seem to be the complaint of a straw man. But in fact some critics of infinite divisibility did invoke the mind-boggling nature of the extremely tiny as a reason to resist infinite divisibility. (See my discussion of Walter Charleton in the appendix to Ch. 1.) The point that experimental results show that there *are* mind-bogglingly small things would be a reply against this (admittedly lame) sort of objection.

6

The Kant–Boscovich Force-Shell Atom Theory

> Whereas Copernicus had to persuade us to believe, contrary to all our senses, that the earth did not stand still, Boscovich taught us to disavow the final 'fixed' thing in the regard to the earth—the belief in 'substance,' in 'matter,' in the little residual earthly clump—the atom. This was the greatest triumph over the senses ever achieved on earth.
>
> Friedrich Nietzsche, *Beyond Good and Evil*

I. Introduction

In this final chapter I examine one of the more ingenious and philosophically intriguing responses to the problems of material structure: the embryonic field theory first suggested by Henry More and subsequently developed as a fully fledged theory of matter by Boscovich and Kant in his younger, pre-critical period. (For brevity's sake, I shall call this account the Kant–Boscovich theory of matter. But the reader should bear in mind that the later, critical period Kant will renounce this theory for a different model of material structure altogether.)

In focusing on this response in particular, I do not mean to imply that it is the only viable approach to the problems of material structure, or even that it is necessarily the most plausible. For all that has been said in this book, certain other responses would still appear equally feasible. Think, for instance, of the potential parts resolution of Aristotle, Hobbes, and the later Kant (see Chapter 1, section IX, 'faction 1'); or of Galileo's system of actual infinities of ultimate parts (see Chapter 1, section IX, 'faction 3'). For a systematic survey of all the logically and conceptually respectable accounts of material structure, see the Conclusion.

I single out the Kant–Boscovich approach for further examination only because this proposal has a particular interest from both a philosophical and a historical point of view. First, it is a response that operates within the framework of the basic conflict explored in this book, the clash between two central tenets of the new world-view of Enlightenment science: the actual parts doctrine on the one hand, and the geometrization of nature on the other. The Kant–Boscovich approach attempts to broker a peace between these two warring doctrines, thereby disarming the antinomy at the heart of the new science. This account thus seeks to preserve both of these core doctrines of the new science, rather than simply jettisoning one of them and thereby stepping outside of the problem space of the main historical debate. Second, the proposed resolution is also of particular historical interest as it introduces the dynamical model of matter and stands as a direct ancestor of the field systems of Faraday and Maxwell. In exploring this account, we will thus have the opportunity to examine various conceptual issues raised by dynamism and the field metaphysic: questions about the ontological status of fields of force; about their structure and occupancy of space; about action at a distance and the relationship between a substance and the forces or powers it projects.

The Kant–Boscovich response to the problems of material structure runs as follows. Space is admitted to be infinitely f-divisible, and, since each body occupies a region of space, bodies are likewise f-divisible ad infinitum. The actual parts doctrine is also accepted: each body is a composite structure, built up from distinct parts as far as it is m-divisible. *If* m-divisibility tracked f-divisibility, each body would then be built out of an infinity of parts, and the classic paradoxes of material structure would beckon (see Chapter 1, sections VII and VIII, for these paradoxes). But the key move that characterizes the Kant–Boscovich resolution is the claim that m-divisibility and f-divisibility can come apart: although bodies are f-divisible ad infinitum, they are only finitely m-divisible. Each body is a construct from a merely *finite* array of extended but m-indivisible first parts: metaphysical atoms that, though extended, cannot be broken apart, even by God.

The immediate challenge facing this approach is obvious: how can these first parts of bodies be extended and f-divisible, and yet at the same time m-indivisible? If a so-called atom is extended and has spatial subparts, then why couldn't God rupture those subparts? Where is the logical impossibility in supposing its spatially differentiable subparts separated and distanced one from another? The Kant–Boscovich solution is radical. The m-indivisible first parts of bodies are no longer the precisely figured and absolutely impenetrable microbodies of traditional corpuscularian theory. Rather, these perfectly solid, sharply defined and—it would seem—m-divisible 'atoms', familiar from the main tradition of Gassendi, Boyle, Locke, and Newton, are each replaced with a diffused shell of force projected by an unextended central *punctum*. Since the core *punctum* is altogether without extension, it cannot be m-divided. And since the surrounding extended atom is simply a field of force thrown out by the core, its spatially differentiable parts cannot be broken apart from the whole. It is f-divisible but m-indivisible: an extended metaphysical atom. The theory raises a host of questions I will need to address. But now we are ready to turn to the texts and to examine the historical development of the doctrine.

I proceed as follows. In section II, I look at Henry More's seventeenth-century version of this model of extended but m-indivisible entities. More uses the model as a picture of spiritual substances rather than material atoms. Nonetheless, his 'indiscerpible' spirits present us with a formal archetype—or at least a precursor—of the Kant–Boscovich force-shell atom. And, as we shall see, More's arguments in defence of his extended but m-indivisible spirits also presage Kant's and Boscovich's arguments in defence of their extended but m-indivisible force-shells.

In the next two sections I introduce and examine the Kant–Boscovich force-shell atom theory. Section III is essentially expository: it sets out Kant's and Boscovich's respective versions of the theory and points out the (relatively minor) ways in which they diverge. Section IV is more critical and analytical in timbre. Here I make some further observations concerning the force-shell theory, noting in particular the ways in which it breaks with the corpuscularian metaphysic and introduces the dynamical or field-theoretic conception of matter. In this section I also respond to some initial objections to the force-shell theory that come from the traditional corpuscularian perspective.

Finally, in section V, I look at what I take to be the most serious objection to the Kant–Boscovich force-shell theory, and indeed to any dynamical account of matter. This is the *hollow world problem*: the worry that any account that identifies matter with fields of force ends up dissolving the material world into a realm of pure potentiality and no actuality.

II. Henry More's 'Indiscerpible' Spirits

Henry More advances this model of f-divisible but m-indivisible entities only as an account the nature of immaterial *spirits*. It is decidedly not intended as an account of the ultimate structure of *matter*, and indeed there are other elements of More's metaphysical system that would rule out any such proposal. Nor can I find any positive textual evidence that More's writings had any direct or immediate influence on either Boscovich or the pre-critical Kant, the two theorists who do ultimately advance the model as an account of the internal architecture of matter. However, there is the strong likelihood of an *indirect* influence here. Boscovich and Kant would certainly have been aware of the fact that Samuel Clarke and other leading eighteenth-century Newtonians allowed extended and f-divisible but m-indivisible spiritual substances—and in this matter Clarke *et al.* were consciously following More's lead.[1] In any case, whether or not a positive influence running from More to Kant and Boscovich exists, More does deserve the credit for pioneering this model of extended but m-indivisible entities a full century before these latter philosophers, and for thereby introducing the rudiments of the field metaphysic.[2]

Originally an enthusiastic disciple of Descartes, the Cambridge divine Henry More (1614–87) became increasingly disenchanted with what he saw as the materialistic and atheistic tendencies of the Cartesian system of nature. Although Descartes himself remained politely unimpressed, the young More pressed him with fervent letters outlining the danger of the Cartesian system collapsing into the bleak materialism of Hobbes. More's decisive break with Descartes was triggered by the latter's famous identification of extension and matter, and the implications of this doctrine for the theory of immaterial substance. In the Cartesian system, everything extended is *ipso facto* material. Immaterial spiritual substances must thus be

---

[1] Samuel Clarke, *A Defence of the Immateriality and natural Immortality of the Soul*, in *The Works of Samuel Clarke*, 4 vols. (London: John and Paul Knapton, 1738; facsimile edn.: New York: Garland, 1978), iii. 762. For commentary, see Ezio Vailati, 'Clarke's Extended Soul', *Journal of the History of Philosophy*, 31 (1993), 387–404. See also George Cheyne, *Philosophical Principles of Religion, Natural and Revealed*, 2 vols. (London, 1725), ii. 5.

[2] I find no mention of More as a precursor of Boscovich's and Kant's force-shell atom theory in the existing secondary literature. Nor has More received adequate recognition for introducing basic concepts that set the stage for the field metaphysic—though an important exception here is Alexander Koyré, *From the Closed World to the Infinite Universe* (Baltimore: Johns Hopkins University Press, 1968), 132.

altogether extensionless. But (More insists) this means that spiritual substances cannot be located in the spatial universe at all and hence can exist nowhere. In denying extension to immaterial spirits, Descartes shuts them out of space, and thus out of existence altogether.[3] (Descartes will of course resist this last inference.) More thus scornfully christens the Cartesian doctrine of the unextended soul 'Nullibilism': a 'Nowhere-ism' that—quite literally—can find no room for spiritual substance, and thus collapses into the 'gross and dirty' materialism of Hobbes.[4]

For More, then, spirits no less than matter must be extended. The difference between the two classes of substance is that matter is extended, impenetrable, and 'discerpible' (i.e. m-divisible), whereas spirit is extended, penetrable, and 'indiscerpible' (i.e. m-indivisible). Thus we have More's doctrine of extended, figured spirits that can flit through matter and through one another, and that are each absolutely m-indivisible: 'utterly *Indivisible* and *Indiscerpible* into real Physical parts'.[5] More's critics will be quick to note the similarities between his spiritual substances and the ghosts of fancy and folklore; perhaps they will remind us of More's own belief in the full phantasmagoria of the spirit-realm, or that his pupil and associate Joseph Glanvill would deploy the theory in his infamous demonology of witches and apparitions, the *Saducismus Triumphatus*. More's apologists, on the other hand, will insist with equal celerity on the similarities between More's immaterial substances and the penetrable fields of modern science; they will remind us that More's notion of space as an infinite, indiscerpible spiritual substance which penetrates and contains all finite substances was adopted wholesale by Newton in the *Principia*; they may also remind us that More's doctrine of extended, penetrable, and indiscerpible *finite* spirits was also taken up as orthodox by the main tradition of eighteenth-century Newtonianism.

The feature of More's extended spiritual substances of interest to us is, of course, their m-indivisibility. How can More reconcile this feature of spirits with their extension? If a finite spiritual substance is extended and

---

[3] '[I]t being of the very essence of whatever *is*, to have Parts or Extension in some measure or another. For to take away all extension is to reduce a thing onely to a Mathematical point, which is nothing else but pure Negation or Non-entity... it is plain that if a thing *be* at all, it must be *extended*', Henry More, *Immortality of the Soul*, ed. A. Jacob (Dordrecht: Kluwer, 1987), 7. See also *The Philosophical Writings of Henry More*, ed. F. I. Mackinnon (New York and Oxford: OUP, 1925), 228.

[4] More, *Philosophical Writings*, 228.

[5] Ibid. 208

thus f-divisible, surely God (at least) could break its spatially differentiated parts apart from one another? The objection crops up again and again in opposition to the system of extended but m-indivisible spirits—for instance in the pamphlets of Anthony Collins against Samuel Clarke, in Isaac Watts's essay *Of the Place and Motions of Spirits*, and in the pamphlet debate between Joseph Priestley and Richard Price. Versions of it are endorsed by Isaac Barrow and Pierre Bayle, and indeed also in the recent secondary literature. The argument is also reported—thought not endorsed—by Kant.[6] More himself phrases the objection as follows:

[I]t is objected...that *Extension* cannot be imagined without *diversity of parts*, nor *diversity of parts* without a *possibility* of *division*, or separation of them...from whence it will follow, that *Indivisibility* is incompateble to a *Spirit*, which notwithstanding we have added in the Definition thereof.[7]

More confesses the objection 'very ingenious, and set on home'; 'it is the chief Edge or Sting of the whole Difficulty, and yet such as I hope I shall with ease File off or Blunt'.[8]

The 'filing off' begins with the reply that the inference from an entity's extension to its m-divisibility is supported by 'nothing but corrupt *Imagination*' and the 'impure Dregs of Fancy'.[9] It is only the pictorial faculty of imagination that encourages us to think that whatever is f-divisible is thereby also m-divisible:

Altho' every Being as a Being is *Extended*, because Extension in its precise Notion does not include any *Physical Division*, but the Mind infected with *corporeal Imagination*, does falsely and unskilfully feign it to be necessarily there. ...
But that one should adjoyn a *Physical* divisibility to such an Extension, surely that must necessarily proceed from the impotency of the *Imagination* which his Mind cannot curb nor separate herself from the dregs and corporeal Foulness thereof;

---

[6] Anthony Collins, *Reflections on Mr. Clarke's Second Defence of his Letter to Mr. Dodwell*, in Clarke, *Works*, iii. 813. Isaac Watts, *Of the Place and Motion of Spirits*, in *The Works of Isaac Watts*, 6 vols. (London: John Barfield, 1810), v. 562. Joseph Priestley, response to Richard Price, in Richard Price, *A Free Discussion of the Doctrines of Materialism and Philosophical Necessity* (New York: Garland, 1978), 62. For Barrow, Bayle, and Kant, and for versions of the argument in the current literature, see the references given in Ch. 2, s. IV, under argument (A2).

[7] Henry More, Appendix to *An Antidote Against Atheism*, in *A Collection of Several Philosophical Writings*, (London, 1712), 186. See also *Philosophical Writings*, 210, 214.

[8] More, Appendix to *An Antidote Against Atheism*, 186, and *Philosophical Writings*, 218.

[9] More, *Philosophical Writings*, 214, 218.

and hence it is that she tinctures and infects this Pure and Spiritual Extension with Corporeal Properties.[10]

But this can be only a beginning. Perhaps it is the pictorial faculty of the imagination that leads us to think that the parts of an extended spirit could be split one from another. But doesn't our ability to imagine an extended spirit breaking apart suggest that such an m-division *is* logically possible? Why think that the imagination misleads us here? Here we come to More's real answer to the objection and his ingenious model of an extended but m-indivisible entity. It will be worth quoting in some detail. First, the 1653 *Antidote Against Atheism*:

> Suppose *a Point of Light* from which rays out *a luminous Orb*, according to the known principles of *Opticks*: This *Orb of light* does very much resemble the *nature of a Spirit*, which is *diffus'd* and *extended*, and yet indivisible. For we'll suppose in this *Spirit* the Center of life to be indivisible, and yet to diffuse itself by a kind of *circumsrib'd Omnipresency*, as the Point of Light is discernible in every point of the Luminous Sphere. And yet supposing the Central lucid Point *indivisible*, there is nothing divisible in all that Sphere of light. For it is ridiculous to think by any Engine or Art whatsoever to separate the luminous rays from the shining Center, and keep them apart by themselves.[11]

And again in the 1659 *Immortality of the Soul*, the same analogy:

> But besides that *Reason* may thus easily apprehend that [extended spirits may be indiscerpible], I shall a little gratifie *Imagination*, and it may be *Reason* too, in offering the *manner* how it is so, in this kind of *Spirit* we now speak of.... Now it is observable in *Light*, that it is most vigorous toward its fountain, and fainter by degrees. But we will reduce the matter to *one lucid point*, which, according to the acknowledged Principles of Opticks, will fill a distance of space with its *rays of light*: Which *rayes* may indeed be reverberated back towards their Centre by imposing some opake body, and so this *Orbe of light* contracted; but, according to the *Aristotelian* hypothesis, it was always accounted impossible that they should be clipt off, or cut from this *lucid point*, and be kept apart by themselves. Those whom dry Reason will not satisfy, may, if they please, entertain their Phansy with such Representations as this, which may a little ease the anxious importunity of their Mind, when it too eagerly would comprehend the manner of how this *Spirit* we speak of may be said to be *Indiscerpible*. For think of any *ray* of this *Orbe of light*, it does sufficiently set out to the *Imagination* how *Extension* and *Indiscerpibility* may consist together.[12]

[10] More, *Philiosophical Writings* 217.
[11] More, Appendix to *An Antidote Against Atheism*, 186.
[12] More, *Immortality of the Soul*, 34–5.

Since the extensionless *punctum* at the heart of the spirit occupies no space, it is neither f-divisible nor m-divisible. But this central core then throws out a shell through the surrounding space 'which we may in some sense call Substance, though but *Secondary* or *Emanatory*'.[13] This extended and f-divisible shell—an 'Orbe' that 'swells out from the Centre of [the] Spirit'—defines the spirit's '*circumscrib'd Omnipresency*':[14] the presence of its receptivity and activity, the defined region where it can both experience and act upon things in space. However, since the shell is simply a sphere of activity thrown out by the m-indivisible core, it too is m-indivisible. The shell is simply the region where the *punctum* exerts an influence—it does not represent the actual presence of any entity, only the 'virtual presence' of its effects. As More puts it, because the sphere's extension arises 'by graduall Emanation from the First and Primest Essence, which we call the *Centre of the Spirit*... we are led from hence to a necessary acknowledgement of perfect *Indiscerpibility* [m-indivisibility] of parts, though not intellectual Indivisibility [f- or i-indivisibility]... For it implies a contradiction that an *Emanative* effect should be disjoyned from the originall.'[15] Just as one cannot cut off and remove the parts of the shell of heat or light projected by a fire, or separate the subparts of a magnetic field thrown out by a lodestone, so likewise one cannot disjoin the parts of a spirit, its extension and f-divisibility notwithstanding. All in all, 'the parts of a *Spirit* can be no more separable, though they be dilated, than you can cut off the *Rayes* of the *Sun* by a pair of Scissors made of pellucid Crystall'.[16]

Taken against the backdrop of the then increasingly dominant corpuscularian metaphysic, More's model of an extended but m-indivisible entity is remarkably original and inventive. But clearly his proposal will raise questions and is open to possible challenges. Does the core's projection of a shell of activity involve action at a distance (an occult anathema to advocates of the mechanist new science)? What is the ontological status of the shell? Can it really be classed as a substance, as More wishes? Moreover, since's More's own principles lead him to reject unextended entities as 'nothing else but pure Negation or Non-entity', how can he then allow the existence of an extensionless core *punctum*, the *sine qua non* of the whole

---

[13] Ibid. 38.
[14] Ibid. 40; More, Appendix to *An Antidote Against Atheism*, 186.
[15] More, *Immortality of the Soul*. 39. See also More, *Collection of Several Philosophical Writings*, 186.
[16] More, *Philosophical Writings*, 12–13.

extended spirit?[17] These are serious issues, and I will return to them. But first let us look at the way Boscovich and Kant mobilize the same model of an extended but m-indivisible entity, this time as an account of the first parts of matter.

III. The Kant–Boscovich Force-Shell Atom Theory

This account of the structure of extended and m-indivisible material atoms is most well-known—today as in the eighteenth century—from Roger Boscovich's *magnum opus*, the 1758 *Theory of Natural Philosophy* (extensively revised and expanded in the 1763 edition). This work ran through many editions and was much fêted in scientific circles, ultimately winning the Croat natural philosopher election to both the French Academy and the Royal Society of London. Its subsequent influence on Priestley, Faraday, and Nietzsche has been well documented.[18]

Kant's remarkably similar and slightly earlier version of the force-shell atom theory, by contrast, is much less well-known. It appears in his little-read 1756 dissertation *Physical Monadology*—a thesis submitted in fruitless application for the position of *professor extraordinarius* in logic and metaphysics at Königsberg. (Kant would eventually be appointed to an equivalent chair a full fourteen years later.) The work is thus somewhat obscure and current commentators tend to neglect it, focusing instead on Kant's quite different matter theory of the later period. One major commentator even speculatively pieces together a version of the force-shell atom theory as a system that the later Kant 'is trying to set up for attack', tentatively suggesting that 'one can believe that some of [Leibniz's] would-be followers entertained it'—all without realizing that Kant is in fact attacking his own previous view from the *Physical Monadology*![19]

Although the pre-critical Kant's and Boscovich's versions of the force-shell atom theory are so similar, I can find no evidence of either philoso-

---

[17] More, *Immortality of the Soul*, 7.

[18] The 1758/63 *Theory of Natural Philosophy* stands as Boscovich's canonical statement of the theory, though it is in some ways prefigured in his 1745 *De Viribus Vivis*. On Boscovich's influence on Priestley and Faraday, see Mary Hesse, *Forces and Fields* (London: Thomas Nelson & Sons, 1961), and Rom Harré and E. H. Madden, *Causal Powers* (Oxford: Basil Blackwell, 1975), 172–5. On his influence on Nietzsche, see George J. Stack, 'Nietzsche and Boscovich's Natural Philosophy', *Pacific Philosophical Quarterly*, 62 (1981), 69–86.

[19] Jonathan Bennett, *Kant's Dialectic* (Cambridge: CUP, 1974), 171.

pher influencing the other (although, as I have said, Samuel Clarke's system of extended spirits, borrowed from More, is a possible mutual influence). In fact, the circumstantial evidence of Kant's and Boscovich's respective larger theoretical programmes seems to suggest that they each arrived at the force-shell model of material atoms independently of one another.

Kant's account was developed in the aftermath of the notorious Berlin Academy dispute over the structure of matter. It stands as an attempt to split the difference between the Newtonian lobby, who had insisted upon the infinite f-divisibility of space (pressing the argument from geometry discussed in Chapter 5), and the Wolffians, who had argued from the actual parts doctrine to a system of elemental simples or monads (via the argument from composition, discussed in Chapter 4).[20] In the *Physical Monadology*, Kant presents the system of extended force-shell atoms, each projected by a core *punctum*, as a way of reconciling the Newtonian and Wolffian positions: the first parts of matter are extended monads that are f-divisible ad infinitum but nevertheless m-indivisible. (The later Kant—of the 1781/7 *Critique of Pure Reason* and the 1786 *Metaphysical Foundations of Natural Science*—will disavow the *Physical Monadology*'s proposed solution. In this later period, Kant will embrace the potential parts metaphysic instead, blaming transcendental realism for the actual parts doctrine and rejecting the Leibnizian–Wolffian case for simples.[21])

The Croat Jesuit and natural philosopher Roger Boscovich (1711–87), by contrast, arrived at the theory of force-shell atoms through a consideration of the problem of impacts within Newtonian dynamics. The traditional perfectly rigid and undeformable atoms of classical Newtonian theory, being completely inelastic, would have to change velocities *instantaneously* in shock impacts. But this is absurd, not least because it would require infinite acceleration and hence infinite force. Boscovich was thus led to replace the system of perfectly solid and rigid Newtonian atoms with elastic shells of resistant force, each of which is projected by a central core *punctum*. Now, Boscovich does note that his theory has the additional 'convenience' of solving the traditional problems of material structure, since it reconciles the system of extended but m-indivisible monadic simples with infinite

---

[20] On the Berlin Academy dispute, see Ch. 4, s. II, esp. n. 10.
[21] See my account and the various references given in Ch. 1, s. IX (under 'faction (1)') and in Ch. 2, s. III.

f-divisibility.[22] But for Boscovich this result is presented as a happy bonus of his system rather than its underlying motivation. While Kant explicitly developed the force-shell atom theory first and foremost as a way out of the threatened antinomies of material structure, Boscovich 'came upon a theory of this kind' from his investigations into the nature of impacts in physical dynamics.[23]

Notwithstanding the different paths that led to their respective versions of the theory, the accounts of the pre-critical Kant and of Boscovich are remarkably similar. For each of them, matter is ultimately built up from elemental first parts, each of which is an extensionless core *punctum* that radiates a surrounding shell of force. A given body is thus composed of a finite number of these force-shell atoms, each of which occupies a region of space with a sphere of resistant force, repelling the approach of neighbouring atoms. (This repulsive force increases exponentially toward the *punctum* but is finite at any distance from the core. So the atoms are not *absolutely* impenetrable: adjacent force-shell atoms may partially overlap if they are compressed together with sufficient force. I return to this issue and its implications for the notion of materiality and the filling of space below.)

In addition to this shell of *resistant* force (which accounts for the atoms' occupancy of space and the nature of elastic impacts), since the atoms must also account for the local cohesion of the parts of bodies and indeed the gravitational attraction between all matter, the *puncta* must also emanate an *attractive* force. Kant thus conceptualizes his *puncta* as projecting two forces, a resistant one which falls off rapidly even at a very small distance, and an overlaid attractive force which decreases much more gradually, in accordance with Newton's inverse square law of gravitation. At a certain microscopic distance—in effect, the 'edge' of the force-shell atom—the

---

[22] 'The [force-shell atom] theory is also convenient for eliminating from Nature all idea of a coexistent continuum—to explain which philosophers have up till now laboured so very hard and generally in vain. Assuming [the force-shell atom theory], no division of a real entity can be carried on indefinitely; we shall not be brought to a standstill when we seek to find out whether the number of parts that are actually distinct and separable is finite or infinite; nor with it will there come in any of those truly innumerable difficulties that, with the idea of continuous composition, have given so much trouble to philosophers...by doing away with all idea of an actual infinity in existing things, truly countless difficulties are got rid of.' Roger Boscovich, *A Theory of Natural Philosophy*, tr. J. M. Child (Cambridge, Mass.: MIT Press, 1966), s. 90; see also ss. 138, 142, 372, and the Synopsis, paragraphs 12 and 13.

[23] Boscovich, *Theory of Natural Philosophy*, 10.

two forces cancel each other out and adjacent atoms will rest at equilibrium with one another.[24] Push the atoms closer together and the resistant force becomes stronger than the attractive force: the two *puncta* will attempt to move apart. Push the atoms apart from the equilibrium distance and there will be an attractive force between them, powerful enough at short distances to hold clusters of force-shell atoms together in tight structures (molecules, and, ultimately, bodies), falling off to become the familiar pull of gravitation at longer distances.

While Kant talks in terms of two overlaid forces, one attractive, one repellent, Boscovich conceptualizes his *puncta* as each projecting one unified force that repels at some distances and attracts at others. This difference would seem to be simply notational: one could think of the 'one' force as the sum of the 'two', or just as easily of the 'two' as aspects of the 'one'. But one genuine difference in their systems is that, where Kant admits only one equilibrium point or 'edge' for each atom as one proceeds out from its core, Boscovich's force-shell atoms each have several such equilibria. The force thrown out by one of his *puncta* becomes exponentially repellent as one approaches the core and then falls off in a wave-like function with distance, becoming attractive, then repellent again, then attractive again—and so on until somewhere outside the level of molecular structures it stays attractive, but ever more weakly as distance increases, in accordance with Newton's inverse square law.[25] Boscovich's undulating force thus allows several equilibria at which *puncta* will hold together in dense coalitions to form material structures like molecules. One could thus think of Boscovich's force-shell atoms as having several nested 'outside' surfaces (in the fashion of a Russian nesting doll), the outermost one being the last equilibrium point from whereon out the force remains (weakly) attractive. This, he hopes, will allow him to build the full range of chemical elements and to explain the divergent structures of gases, solids, and liquids, all as constructs from perfectly homogeneous force-shell atoms, locked together into different arrays around their various different equilibrium points.[26]

---

[24] '[T]here must be some point on the diameter where attraction and repulsion are equal. This point will determine the limit of impenetrability and the orbit of external contact; that is to say, it will determine the volume; for the repulsive force, once it has been overcome by attraction, ceases to act any further.' Kant, *Physical Monadology*, in *The Cambridge Edition of the Works of Immanuel Kant: Theoretical Philosophy 1755–1770*, tr. and ed. David Walford and Ralf Meerbone (Cambridge: CUP, 1992), 1: 485.

[25] Boscovich, *Theory of Natural Philosophy*, ss. 9–12.   [26] Ibid., s. 79.

248 / The Force-Shell Atom Theory

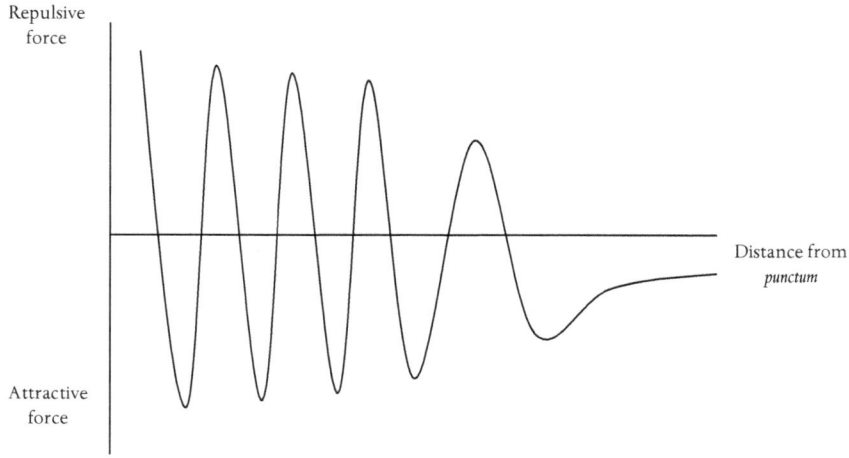

Figure 2 Force projected by Boscovich's *puncta*

By now Kant's and Boscovich's resolution of the problems of material structure should come as no surprise. As with More's extended spirits, their force-shell atoms combine extension and f-divisibility with m-indivisibility. They occupy space through their shell of repulsive force. But this shell is simply 'an orbit of activity', a sphere of force projected by the extensionless f- and m-indivisible core. So the parts of the shell cannot be ruptured and separated one from another, any more than the parts of a 'shell' of illumination can be separated from each other or from the light source from which they are emanated: force-shell atoms are m-indivisible.[27] Any given body is thus an aggregate of an unparadoxical finite number of these extended but m-indivisible atoms. The actual parts metaphysic is thus squared with the infinite f-divisibility of space.

IV. Further Observations Concerning the Kant–Boscovich Force-Shell Atom Theory

The force-shell atom theory constitutes a radical break with the traditional corpuscularian model and a bold first step towards the modern physicist's dynamical account of matter. It is certainly worth stressing the formal

---

[27] Kant writes that 'The monad...which is the fundamental element of the body, in so far as it fills space, certainly has a certain extensive quantity, namely an orbit of activity. You will

similarities between the Kant–Boscovich atomic theory and the tradition of Rutherford, Bohr, Heisenberg, and Gell-Man, the pioneers of the current 'standard model' of particle physics. This standard model also conceptualizes the fundamental particles as unextended *puncta*, each of which projects a shell of influence around it. As the physicist Brian Greene puts it, 'the standard model views the elementary constituents of the universe as point-like ingredients with no internal structure'; it envisages 'all matter particles and all force particles to be pointlike objects with literally no spatial extent'.[28] (This standard model can be contrasted with the anti-atomistic approach of Faraday and Einstein, who suggest that fields are more basic than so-called particles, conceptualizing these latter as simply focal concentrations or pulses within fields.[29] It can also be contrasted with string theory, which replaces the standard model's extensionless point-particles with miniscule but nevertheless extended loops of vibrating thread.)

In this section I want to make a few further observations concerning the novel features of Kant's and Boscovich's account, all the while keeping one eye on the ramifications of these issues for the dynamical approach to matter in general. These observations will also lead me to touch on some possible objections to the Kant-Boscovich force-shell theory. (The objections considered in this section are ultimately ineffective, as I shall show. I leave consideration of the most serious objection to section V below.)

---

not find in this orbit of activity a plurality of things, of which each one, existing on its own and in isolation from the others, would have its own permanence. For what found is in the space [occupied by one half of the force-shell] cannot be separated from what is present in the space [occupied by the other half of the force-shell] so that each existed on its own, for each is nothing but an external determination of one and the same substance; but accidents do not exist independently of their substances.' *Physical Monadology*, in *Works*, 1: 481–2. Likewise Boscovich: 'in my Theory, in which the primary elements of matter [he means the *puncta*, not their projected force-shells] are simple and nonextended, it is easily seen that there can be no divisibility of the elements. Also masses in so far as they actually exist, are to me merely sets of points finite in number. Hence these sets of points can at any rate be divided into parts, but not into a greater number of points than that given by the number of points constituting the mass, since no part can contain less than one of these points. Nor do geometrical arguments prove anything as far as my Theory is concerned, in favour of divisibility beyond this limit; for as soon as we reach intervals that are less than the distance between any two points we cut these empty intervals and not matter.' *Theory of Natural Philosophy*, s. 393.

[28] Brian Greene, *The Elegant Universe* (New York: Random House, 2000), 127, 157; see also 158.

[29] On Faraday's approach, see s. V below. On Einstein's approach, see James W. Ellington, introduction to Kant, *Metaphysical Foundations of Material Nature*, in Kant, *Philosophy of Material Nature*, tr. James W. Ellington (Indianapolis: Hackett, 1985), p. xxvii.

The observations I will make concern: (1) the ontological status of the force-shell; (2) the force-shell theory's admission of action at a distance; (3) the way in which the shell fills space to various degrees of intensity; and (4) the replacement of the corpuscularian's space-filling categorical property of solidity with a dispositional property or power of repulsion.

1. The Projected Force-Shell of a Kant–Boscovich Atom is Conceptualized as a Property, Not a Substance

The first point concerns the ontological status of Kant's and Boscovich's force-shells. Recall that More wanted to preserve the notion that the extended sphere of activity of his spirits (and not just their core *punctum*) is properly classed as a substance. Despite the fact that it arises 'by graduall Emanation from the First and Primest Essence, which we call the *Centre of the Spirit*', More claims that 'this *Secondary* or *Emanatory* Substance may be rightly called *Substance*'; 'Nor is there any incongruity, that one Substance should cause something else which we may in some sense call Substance, though but *Secondary* or *Emanatory*'.[30] But it must be said that More's shells of spiritual activity are poor candidates for substancehood. They fall short of each of the two criteria for substancehood that are standard (and often confusedly run together) in early modern metaphysics. First, they do not seem to be *ultimate hosts for properties*—subjects of properties that are not themselves also properties—since it seems perfectly possible to conceptualize the whole shell as a property of the core: it is just the core's power to act on things in the surrounding region of space. Second, they seem to lack *ontological independence* or *self-sufficiency*, since (as More admits) they are produced from the '*Primary Substance*'—the core *punctum*—'by Emanative causality'.[31]

In any event, Kant and Boscovich reject this Morean idea that projected shells of activity qualify as substances. Kant is clearest about this, since he

---

[30] More, *Immortality of the Soul*, 39.

[31] Ibid. 38. In fact More argues that his projected shells of spiritual activity do satisfy both criteria of substancehood (More runs the criteria together somewhat, as is quite common in his time): '[T]his *Secondary* or *Emanatory* Substance may rightly be called *Substance*, because it is a Subject imbued with certain powers and activities, and that it does not inhere as an *Accident* in any other Substance or Matter, but could maintain its place, though all Matter or what other Substance soever were removed out of that space it is extended through, provided its *Primary Substance* be but safe' (ibid. 39). But these arguments are very thin. *Vis-à-vis* the first criterion

readily employs the familiar jargon of early modern substance metaphysics. The core *punctum* at the heart of one of his force-shell atoms is indeed a 'simple substance, which is also called a monad'.[32] However the shell of force—and indeed each part of that shell—is then conceptualized as an 'external determination' or 'relational determination'—that is to say (for Kant is quite explicit), as an 'accident' of the substantial *punctum*.[33]

It is a little harder to pin down Boscovich's answer to the question of how his force-shells fit the traditional taxonomy of substance and accident. Influenced, no doubt, by the modish positivism that dominated natural philosophy in the mid-eighteenth century, his text eschews such traditional metaphysical language, presenting itself instead as a hard-headed mathematico-physical hypothesis driven by practical issues in dynamics rather than the baroque quandaries of the metaphysician. But his system does (of course) betray implicit metaphysical attitudes. When the chips are down, bodies are ultimately conceived of as coalitions of the monadic point-centres: 'matter is unchangeable and consists of points that are perfectly simple, indivisible, of no extent, and separated from one another'; 'the primary elements of matter are in my opinion perfectly indivisible and non-extended points'.[34] These simple *puncta* are treated as the fundamental existents of the system, grounding larger bodies and serving as the basic elements for a relational system of space and time.[35] (As Boscovich notes, his *puncta* are thus far parallel to 'those simple and non-extended elements upon which is founded the theory of Leibniz'.[36]) So it seems that these point cores are naturally handled as substances.

Each of these points then 'has ...a mutual acting force', they are 'subject to a determination to approach one another at some distances and to recede from one another at other distances'.[37] Now it is true that Boscovich

---

(ultimate subject of properties), this passage simply asserts that the shell does not 'inhere as an *Accident* in any other Substance'. But no real reason has been given to think that it could not be conceptualized as a property of the core. Vis-à-vis the second criterion (ontological independence), More claims that his shells are independent entities, since they can endure even when 'all Matter or what other Substance soever' is removed from their locale. But his concession in the last few lines ('provided its *Primary Substance* be but safe') gives the game away: the projected shells are dependent on their cores, and a being that requires the another created being as its sustaining cause is not a substance by this traditional criterion.

[32] Kant, *Physical Monadology*, in *Works*, 1: 477; see also 1: 481.   [33] Ibid. 1: 482.
[34] Boscovich, *Theory of Natural Philosophy*, Synopsis 10; s. 7.
[35] Ibid., Supplement 1, 197–202.   [36] Ibid., s. 2.
[37] Ibid., Synopsis 10; s. 9.

is cagey about the origin and metaphysical status of this force: he wants simply to present a mathematical model of the phenomena; concerning its origins, *hypotheses non fingo* remains the dominant attitude.[38] But since 'matter' is being identified with the point centres, and since the space between any two such *puncta* is described simply as a 'vacuum' or 'empty interval', it seems that Boscovich is effectively treating the *puncta* as substances while rejecting the idea that the force-shell around each *punctum* qualifies as a substance.[39] The shell of force thus seems to be a relational or dispositional property that holds between the core *punctum* and the various other *puncta* in the system.

If this reconstruction is fair, then it seems that Kant and Boscovich are both conceptualizing the *puncta* as substances while treating the force-shells as properties of these point-substances. They do then preserve the traditional notion of a substance at the heart of matter and as the base of its relational properties. But in equating the space-filling part of matter—the projected force-shell—with 'relational determinations', these two philosophers have taken a bold first step towards the rejection of the substance metaphysic and its replacement with dynamism, the exclusive identification of matter with fields of force.

2. The Kant–Boscovich Theory Admits Action at a Distance

The Kant–Boscovich matter theory's admission of action at a distance constitutes a further radical break from the traditional corpuscularian–mechanist world-view of seventeenth-century natural philosophy. Perhaps the most famous principle of the mechanistic science of the seventeenth century is the doctrine that all natural causation proceeds via direct contact-action. For advocates of this mechanistic world-system, explanations of causal interaction in physics were acceptable only if they bottomed out in immediate contact between the parts of matter. The only adequate explanations of the activity of physical systems were those that appealed to a mechanistic or hydraulic model of causation, the behaviour

---

[38] Concerning the *puncta*'s inertia, Boscovich writes that 'whether this [propensity] is dependent on an arbitrary law of the Supreme Architect, or on the nature of the points themselves, or on some attribute of them, whatever it may be, I do not seek to know; even if I did wish to do so, I see no hope of finding the answer; and I truly think that this also applies to the law of forces'. *Theory of Natural Philosophy*, s. 8; see also ss. 101, 102.

[39] Ibid., Synopsis, ss. 7 and 393.

of the whole arising from the complex contact-interaction of the various parts—the pressure, displacement, impulse, friction, and tension of the material elements as they act on one another through immediate contact.

The motive behind this explanatory principle appears to have come from the doctrine that, of all putative varieties of causation, action through direct contact presents the only case where one can actually 'see' the causal influence (perhaps immediately in experience, or perhaps with the ancillary use of reason): direct contact-action is the only case where a necessary connection between cause and effect is intuitively evident. This mechanist conceit was then paired with the assumption that nature is fundamentally intelligible, and thus that our ability to understand direct-contact causation but not other putative sorts of causation mirrors the fact that direct-contact causal necessities exist in nature whereas other sorts do not. Thus we have the conclusion that all causation in nature must ultimately arise in mechanical fashion from the immediate contact of the parts of matter. In the language of the day, causal influence requires 'substantial presence': the immediate contact of the acting substance. 'Virtual presence' without substantial presence'—the influence of a substance's power or causal efficacy without the immediate contact of that acting substance—cannot occur in nature. Some argument like this, coupled perhaps with an appreciation of the great successes of the mechanist programme in experimental physics, seems to underpin the corpuscularian tradition's rejection of action at a distance. In many writers the rejection even rises to the charge that action at a distance is *inconceivable* or an outright *contradictio in adjecto*.

Descartes, Hobbes, and Leibniz stand as obvious advocates of this mechanistic view of causation.[40] But even the Newtonians, with their gravitational attraction between distant bodies, held to the mechanistic explanatory principle and sought for a medium that could transmit gravitational influence through a system of direct contact (the aether, perhaps,

---

[40] See Descartes's laws of motion in the *Principles of Philosophy*, part 2, articles 37–55, in *The Philosophical Writings of Descartes*, tr. and ed. John Cottingham, Anthony Kenny, Dugald Murdoch, and Robert Stoothoff, 3 vols. (Cambridge: CUP, 1984), i. 240–6. Hobbes: 'There can be no cause of motion, except in a body contiguous and moved', *De Corpore*, in Hobbes, *The Collected Works of Thomas Hobbes*, ed. W. Molesworth, 11 vols. (London, 1839–45; facsimile edn.: London: Routledge Thoemmes Press, 1992), i, ch. 9, s. 7. Leibniz: 'A body is never moved naturally, except by another body which touches and pushes it', in Samuel Clarke and G. W. Leibniz, *The Clarke–Leibniz Correspondence*, ed. H. G. Alexander (Manchester Manchester University Press, 1956), 66.

or God's omnipresence). Newton himself is quite clear about this, for instance in the following famous passage from his correspondence with the theologian Richard Bentley:

That Gravity should be innate, inherent and essential to Matter, so that one Body may act upon another at a Distance thro' a *Vacuum*, without the Mediation of any thing else, by and through which their Action and Force may be conveyed from one to another, is to me so great an Absurdity, that I believe no Man who has in philosophical Matters a competent Faculty of thinking, can ever fall into it.[41]

And likewise Newton's myrmidion Samuel Clarke:

That one body should attract another without any intermediate influence, is indeed not a miracle, but a contradiction: for 'tis supposing something to act where it is not.[42]

The Kant–Boscovich system is in clear conflict with this rejection of action at a distance. The forces of repulsion and attraction thrown out by a Kant–Boscovich *punctum* present a paradigm case of virtual presence without substantial presence—of a causal power transmitted across a distance without the immediate contact of the active substance. Each *punctum* is supposed to project this influence across regions of space without itself being extended into those regions: indeed this is precisely how it dominates a volume without compromising its m-indivisibility. In fact any given *punctum* simultaneously acts at a distance on *all* other *puncta* in the world, attracting those that are far away and repelling those that come excessively close. And all this occurs without the immediate contact of any two *puncta*. So Kant's and Boscovich's system is in obvious violation of mechanism's prohibition on action at a distance. By the lights of the mechanist's explanatory principles, it reintroduces an unintelligible and occult form of

---

[41] Isaac Newton in correspondence with Richard Bentley, in *Isaac Newton's Papers and Letters on Natural Philosophy and Related Documents*, ed. I. Bernard Cohen (Cambridge Mass.: Harvard University Press, 1978), 25–6. In the 1687 *Principia*'s General Scholium, this general principle that all action requires substantial presence is invoked to demonstrate that God's omnipresence involves his literal extension through all space: 'He is omnipresent not *virtually* [i.e. through influence] only, but also *substantially*; for virtue cannot subsist without substance.' (Newton, *Principia Mathematica*, tr. Andrew Motte (Amherst, NY: Prometheus Books, 1995), 441.) For other 17th-century examples of this reasoning, see Edward Grant, *Much Ado About Nothing: Theories of Space and Vacuum from the Middle Ages to the Scientific Revolution* (New York: CUP, 1981), 230.

[42] Samuel Clarke, in *Clarke–Leibniz Correspondence*, 53. Also 21: 'Nothing can any more act, or be acted upon, where it is not present, than it can be where it is not.'

causation into physics—a nonsense that was rightly cleansed from natural philosophy by the mechanist world-model of the new science.

(Interestingly, the later, critical period Kant seems to utilize this point in an argument against his own earlier *Physical Monadology* force-shell theory.[43] This argument must, however, also be set alongside another background reason that the later Kant has to resist the force-shell view: his critical period adoption of the potential parts doctrine, which undercuts the whole case for atomic simples or first parts of matter.[44])

But the mechanist's antipathy to action at a distance does not present a serious objection to the Kant–Boscovich theory. The mechanist argument against action at a distance required the claim (i) that contact-action stood alone among putative forms of causation as an intelligible, transparent

---

[43] The force-shell theorist claims that 'matter consists of physical points each of which... has no movable parts but yet fills a space by mere repulsive force'; 'he would compound matter from physically indivisible parts and yet allow it to occupy space in a dynamical way'. But this is 'completely impossible'. '[I]n a filled space there can be no point that does not itself on all sides repel in the same way as it is repelled, i.e., as a reacting subject, of itself movable, existing outside every other repelling point... In order to make this fact... intuitable, let it be assumed that A is the place of a monad [i.e. the force-shell theorist's *punctum*] in space, that *ab* is diameter of the sphere of its repulsive force, and hence that *aA* is the radius of the sphere. Thus between *a*, where the penetration of an external monad into the space occupied by the sphere in question is resisted and A, the centre of the sphere, a point *c* can be specified (according to the infinite divisibility of space). Now, *if A resists whatever endeavours to penetrate into a, then c must resist both the points a and A, for f this were not so, they would approach each other unimpeded; consequently, A and a would meet in the point c, i.e., the space would be penetrated.* Therefore, there must be something in *c* that resists the penetration of *a*, and thus repels the monad A as much as this something is repelled by the monad. Now, since repulsion is a motion, *c* is something movable in space, i.e. matter; and the space between A and *a* could not be filled by the sphere of activity of a single monad, neither could the space between *c* and A, and so on to infinity.' (*Metaphysical Foundations of Modern Science*, 50–1; my emphasis.) Notice that the key move—the portion I have italicized—seems to tacitly assume that there can be no action at a distance. If there were action at a distance, then A could resist the penetration of *a* without any point between the two of them (such as *c*) being an independent source of resistance. The argument (so interpreted) is quite puzzling, however, since Kant explicitly allows remote action elsewhere in the *Metaphysical Foundations* (ibid. 61–2). For commentary, see Eric Watkins, 'Kant's Justification of the Laws of Mechanics', *Studies in History and Philosophy of Science*, 29 (1998), 539–60: 550–1.

[44] Immanuel Kant, *Critique of Pure Reason*, tr. Werner S. Pluhar (Indianapolis: Hackett, 1996), B333, B552 *Metaphysical Foundations of Modern Science*, 53–5. Michael Friedman suggests that Kant has another critical period argument against force-shell atoms, which proceeds from conservation laws governing the quantity of material substance. 'Matter and Material Substance in Kant's Philosophy of Material Nature: The Problem of Infinite Divisibility', in Hoke Robinson (ed.), *Proceedings of the Eighth International Kant Congress*, 5 vols., i/2 (Milwaukee: Marquette University Press, 1995), 595–610.

form of necessary connection, and the claim (ii) that the physical world mirrors our ability to apprehend it with the result that only this intuitively evident contact-action form of causation exists in nature. But each of these premises is questionable. Particularly in the light of Hume's famous critique of the idea that immediate contact-action is any more logically necessary, intuitively evident, or intelligible in its inner nature than other cases of constantly conjoined motions, I think that Kant and Boscovich can safely reject this part of the mechanist world-picture.[45] Furthermore, the mechanist doctrine that all action proceeds from immediate contact is itself plagued by an array of serious problems. First, it seems unable to explain the cohesion of solid bodies (the old appeal to hooks on corpuscles that lock them together simply pushes the question back: how do the parts of the corpuscles—including their hooks—themselves cohere?). Second, it has problems explaining the apparent action at a distance in cases of magnetic, electrical, and gravitational attraction (witness the Cartesians' baroque system of vortices). Finally, as we have seen, impulse itself—the most basic form of mechanistic contact-action—seems to require an elasticity in matter that is inexplicable in mechanical terms.

In the following quotation from Boscovich's *Theory of Natural Philosophy*, I think we can see both the Humean point that contact-action is no more intelligible in its intrinsic nature than action at a distance, and—at a stretch—the point that the force-shell atom theory offers an elegant, unifying solution to the various problems within mechanism.

To my mind indeed it is clear that motions produced by these forces depending on the distances are not a whit more mysterious, involved or difficult of understanding than the production of motion by immediate impulse as it is usually accepted; in which impenetrability determines the motion, and the latter has to be derived just the same either from the nature of solid bodies, or from an arbitrary law of the founder of the universe.

---

[45] Of course, this is not to say that there could be no evidence for the claim that there is no action at a distance in nature. For instance, if all transmissions of causal influence took time to reach from the agent to the patient, this would seem to suggest that that influence is propagated continuously across an intervening medium rather than acting at a distance. Or again, if all such transmissions could be affected by intervening objects, this would also seem to suggest that the influence does not act at a distance. (Here it is interesting to contrast electromagnetic attraction, which can be blocked by insulation, with gravity, which cannot.) My claim here is simply that the mechanist has failed to show that there is anything conceptually incoherent in the Kant–Boscovich model of *puncta* that act on one another at a distance.

Now, that the investigation of the causes and laws of motion are better made by my method, than through the idea of impulse, is sufficiently indicated by the fact that, where hitherto we have omitted impulse and employed forces depending on the distances only, in this way everything has been accurately defined and determined, and when reduced to calculation everything agrees with the phenomena with far more accuracy than we could have possibly expected.... [I]n other parts of physics most of the explanations are independent of, and disconnected from one another, being based on several subsidiary principles. Hence we may now conclude that if, relinquishing all idea of immediate impulses, we employ a reason for the action of Nature that is everywhere the same and depends on the distances, the remainder will be explained with far greater ease and certainty; and indeed it is altogether successful in my hands, as will be evident later, when I come to apply the Theory of Nature.[46]

3. In the Kant–Boscovich Theory, Matter Fills Space by Degrees

I have already noted that Kant–Boscovich force-shell atoms 'fill in' regions of space with materiality by means of a projected repulsive force. The region where such an atom resists other encroaching atoms marks the space it fills: the equilibrium point at which attractive force and repulsive force cancel each other out marks the effective edge and radius of the force-shell atom. (Of course, Boscovich's atoms each have several such equilibrium points, and—in effect—several nested 'outside' surfaces.) I have also noted that force-shell atoms are not absolutely impenetrable: their surface marks the point at which another approaching atom will start to experience a repellent force; it does not mark an absolutely impenetrable border. It is only towards the very centre of the atom and its core *punctum* that repulsive force increases exponentially, and even here, at any finite distance from the core one can always approach a little closer, given sufficient force to overcome the increasing resistance. Force-shell atoms can thus partially overlap when they impact one another (this explains the elasticity of their collisions) or when they are compressed together.

This new analysis of matter's filling in of space, defined as the domination of the filled-in region by a repulsive force, constitutes a further

---

[46] Boscovich, *Theory of Natural Philosophy*, ss. 102, 103.

dramatic break with the traditional system of corpuscularian metaphysics. For corpuscularians such as Locke and Newton, matter filled space through its property of *absolute* solidity. A region of space was thus either completely empty (freely penetrable) or completely full (host to a perfectly solid and absolutely impenetrable corpuscle of matter). True matter—the solid corpuscle—could be neither compressed nor penetrated. If a larger body could be compressed or penetrated, this was only because it contained 'interstices' of empty space between its corpuscles that could be squeezed together or filled in by smaller corpuscles. Kant's and Boscovich's new system, on the other hand, allows for 'filled' regions of space to be penetrated and compressed, and indeed for material presence in space to come in a spectrum of intensity, from weakly repellent fields to stronger ones. Newton's all-or-nothing, binary system of absolutely solid atoms and absolutely empty void-space is thus replaced with the more supple and gradualistic field-theoretic notion of *degrees* of material presence. Spaces that are occupied by matter are now full to varying degrees: materiality comes in intensive as well as extensive magnitudes.[47]

One corollary of this is that the new concept of material presence and the filling of space introduced by Kant and Boscovich allows for material entities to have fuzzy rather than sharply defined surfaces. Although I have spoken of an exterior 'edge' of the force-shell atom—the outside surface that marks the place where repellent force is cancelled out by attractive force—it is important to appreciate that this edge point need not mark a discontinuous or sharp break. The repulsive force experienced as one passes through this surface towards the interior of the atom might be extremely weak at first, and only increase to significant levels towards the core (perhaps in a continuous function from zero). This can once again be contrasted with the abrupt borders of the corpuscularian's precisely figured and sharply defined atoms.

4. In the Kant–Boscovich Theory, Matter Fills Space through its Powers

I now turn from the issue of whether or not the space-filling property of matter comes in a range of intensity to the related issue of whether

---

[47] Kant famously develops this point in the critical period in the Anticipations of Perception in the *Critique of Pure Reason*, B207–18. For useful commentary, see Jay F. Rosenberg, *One World and our Knowledge of it* (Dordrecht: Reidel, 1980), 26, 27, and Rae Langton, *Kantian Humility* (Oxford: OUP, 1988), 168–72.

that property is dispositional or intrinsic. (In what follows I use 'disposition' and 'power' interchangeably to stand for a thing's propensity to affect other things in certain ways. This relational sort of property can be contrasted with non-relational, 'intrinsic' or 'categorical' properties: properties that tell us what a thing *is* in itself, rather than how it affects other things.[48])

Corpuscularians notoriously wavered on this question. By the turn of the eighteenth century, the dominant corpuscularian view was that matter, in addition to its purely geometrical properties, must have a further essential property through which it fills space.[49] This space-filling property was known as 'solidity' and was thought to come in an all-or-nothing polarity of absolute 'repletion' on the one hand, or complete absence on the other—thereby establishing the stark contrast between absolutely solid matter and absolutely empty void-space. But what is the nature of this solidity? Is it a categorical or dispositional property, an intrinsic feature or a power?

On the one hand, corpuscularians tended to equate solidity with *impenetrability*: a power to resist the entry of other things into the occupied region of space, and a paradigmatically relational and dispositional property. On the other hand, corpuscularians also wanted solidity to play the role of an intrinsic, non-relational and non-dispositional property: something that tells us what matter *is*, not just how it will affect such-and-such another thing in such-and-such a circumstance. Thought of in this second way, this solidity-qua-intrinsic-property is then considered the ground or cause of impenetrability, but it is not itself that power.

Locke presents perhaps the clearest example of this Janus-faced attitude, if only because he is more self-aware than most corpuscularians and thus goes out of his way to point out solidity's moonlighting double role (a double role most corpuscularians overlook entirely). First, solidity as the disposition of impenetrability: 'the *Idea* of *Solidity*... arises from the resistance which we find in Body, to the entrance of any other Body into the

---

[48] My use of the term 'disposition' is thus somewhat narrower than another common usage. I will be using it to refer only to a thing's propensities to affect *other things* in certain ways—rather than the broader usage in which it covers a thing's propensities to affect other things *and* its propensities to affect or alter itself.

[49] Here I am thinking of the increasingly dominant tradition of Gassendi, Boyle, Locke, and especially Newton, which can be contrasted with the Cartesian branch of corpuscularianism (which held that matter's properties were exclusively geometrical).

Place it possesses'; 'Solidity consists in repletion, and so an utter Exclusion of other Bodies out of the space it possesses'; '[I]f anyone think better to call [solidity] *Impenetrability*, he has my Consent'.[50] Such passages certainly seem to suggest that Locke identifies solidity with the disposition or power of impenetrability. But then Locke also feels a pull towards treating solidity as a categorical property that tells us what matter *is*—an intrinsic feature that somehow stands behind the merely dispositional property of impenetrability. Having granted the 'Consent' just quoted, Locke immediately adds 'Only I have thought the term *Solidity*, the more proper to express this *Idea*... because it carries something more positive in it, than *Impenetrability*, which is negative, and is, perhaps, more a consequence of *Solidity*, than *Solidity* it self.'[51] This underlying 'more positive' solidity is then subsequently contrasted with the powers of *hardness* (a disposition to affect our tactile sense and a secondary quality) and *repulsion* (a disposition to affect other bodies and a tertiary quality)—powers which 'depend' on solidity, but are not to be identified with it.[52]

One can, I think, reconstruct the underlying reasons for this hedging and ambiguity. On the one hand, we can only be acquainted with this space-filling feature of matter if it is a power. We can only *directly* experience it in so far as it immediately affects us; and we can only be *indirectly* acquainted with it in so far as it affects other things, from whence we can infer back to it. This would suggest that the space-filling property—at least in so far as we can ever be acquainted with it—must be a power to affect either ourselves or other things. But then, on the other hand, there is also a common-sense intuition that matter does not fill space simply through a capacity or potentiality, a disposition or power. Matter's space-filling property should be real, actual, categorical, and intrinsic—not simply a potentiality to affect such-and-such another thing in such-and-such a way in such-and-such a situation. It should be concrete: the ground of dispositions and relations, perhaps, but not itself a disposition or relation. This common-sense intuition thus leads us to think of the space-filling feature of matter as an intrinsic, categorical property—and not simply as the relational, dispositional power of impenetrability.

---

[50] John Locke, *An Essay Concerning Human Understanding*, ed. P. H. Nidditch (Oxford: OUP, 1975), II. iv. 1, 4.
[51] Ibid. II. iv. 1.
[52] Ibid. II. ii. 4, 5.

What Kant and Boscovich do, in effect, is to call a halt to this ambiguity endemic in the corpuscularian tradition and come down squarely on the side of the view that matter fills space simply through a dispositional property or power. This power is not the *absolute* impenetrability of the corpuscularians, but is at least a linear descendant of it: it is what we might call the *relative* impenetrability of repulsive force. Regions of space that are occupied by bodies are so occupied in virtue of a force (though not necessarily an overwhelming one) that resists the movement of other bodies into that region. What *is* completely rejected is the notion that matter requires a categorical, non-dispositional, and non-relational property to fill space. Matter simply fills space through its dispositions and powers, occupying a region (as Kant says) purely 'by a certain activity which prevents other bodies from penetrating it'.[53] Fields of repulsive force—which are certainly relational, dispositional, and non-categorical in nature—are all that is required to fill a region of space; and the stronger the force, the greater the degree of intensity to which the space is filled.[54]

But what of the common-sense intuition that matter must fill space with an intrinsic, non-relational and non-dispositional property? Kant and Boscovich do not explicitly address this thought; they simply seem unmoved by it. And in fact there are good reasons to be suspicious of this sort of desire for a categorical base property.[55] First, one can raise worries about the relationship between the posited categorical property and the dispositional property of impenetrability that it supposedly underpins. If

---

[53] Kant, *Physical Monadology*, in *Works*, 1: 482.

[54] One recent commentator finds this analysis of the filling of space objectionable: 'The idea that an extended whole could be composed of finitely many unextended elements is generally thought plausible only in conjunction with the hypothesis, originated by Boscovich, that each of them exerts a force that prevents anything else from entering its sphere of influence... Points endowed with force may constitute an impenetrable field, but I do not see how a finite number of them can constitute an extended object. A region populated by a scattered colony of Boscovichean points is a region of which no subregion, however small, is filled; and if no subregion of a given region is filled, it is surely incorrect to say that the region contains anything that is extended. What Boscovichean points give us is the *appearance* of extension, not extension itself.' (James Van Cleve, 'Reflections on Kant's Second Antinomy', *Synthese*, 47 (1981), 490.) Van Cleve does not say *why* such a repulsive force dominating a region couldn't thereby fill it, but perhaps he is moved to this conclusion by the common-sense intuition that an intrinsic, non-dispositional property is required to fill space.

[55] Here I draw lessons from the following discussions: Howard Robinson, *Matter and Sense: A Critique of Contemporary Materialism* (Cambridge: CUP, 1982), 109–11. Simon Blackburn, 'Filling in Space', reprinted in Blackburn, *Essays in Quasi-Realism* (New York: OUP, 1993), 255–8. Langton, *Kantian Humility*, 172–6.

there are really two distinct properties here, then for all logic tells us they could be found apart from one another: logically speaking, the connection between the categorical property and the dispositional property is perfectly contingent. But this then allows the spectre of bodies that fill space with the categorical property of solidity, but are freely penetrable for all of that. It is difficult to see what explanatory role such a categorical property could play, or what motivation there could be for positing it. Second, as Simon Blackburn has noted, if the idea is that the underlying categorical property somehow gives rise to the dispositional property in virtue of a law of nature (so there is a nomic necessity binding the properties together, if not a logically necessity), this would seem to impute a power or disposition to the so-called categorical property, whereby it generates the further dispositional property. If this new power then needs a further categorical property underlying it—as the common-sense intuition, with its adversity to bare dispositions, would suggest—then we are off on an infinite regress. Third, the supposedly underlying categorical, non-dispositional property of solidity would be entirely inscrutable. We can only ever be acquainted with the powers of an object: we can only be acquainted with it in so far as it affects us or other entities that in turn affect us. If there is some further property behind these, such as solidity-qua-categorical-nature, a property that is not itself an active power, then this feature will remain forever unobservable, a something-we-know-not-what that always eludes us. The common-sense intuition, in demanding an intrinsic, non-dispositional property with which matter fills space, is asking us to posit a feature forever beyond our ken and whose explanatory value is nil.[56]

Given these qualms about the common-sense intuition, I think we can see the appeal of the Kant–Boscovich alternative. They simply jettison the idea that a categorical, non-relational property is needed if matter is to fill space. Rather, matter can occupy a region of space simply through its repulsive force. Granted, this means that matter fills space with properties that are fundamentally relational and dispositional: with its powers rather than an intrinsic nature. But if the only objection to this comes from the common-sense intuition, then we need not worry. For all its gut appeal and all its influence on the corpuscularian tradition, that intuition is itself quite naïve. Perhaps we *do* have a stubborn, possibly inborn instinct to

---

[56] On these points, see Blackburn, 'Filling in Space', 256–7. Langton, *Kantian Humility*, 176. John Foster, *The Case for Idealism* (London: Routledge & Kegan Paul, 1982), 59–67.

believe in what Nietzsche (in the quotation that heads this chapter) calls 'a final "fixed" thing... [a] little residual earthly clump' underlying the powers of matter. But that belief is quite unsupported. The posited clump of earth proves ultimately inscrutable and lacks any genuine explanatory value. Science can encounter only the powers of matter, not the non-relational internal nature that supposedly stands behind those powers. And in the light of all this, it is easy to see the attraction of the Kant–Boscovich matter theory, which (along with the subsequent course of modern physics) simply dissolves this evanescent clump of earth into a haze of force.

V. The Hollow World Problem

In this final section I examine what I take to be the most serious objection to the Kant–Boscovich matter theory. The objection is best introduced in two distinct stages. First, it begins with a challenge to the notion of the substantial *punctum* posited at the heart of a Kant–Boscovich force-shell atom. According to the objection, this point-substance is an unwarranted metaphysical excrescence. There is simply no content to the suggestion that the *punctum*-core has any distinct existence over and above the shell of force manifested in that region of space. Properly understood, the Kant–Boscovich theory then resolves simply to a system of mutually influencing fields of force which alternately attract or repel one another. The notion of independently existing *puncta* over and above these fields is a confusion.

The second stage of the objection then asserts the emptiness of this leftover system. Once we (quite rightly) abolish the *puncta*-substances, this unmasks a vicious regress or circularity that all along hid at the heart of the whole force-shell atom theory. The objection is that Kant and Boscovich are offering us a fundamentally vacuous system where material entities are defined in terms of powers to attract or repel other material entities, where these latter entities are in turn defined in terms of powers to attract or repel yet further material entities.... This is a hollow, empty world where concrete materiality disappears in an endless regress. (This second stage of the objection generalizes to challenge all systems that identify matter with forces or powers.)

First, the case against the independent existence of the *puncta*. I suppose one might, like More and Hobbes, resist the idea that there could be point-existents on the grounds that 'whatsoever is, is of necessity *extended*'—that 'to take away all *Extension* is to reduce a thing onely to a Mathematical point, which is nothing else but Negation or Non-entity'.[57] Personally, however, I can see no such conceptual incoherence in the bare idea of an unextended, point-like existent. There is in any case a much more compelling argument available against the independent existence of Kant's and Boscovich's core *puncta*.

Recall the argument (in section IV (4) above) against the common-sense intuition that we should posit a space-filling property that stands behind matter's merely dispositional features and that is in itself intrinsic, categorical, non-relational, and non-dispositional. The argument was that such a property lacks any possible explanatory value and would be entirely inscrutable. At that stage of the argument I used this point to uphold the idea that matter could fill space simply through a dispositional property such as its repulsive force, and to discredit the claim that an intrinsic, categorical property is required. But we can now extend that same argument to attack not merely the notion of a categorical, non-dispositional space-filling property, but also the idea of an independent core *punctum* that projects the space-filling dispositional properties of the force-shell atom. Whether we are thinking of this supposed point centre in terms of various categorical, non-dispositional properties, or in terms of a substratum that stands behind all properties whatsoever, precisely the same argument applies. The *punctum*-core that allegedly underpins the relational and dispositional properties of the force-shell lies forever beyond our ken and is without any genuine explanatory value.

Michael Faraday made just this point in 1844. Supposing we distinguish a particle a, from the powers or forces m which it is said to project, the only properties we can ever encounter are

the properties or forces of the m, not those of the a, which, without the forces, is conceived of as having no powers. But then surely the m is the *matter*?... To my mind, therefore, the a or nucleus vanishes, and the substance consists of the powers, or m; and indeed what notion can we form of the nucleus independent of its powers? all our perception and knowledge of the atom, and even our fancy, is limited to ideas of its powers: what thought remains on which to hang the imagination of an a independent of the acknowledged forces?... why then assume

---

[57] More, *Philosophical Writings*, 228; *Immortality of the Soul*, preface, 7.

the existence of that of which we are ignorant, which we cannot conceive, and for which there is no philosophical necessity?[58]

Anthony Quinton prosecutes a version of this objection explicitly against Boscovich: 'The [Boscovichean] point is parasitic on its activities in the sense that the idea of a wholly inert material point seems to be entirely vacuous. The atom might just as well be identified with the extended system of forces it is said to be the source of.'[59] If a *punctum* did not project a force-shell, there could be no possible reason to posit its existence. But if it does manifest itself with the projection of such a shell, then there is simply no cash value to the suggestion that it is in any way distinct from that extended field of force. The idea of a distinct *punctum*-substance thus goes the way of that other unwarranted and vacuous posit, the corpuscularian's space-filling but non-dispositional property. A *punctum* is at best merely the focal point of the force-shell: it is not an independent or free-standing entity in its own right. Properly cleansed of this vacuous something-I-know-not-what, the Kant–Boscovich system would admit only the existence of fields of force which shimmer around a central concentration or singularity, a knot or pulse in the field that is not itself anything more than a mode of that field.[60]

Thus far the objection might simply seem to warrant a minor adaptation of the Kant–Boscovich model of matter. Perhaps Kant and Boscovich *should* jettison the idea that the core *punctum* is an independent entity. (For all Nietzsche's effusive praise in the epigraph heading this chapter, this may be Boscovich's own 'little residual earthly clump', a last parting homage to the metaphysical category of substance, or to the idea that matter need have an intrinsic, non-relational property at core.) So far this seems to leave their system pretty much intact: it is just that force-shells are now conceptualized as having central focal points rather than ontologically prior point centres.

But this brings us to the second stage of the objection. Once we abolish the *puncta* as distinct entities, we can then see clearly a problem that all along hid at the heart of the force-shell theory, and indeed that plagues

---

[58] Michael Faraday, 'A speculation touching Electrical Conduction and the Nature of Matter', in *Experimental Researches in Electricity* (London: Richard and John Edward Taylor, 1844), ii. 290–1; quote and reference from Langton, *Kantian Humility*, 181.

[59] Anthony Quinton, *The Nature of Things* (London: Routledge & Kegan Paul, 1973), 87.

[60] Similar points are made in Rom Harré and E. H. Madden, *Causal Powers* (Oxford: Blackwell, 1975), 266; and in Langton, *Kantian Humility*, 181.

dynamism and field systems *tout court*. The problem is that we are left with nothing but dispositions or powers with which to populate the universe. A material atom turns out to be a shell of force, where this is just a power or disposition to repel things in such-and-such situations and to attract other things in such-and-such other situations. But what are these other things that are alternately either repelled or attracted? They too are merely shells of force, located powers or dispositions to repel or attract yet further things in such-and-such yet further situations. What we have here is the world as an endless sequence of promissory notes. Everything we thought of as concrete dissolves into a power to affect further things we also once thought of as concrete; but on inspection these further things themselves dissolve into powers to affect yet further things we also once thought of as concrete... And this regress looks vicious. As Russell famously puts the problem, 'There are many possible ways of turning some things hitherto regarded as "real" into mere laws concerning the other things. Obviously there must be some limit to this process, or else all the things in the world will merely be each other's washing.'[61]

It is sometimes thought that we can draw the teeth of this problem by pointing out that various of the powers or dispositions in the system will be presumably be manifested at any given moment. And such manifested powers look like they may block the regress or circularity, restoring concrete materiality to the overall system. Manifested powers or dispositions exist all right (after all, they are *manifested*) and these manifestations will be spatially and temporally located—so don't they provide us with the thing-like concreta the system seemed to be missing? But in fact, although the

---

[61] Bertrand Russell, *The Analysis of Matter* (London: Kegan Paul, 1927), 325. Quoted in Blackburn's 'Filling in Space', 257. For Hume's version of this problem, see David Hume, *A Treatise of Human Nature*, ed. David Fate Norton and Mary J. Norton (Oxford and New York: OUP, 2000), 1. 4. 4. 9. Additional references abbreviated 'SBN' give the corresponding page numbers in David Hume, *A Treatise of Human Nature*, ed. L. A. Selby-Bigge, 2nd edn. with text revised and notes by P. H. Nidditch (Oxford: OUP, 1978). Here the SBN reference is 228–9. Hume raises an objection to taking the disposition of impenetrability as the defining property of matter. The very idea of impenetrability, he insists, already presupposes some idea of the material entities that cannot move into each other's space: one has to already know what exactly it is that is being occluded. But then impenetrability cannot be the defining property of matter, 'For that would be to run in a circle, and make one idea depend on another, while at the same time the latter depends on the former.' (Hume actually speaks of 'solidity' here, but he certainly means solidity-qua-the-power-of-impenetrability, not solidity-qua-a-categorical-property: 'The idea of solidity is that of two objects, which being impell'd by the utmost force, cannot penetrate each other' (*Treatise*, 1. 4. 4. 9; SBN 228).) For a more recent statement of the problem, see Foster, *Case for Idealism*, 67–72.

possibility of manifested powers is certainly a reprimand to any lazy attempt to equate powers and dispositions exclusively with unactualized potentialities, I don't think that their invocation will solve the current problem. Even if a given power is manifested at a given moment, it will nonetheless still simply be a (manifested) power to affect further powers, which in turn are simply (manifested or unmanifested) powers to affect yet further (manifested or unmanifested) powers. So even a manifested power will never engage anything concrete or thing-like in this system; in Russell's terms we are still stuck with nothing but (particular manifestations of) laws concerning other things, where these other things turn out to be themselves further (particular manifestations of) laws concerning yet other things, and so on, washing all the way round. So the regress and circularity problems are not dissolved simply by the reminder that some of the powers may be manifested at a given moment.

To sum up. The first stage of the objection maintains that the Kant–Boscovich *puncta* are not really independent thing-like entities; rather they are just the focal centres of fields of force. So, according to this part of the objection, the Kant–Boscovich theory, properly understood and cleansed of vacuous posits, ultimately identifies matter exclusively with forces or powers. The second stage of the objection then maintains that, once we have a system that identifies matter exclusively with forces or powers, we are left with a grand circle or infinite regress, a great network of dispositions that interrelate with one another but that never truly engage anything actual. Lacking any concrete non-force, non-power, non-dispositional, thing-like entities for this system of dispositions to latch onto, the world is left a miasma of shifting relations that never achieves any concrete materiality. (Notice that analogous problems would seem to apply to all systems that build the material realm entirely from powers or dispositions. This is particularly worrying, since—as we have seen—there are good reasons to think that we cannot ever discover intrinsic properties or property-less substrata: powers are all that science can ever show us.[62])

---

[62] As Blackburn and Robinson note, the only categorical properties we encounter are the sensations of first person subjective experience (colours, sounds, pain, and so on); categorical properties do not characterize the objective material world as given in the scientific image (excepting spatial properties such as shape and size, which give us structure but not stuffing). Robinson, *Matter and Sense*, 117. Simon Blackburn's 'Filling in Space', 257. Compare also Foster, *Case for Idealism*, 64–72. See also Langton, *Kantian Humility*, 164–5, where she suggests that one might hold that *all* the properties of the scientific image of the world are dispositional rather than categorical, spatial properties such as size and shape *included*.

I conclude by considering two types of response to the hollow world problem.

Response (1). Is the Circularity or Regress of Powers on Powers Really Vicious?

This response attempts to disarm the second stage of the argument. Perhaps the first stage is correct: the Kant–Boscovich theory does resolve to a system in which matter is exclusively identified with powers or dispositions, leaving us with a material realm that is purely a network or web of dispositions to affect dispositions to affect dispositions... If one is looking for a thing-like entity to give this fabric of interrelated powers something concrete to affix itself to, none is to be had. But are these endless regresses and circular feedback loops of dispositions upon dispositions altogether empty? Perhaps we might take a lead from coherentist models in epistemology, where similar web-like structures are accepted as non-vicious, their ultimate circularity notwithstanding. According to coherentism, suitably elaborate networks of interrelated beliefs—beliefs that ultimately depend on one another in circular feedback loops—*can* give rise to a form of justification. And, following this analogy, mightn't a (suitably elaborate) web of dispositions bootstrap itself (so to speak) into a form of materiality?[63]

Whatever one thinks of coherentism or holism in epistemology, I don't think that this sort of trick will work here. As Howard Robinson has noted, the response trades on a certain vagueness.[64] We are told to think about the web as a whole and that, once thought of in this way, the hollow world regains its materiality. We are not to worry about the regresses that we encounter when attempting to characterize any particular power, but rather to shift our focus to the entire holistic structure. But why should we do this? The original problem was stated in terms of the regresses that we encounter when attempting to give a concrete specification of a particular power, and when we look for an actual thing-like entity for these regressive structures to latch onto. And to be told that a particular power P is just the power to affect a power Q in such-and-such

---

[63] A version of this response appears in Richard Holton, 'Dispositions All the Way Round', *Analysis*, 59 (1999), 9–14.

[64] Robinson, *Matter and Sense*, 116.

a way, where power Q is just the power to affect a further power R in such-and-such a way, where power R is just...and so on all the way round is not to be told anything. What does P actually amount to in concrete terms? We are never told. And no good reason has been given for thinking that this precisely specified problem evaporates once we look to the whole system of interrelated dispositions in its entirety. We might lose our focus on this precise problem as we make the Zen-like move to apprehend the whole structure. But I don't see that this makes the problem go away so much as simply shifts our attention away from it.

Response (2). Reintroduce the *Puncta*, Since the Hollow World Problem Shows that they do Have Explanatory Value After All

It is clear that the Kant–Boscovich *puncta*, conceived of as independent entities distinct from their force-shells, are inscrutable. It is also clear, I think, that they can have no intrinsic, non-dispositional properties that could in any way help to explain the powers and dispositional properties manifested in their surrounding fields. These were the reasons given in the first stage of the argument for dismissing the independent existence of the *puncta* over and above the shells of force.

But, one might argue, now we *do* have a reason to posit their existence after all. Even though *puncta*-substances are inscrutable and can have no properties that explain their surrounding fields of power, they are not altogether without explanatory value. The very fact that their abolition plunges us into the hollow ghost-world of pure dispositions shows us that we do have reason to retain them after all: a metaphysical reason if not an empirical or scientific one. Without them, the material universe becomes an empty web of dispositions. With them, the system of dispositions and potentialities has a concrete linchpin that restores its materiality. The *puncta* then stand as the thing-like entities that ground the system of powers, restoring an actuality to a world where there seemed to be nothing but dispositions. The response can be made directly to Faraday's attack on the *punctum*. Recall his tripartite challenge: 'why...assume the existence of that of which we are ignorant, which we cannot conceive, and for which there is no philosophical necessity?' According to the current response, we may be ignorant of the *puncta* (in that they are inscrutable), and we may not be able to conceive of them (in that we can form no positive conception of their properties behind the manifested shell of powers). But there *is* a 'philosophical necessity' for them. They are the *sine qua non* of the material world.

There is something desperate about this response. Our original arguments showed us that the supposed *punctum*-core is inscrutable, and that we can form no positive conception of its intrinsic nature. Nor can we characterize the *punctum* in terms of its purported *role* in grounding powers—for no such entity, characterized purely in terms of intrinsic, non-dispositional properties, can begin to explain the production of powers or dispositions.[65] (In the familiar jargon of Locke and Berkeley, we then have neither a 'positive idea' of the *punctum* that tells us what it *is*, nor even a 'relative idea' that fixes our reference to it by adequately specifying its relations to other known entities or properties.[66]) So here we have the epistemological point that we cannot positively specify what a *punctum* is in terms of its own categorical nature, or even triangulate in on it via its alleged role in supporting powers. In addition we have the metaphysical point that it is explanatorily otiose: it cannot help to account for the production of fields of force. What then are we supposing the play of dispositions and powers is latching onto? What is this 'linchpin', this '*sine qua non*' of the material world? I think that there is, at this point, a very strong temptation to think of the posited *punctum*-substance as a miniature material particle, a tiny corpuscle-like entity—as (one might even say) a 'little residual earthly clump'. But of course we are not entitled to think of the *punctum* this way, for this is surreptitiously to credit it with all manner of dispositions and manifested powers. We are supposed to be thinking of the *punctum* stripped of all its powers: we are to think of it either in terms of non-dispositional, non-power, intrinsic properties, or perhaps even in terms of a substratum that stands behind all properties whatsoever. But the idea of such a *punctum* seems simply vacuous, lacking any positive content or even any genuine role for it to play. And if there is no content to the notion of the *puncta*, then even a generous sprinkling of these I-know-not-what vacuities can hardly return the hollow world to material existence.

---

[65] See section IV(4) above for this argument. For an attempt to locate intrinsic, non-dispositional natures via their roles, see A. D. Smith, 'Of Primary and Secondary Qualities', *Philosophical Review*, 99 (1990), 221–54, esp. 250–1.

[66] See, for instance, George Berkeley, *Principles*, part 1, s. 16, in *The Works of George Berkeley*, ed. A. A. Luce and T. E. Jessop (London: Thomas Nelson & Sons, 1948–9), ii. 47.

We can generalize this lesson. The proposed response was the suggestion that *puncta* be reintroduced to serve as the thing-like entity missing from a realm of pure dispositions. But as we have just seen, the *puncta*, conceived as thing-like entities independent of their powers, cannot fulfil this role. Now we can extend this point. Just as the *puncta* cannot solve the hollow world problem, nor can the introduction of any other intangible, invisible entity thought of as independent of all powers and dispositions. For instance, consider the introduction of an aether-like fluid medium across which the dispositions and powers of the interacting fields of force ripple back and forth. This medium (let us suppose) is to give the shifting play of dispositions and powers something concrete to engage with: it is the thing-like entity that restores materiality to the hollow world. But here again we have the same dilemma. If we characterize the aether in terms of various powers and dispositions, then it too is simply part of the web of potentialities: it is part of the problem, not part of the solution. If we then try to characterize the aether in terms of intrinsic, non-dispositional, non-power, categorical properties, or as a substratum without properties altogether, then it seems once again simply a vacuous something-I-know-not-what. And what is said here of the *puncta* and the aether likewise goes for any other entity (such as Descartes's plenum, Newton's absolute space, or extended corpuscles characterized in terms of categorical solidity) that might be posited to serve as the thing-ish, non-dispositional linchpin of the material world. And this means that the current problem of characterizing the space-filling feature of matter in terms of categorical or dispositional properties in fact applies to all accounts of material structure. It may come out particularly vividly in the case of Kant–Boscovich force-shell atoms, where the space-filling aspect of matter is explicitly characterized in terms of dispositions. But in fact it applies no less to theories of infinitely m-divisible atomless gunk like Hobbes's and Reid's, to Galileo's actually infinite aggregates of point atoms, or to theories of finitely m-divisible granular structure like Charleton's and Hume's. Whatever one's theory of the structure, divisibility, and composition of material bodies, in the end one must face this problem of whether their space-filling 'stuffing' is defined in terms of categorical or dispositional properties. The force-shell theory is no worse off on this question than these rival accounts of matter's internal form or architecture.

What are we to make of the hollow world problem? I think we have reached a serious difficulty in our understanding of the physical here.

Scientific investigation into the nature of matter can only ever lead us to powers: to relational and dispositional properties. It cannot lead us to categorical or intrinsic properties, still less to their equally inscrutable ancestor, a quality-less substratum that stands behind all properties whatsoever. We never encounter a non-dispositional, non-relational, categorical property in the physicist's material world. And even if we *could*, it would not help us explain the powers or dispositional properties we do encounter. All this suggests that we must conceive of the material realm in dynamical terms: in terms of forces, powers, and dispositions.

But nor can we make sense of the world where all is powers and dispositions. This is Russell's world where everything is everything else's washing, where matter is defined in terms of powers to affect the material. And, putting aside all worries about formal circularity, such a world seems simply an empty realm that is all promissory note and no payoff, all potentiality and no actuality, not the world of stubborn concrete materiality we think we know.

Conclusion

Modern physics has abandoned the dominant early modern picture of material microstructure. The corpuscularian model of rigid, sharply defined chunks of completely solid material has been rejected in favour of a dynamical image: following Boscovich and Kant, we are now to think of the space-filling 'stuffing' of matter in terms of diffused fields of force that range in intensity and have fuzzy, interpenetrating borders. Physicists disagree over whether we should think of these space-filling fields as undergirded by a system of ontologically prior point particles, or whether we should think of the so-called particles as nothing more than focal concentrations or pulses propagated in fields. But in either case, I hope it is clear that the same basic framework of questions that plagued the early moderns still arises for the current conception of matter. Whether we think of a piece of matter as a cloud of point particles throwing out interlocking fields of force, or as a particle-free distribution of pure force across a region of space, either way we can still ask: how far forth can we rupture and separate the spatially distinct parts of this field (or fields, or fields-plus-particles)? Is this piece of matter finitely or infinitely divisible? Are the parts (smaller fields, or fields-plus-particles) into which it can be divided distinct beings, even prior to division? Or are they merely potential entities? Perhaps certain issues that faced the corpuscularian no longer arise—in particular, problems that presuppose sharp boundaries and immediate contact between rigid atoms. But the main questions that structure the early modern debate remain with us.

What are the philosophical morals of the early modern controversy over the internal architecture of material body? First, we should certainly thank the early moderns for focusing our minds on the problem. Barrow's talk of 'Labyrinths, Difficulties and Inconveniences', Boyle's fear of 'truths *unsociable*', Bayle's comprehensively documented bafflement, and Kant's warning about 'the scandal of the apparent contradiction of reason with itself' all serve to draw our attention to the paradoxes that lurk even in such an apparently innocuous and straightforward question as the division of a lump of matter.

But what are the more concrete lessons of the Enlightenment controversy? To begin with, I hope I have shown that the historical debate does indeed turn on metaphysical issues, particularly (of course) the question of actual versus potential parts. While post-Enlightenment developments in mathematics may indeed help to disarm the paradoxes of material structure, any response to the debate must also take account of its metaphysical underpinnings, and take a stand on these fundamental issues of ontology and individuation. Do bodies have actual or potential parts? If bodies have actual parts, are we free to embrace the system of atomless gunk? Or does the actual parts doctrine entail that bodies resolve to metaphysical atoms? (Is such metaphysical atomism compatible with infinite divisibility?) On the other hand, if we embrace the potential parts account, what sort of division is required to actualize parts? What are the divisible-but-simple units of the system? Do we need to endorse the theory of substantial forms here, or is there some alternative account of the unified nature of divisible but simple bodies?

Did successful resolutions of the paradoxes of material structure in fact emerge in the course of the early modern debate? I think that the early moderns did succeed in forging several distinct logically respectable responses, notwithstanding the scepticism of Bayle *et al*. This seems to me our real inheritance from the Enlightenment debate: a road map of theoretical responses to the threatened antinomy, some of them quite surprising and innovative, and each tried and tempered in the fire of the historical controversy. (My positive assessment as to the logical viability of these proposals is admittedly more tentative in some cases than in others.)

Here I shall review the main options that emerge in the course of the early modern debate. First, let us assume, with most early moderns, that bodies have actual parts as far forth as they are m-divisible. Enlightenment philosophers showed that this actual parts account entails that each material body has a fixed and determinate number of parts, in which case we can immediately dismiss the received Aristotelian–scholastic suggestion that they each contain an ever-increasable potential infinity of parts. A material body must then have either a given finite number of parts, or an actually infinite number (a completely given, greater-than-finite set). If the former, then we need an account of how extended bodies can be constructed from a finite base of m-indivisible parts. Here the 'ungeometrical philosophers' (including Charleton, Berkeley, and Hume) take refuge in a granular model of space, with each m-indivisible material atom occupying one such f-indivisible space-granule. For all its initial unfamiliarity and prima-facie strangeness, I have maintained that this view stands unrefuted, and in particular that it can resist the geometrical and conceptual arguments for infinite divisibility (see Chapter 5). So this is one position that emerges unscathed. Another possible account that marries actual parts and finite divisibility is the force-shell theory of Boscovich and the younger Kant. According to this system, a finite array of m-indivisible, extensionless *puncta* can build an extended body, given that each dominates a region of space by projecting a sphere of repulsive force. This approach certainly makes the problem of characterizing the space-filling feature of matter in terms of either dispositional or categorical properties particularly vivid (see Chapter 6). But as we have seen, this is a difficulty for all accounts of material structure, not just for the force-shell theory.

Still working within the system of actual parts, let us now turn to address theories of infinite divisibility. Could a body with actual parts be m-divisible ad infinitum? Only if one is prepared to face down the classic paradoxes of infinite divisibility, maintaining both that we can make sense of a physically realized greater-than-finite collection, and that a finite body could be built up from this sort of collection of parts. Most early moderns were deeply sceptical about both of these claims: the standard view was that an actually infinite collection was incoherent in concept, physically unrealizable, and in any case could not serve as the basis for constructing an extended body on pain of Zeno's metrical paradox. (Armed with the developments of modern set theory, most current metaphysicians are more sanguine here. But the issues remain controversial: a sizeable minority holds that, even if actual infinities are consistent in mathematical formalism, they cannot be physically instantiated on pain of paradox, or at least that they cannot serve as the basis for extended continua.)

Let us press on with those early modern actual parts theorists who *are* prepared to countenance an actual infinity of parts. There are two possible views of material structure we can distinguish at this juncture. On the one hand, one might think that bodies are made up of an actual infinity of *ultimate* parts (extensionless point atoms or perhaps infinitesimals). This is the view developed by Galileo, and formalized in our own era by Adolf Grünbaum. On the other hand, one might maintain that bodies have an actual infinity of ever-smaller actual parts, but no ground floor of ultimate parts or metaphysical atoms. Reid endorses this combination of the actual parts doctrine and the theory of atomless gunk. But for most early moderns these two doctrines were quite inconsistent, since the actual parts doctrine was thought to entail metaphysical atomism, either by way of the argument from the definiteness of parts, or by way of the argument from composition. (The former of these arguments fails, or so I argued in Chapter 3. The latter argument can also be resisted, but only if one is prepared to countenance an infinite regress of ontological parasites with no ultimate hosts (see Chapter 4).)

What if one resists the new philosophers' actual parts system and endorses the potential parts metaphysic instead? Here the paradoxes of material structure rapidly dissolve, since we now have a top–down account where parts are only created as division proceeds, and there is no difficulty in accommodating infinite divisibility by means of an unparadoxical potential infinity of parts: a number of parts that can always be increased further, but is always finite at any given stage. This is the traditional solution of the scholastics; it is also adopted in the early modern period by Digby, Hobbes, and the later Kant. Notice however that, while the potential parts doctrine disarms the paradoxes of infinite divisibility, and while all early modern potential parts advocates took this as a virtue of their system, the logic of the doctrine also leaves open the option of endorsing one of the systems of finite divisibility instead. If one thought that there were other reasons to adopt one of the systems where matter is only finitely divisible, a potential parts theorist could presumably just as well endorse such an account as an actual parts theorist. It is simply that they would then declare that the last, ultimate parts (granular *minima*, perhaps, or individual force-projecting *puncta*) are actualized by division rather than unveiled. Prior to this last step in the finite series of division—each step of which *creates* parts—they are merely potential existents.

Finally, what of the respective merits of the actual parts and potential parts doctrines, our two rival accounts of the ontology of material parts and wholes? For all their mutual inconsistency, and for all the ferocious polemics from each side of the debate, I maintain that both accounts emerge unscathed. At least if we restrict ourselves to assessing the arguments raised in the course of the historical debate, it seems that neither theory is refuted. Each doctrine seems to be logically and conceptually respectable: each is internally consistent, conflicts with no other mutually admitted principle, and each articulates a (distinct) cluster of familiar intuitions about division, parthood, and individuation (see Chapter 2).

Whether or not I am right about the logical situation here, the actual parts doctrine did of course dislodge the potential parts theory as the orthodox account during the early modern period. The success of the actual parts doctrine in permeating early modern thought can be seen both where it is simply implicitly assumed (as in much eighteenth-century natural philosophy) and, somewhat ironically, also in the explicit arguments that are given for the doctrine (by such seventeenth-century figures as Charleton, More, and Bayle). In Chapter 2 I suggested that these explicit arguments each beg the question. They presuppose actual parts premises that potential parts theorists are given no real reason to accept: they simply lay stress on our familiar actual parts intuitions, while appearing blind to the competing (and, it seems to me, equally familiar) potential parts intuitions. This in itself is suggestive of the status of the actual parts doctrine with most early modern thinkers: it is really a more or less self-evident first principle, a banal truism that could only be rejected by someone in thrall to the supposed authority of the scholastic tradition. Perhaps one can give arguments for such a truism, but since the doctrine is so basic it is hardly surprising if these arguments take the underlying actual parts intuitions as obvious—or so actual parts theorists may have reasoned.

But why did the actual parts metaphysic succeed in displacing the potential parts theory, given the weakness of the various official arguments? Part of the answer seems to me simply sociological: rightly or wrongly, the potential parts system was linked in the thinking of early moderns to the dying theory of substantial forms, whereas the actual parts system was associated with the project of the new science and the successes of corpuscularian explanation. There was in fact a substantial *de facto* coincidence of theoretical commitments here, with philosophers tending to endorse one of two doctrinal packages: either the potential parts doctrine and the system of substantial forms on the one hand, or the actual parts doctrine and some version of corpuscularianism on the other. (Of course, these pairings had historical roots in the Aristotelian tradition on the one hand and the Galilean and neo-Epicurean traditions on the other.) And this association of the potential parts theory with the system of substantial forms may account for its decline in popularity. If partisans of the new philosophy assumed that the two theories went hand in hand with one another, then they would have ceased to take the potential parts doctrine seriously when they rejected the system of substantial forms. This may well account for the early moderns' dismissive attitude towards the potential parts theory, which they seem to have regarded as something of a hangover from the Aristotelian-scholastic system.

But this *de facto* association of doctrines has no real *de jure* basis, or so I have argued. The actual parts–potential parts conflict is quite orthogonal to the merits of corpuscularianism and the program of the new science, as indeed one or two of the more discerning early moderns such as Hobbes and Kant appreciated. In particular, the potential parts doctrine need not be married to the system of substantial forms. Substantial forms might offer one possible account of the nature of the potential parts theorist's divisible-but-simple units or individuals, but (as we have seen) there are certainly other rival criteria for the status of simple, noncomposite (but m-divisible) substance: p-indivisibility perhaps, or maximal continuity, or internal qualitative homogeneity, or perhaps some mind-dependent or observer-relative criterion (see Chapter 2).

None of these reservations about the case for actual parts should be taken as positive recommendations of the potential parts system. As we saw, the historical arguments for potential parts are no less question-begging than those for actual parts. Still, if the early moderns' adoption of the actual parts metaphysic and rejection of potential parts was rooted more in unreflective prejudice than well-grounded argument, it is surely worth considering whether the current presumption among metaphysicians in favour of actual parts is itself justified. Reaching beyond the purview of this historical study, perhaps we should ask whether the affinity for actual parts systems in current analytic mereology and metaphysics is itself simply a prejudice inherited from our early modern forebears.

Writing in the persona of an anonymous reviewer in his 1740 *Abstract*, David Hume laments the state of the controversy over material structure and speaks ruefully of its implications for the rational intelligibility of nature: "Twere certainly to be wished, that some expedient were fallen upon to reconcile philosophy and common sense, which with regard to the question of infinite divisibility have waged most cruel wars with each other.'¹ This study has surveyed the battles and campaigns of those wars as they played out across two centuries, from Galileo's revolt against Aristotelian accounts of material structure through to the later Kant's defence of a renewed potential parts model. I have presumed to pass judgement on the outcome of certain of the major battles, distinguishing decisive clashes from standoffs, advances from retreats, and victors from losers. In my view, Hume's wished-for reconciliation of philosophy and common sense was in fact effected: consistent responses to the paradoxes of material structure did emerge in the seventeenth and eighteenth centuries. But while there are early modern accounts of material structure that do seem to me logically and conceptually respectable, these accounts turn on doctrines that remain controversial today. In the end, as Hume says, 'this question must be left to the learned world to judge'.²

[1] David Hume, *An Abstract of a book lately published, entitled, A Treatise of Human Nature, &c. wherein the chief argument of that book is farther illustrated and examined* (1740), paragraph 29, in David Hume, *A Treatise of Human Nature*, ed. David Fate Norton and Mary J. Norton (Oxford and New York: OUP, 2000), 415.

[2] Ibid.

BIBLIOGRAPHY

Primary Sources

Arnauld, Antoine, and Nicole, Pierre, *Logic, or The Art of Thinking*, tr. and ed. Jill Vance Buroker (Cambridge: Cambridge University Press, 1996).

Aristotle, *The Complete Works of Aristotle*, ed. Jonathan Barnes, 2 vols. (Princeton: Princeton University Press, 1984).

Barrow, Isaac, *The Usefulness of Mathematical Learning Explained and Demonstrated: Being Mathematical Lectures Read in the University of Cambridge*, tr. John Kirkby (1734; facsimile edn.: London: Frank Cass, 1970).

Bayle, Pierre, *Historical and Critical Dictionary Selections*, tr. and ed. Richard H. Popkin (Indianapolis: Hackett, 1991).

——— *Œuvres Diverses*, 5 vols. (1731; facsimile edn.: Hildesheim: Georg Olms Verlag, 1968).

——— *Various Thoughts on the Occasion of a Comet*, tr. and ed. Robert C. Barlett (Albany, NY: State University of New York Press, 1999).

Bergerac, Cyrano de, *Voyages to the Moon and the Sun*, tr. Richard Aldington (London: Routledge & Sons, 1923).

Berkeley, George, *The Works of George Berkeley*, ed. A. A. Luce and T. E. Jessop, 9 vols. (London: Thomas Nelson & Sons, 1948–9).

Boscovich, Roger J., *A Theory of Natural Philosophy*, tr. J. M. Child (Cambridge, Mass.: MIT Press, 1966).

Boyle, Robert, *Selected Philosophical Papers of Robert Boyle*, ed. M. A. Stewart (Manchester: Manchester University Press, 1979).

——— *The Works of the Honourable Robert Boyle*, 6 vols. (London: J. and F. Rivington, 1772).

Cartaud de la Vilate, Francois, *Pensées Critiques sur les Mathématiques* (Paris, 1733; facsimile edn.: Geneva: Slatkine Reprints, 1971).

Cavendish, Margaret, *Grounds of Natural Philosophy* (West Cornwall Conn.: Locust Hill Press, 1996).

Charleton, Walter, *Physiologia Epicuro-Gassendo-Charletoniana* (London, 1654; facsimile edn.: New York: Johnson Reprint Co., 1966).

Cheyne, George, *Philosophical Principles of Religion, Natural and Revealed*, 2 vols. (London, 1725).
Clarke, Samuel, *A Demonstration of the Being and Attributes of God, and Other Writings*, ed. Ezio Vailati (Cambridge: Cambridge University Press, 1998).
—— *The Works of Samuel Clarke* (London: John and Paul Knapton, 1738; facsimile edn.: New York: Garland, 1978).
—— and Leibniz, G. W., *The Leibniz–Clarke Correspondence*, ed. H. G. Alexander (Manchester: Manchester University Press, 1956).
Collier, Arthur, *Clavis Universalis: A Demonstration of the Non-Existence, or Impossibility, of an External World*, ed. Samuel Parr (1713; facsimile edn.: New York: Georg Olms Verlag, 1974).
Collins, Anthony, *A Letter to the learned Mr. Henry Dodwell, containing some remarks on a (pretended) Demonstration of the Natural Immortality of the Soul* (1709), in Samuel Clarke, *The Works of Samuel Clarke*, 4 vols. (1738; facsimile edn.: New York: Garland, 1978), iii. 749–53.
—— *A Reply to Mr. Clarke's Defence of his Letter to Mr. Dodwell* (1710), in Samuel Clarke, *The Works of Samuel Clarke*, 4 vols. (1738; facsimile edn.: New York: Garland, 1978), iii. 764–79.
—— *Reflections on Mr. Clarke's Second Defence of his Letter to Mr. Dodwell*, in Samuel Clarke, *The Works of Samuel Clarke*, 4 vols. (1738; facsimile edn.: New York: Garland, 1978), iii. 800–21.
Condillac, Etienne Bonnot de, *Les Monades*, in *Studies on Voltaire and the Eighteenth Century*, 187 (Oxford: Cheney & Sons, 1980), 109–211.
Conway, Anne, *Principles of the Most Ancient and Modern Philosophy* (Cambridge: Cambridge University Press, 1996).
Cordemoy, Gerauld de, *Œuvres Philosophiques*, ed. Pierre Clair and François Girbal (Paris: Presses Universitaires de France, 1968).
Cudworth, Ralph, *The True Intellectual System of the Universe*, 3 vols. (London: Thomas Tegg, 1845).
Descartes, René, *Œuvres de Descartes*, ed. Charles Adam and Paul Tannery, 12 vols. (Paris: Vrin, 1964–76).
—— *The Philosophical Writings of Descartes*, tr. and ed. John Cottingham, Anthony Kenny, Dugald Murdoch, and Robert Stoothoff, 3 vols. (Cambridge: Cambridge University Press, 1991).
Digby, Kenelm, *Two Treatises, in the one of which the Nature of Bodies is expounded; in the other, the Nature of Man's Soule; is looked into* (1644; facsimile edn.: Stuttgart: Friedrich Fromman Verlag, 1970).
Drummond, William, *Academical Questions* (1805; fascimile edn.: Delmar, NY: Scholars' Facsimiles and Reprints, 1984).

Epicurus, *The Epicurus Reader*, tr. and ed. L. P. Gerson and Brad Inwood (Indianapolis: Hackett, 1994).

Euler, Leonhard, *Letters on Different Subjects in Natural Philosophy*, tr. Henry Hunter, ed. David Brewster (New York: J. and J. Harper, 1833).

Galileo, Galilei, *Dialogues Concerning Two New Sciences*, tr. Henry Crew and Alfonso de Salvio (New York: Macmillan, 1914).

Gassendi, *Opera Omnia* (Stuttgart-Bad Canstatt. Friedrich Fromann, 1964).

—— *The Selected Works of Pierre Gassendi*, tr. and ed. Craig B. Bush (London: Johnson Reprint Co., 1972).

Glanvill, Joseph, *The Vanity of Dogmatizing* (1661; facsimile edn.: Hildesheim: Georg Olms Verlag, 1970).

Hobbes, Thomas, *The Collected Works of Thomas Hobbes*, ed. W. Molesworth, 11 vols. (London: 1839–45; facsimile edn.: London: Routledge Thoemmes Press, 1992).

—— *Dialogus Physicus de Natura*, tr. Stephen Shapin, in Stephen Shapin and Simon Shaffer, *Leviathan and the Air-Pump: Hobbes, Boyle and the Experimental Life* (Princeton: Princeton University Press, 1988), 345–91.

—— *Opera Philosophica Quae Latine Scripsit*, ed. William Molesworth, 5 vols. (Aalen: Scientia, 1961).

—— *Thomas White's 'De Mundo' Examined*, tr. Harold Whitmore Jones (London: Bradford University Press, 1976).

Hume, David, *Enquiries Concerning the Human Understanding and the Principles of Morals*, ed. L. A. Selby-Bigge, 3rd edn. revised by P. H. Nidditch (Oxford: Oxford University Press, 1975).

—— *A Treatise of Human Nature*, ed. with an analytical index, by L. A. Selby-Bigge (Oxford: Oxford University Press, 1978).

—— *Dialogues Concerning Natural Religion*, ed. Richard H. Popkin (Indianapolis: Hackett, 1980)

—— *A Treatise of Human Nature*, ed. David Fate Norton and Mary J. Norton (Oxford and New York: Oxford University Press, 2000).

Jackson, John, *A Dissertation Concerning Matter and Spirit* (1735; facsimile edn.: Bristol: Thoemmes Press, 1994).

Kant, Immanuel, *Philosophy of Material Nature*, tr. and ed. J. W. Ellington (Indianapolis: Hackett, 1985).

—— *Critique of Pure Reason*, tr. Werner S. Pluhar (Indianapolis: Hackett, 1996).

—— *The Cambridge Edition of the Works of Immanuel Kant: Theoretical Philosophy 1755–1770*, tr. and ed. David Walford and Ralf Meerbone (Cambridge: Cambridge University Press, 1992).

—— *The Metaphysical Foundations of Natural Science*, in Immanuel Kant, *Philosophy of Material Nature*, tr. and ed. J. W. Ellington (Indianapolis: Hackett, 1985).

Keill, John, *An Introduction to Natural Philosophy* (London: Senex, Innys, Manby, Osborn & Longman, 1733).

Law, Edmund, *An Enquiry into the Ideas of Space, Time, Immensity and Eternity* (Cambridge: 1734; facsimile edn.: New York: Garland, 1976).

Lee, Henry, *Anti-Scepticism, or Notes upon each chapter of Mr. Lock's Essay concerning human understanding* (Hildesheim: Georg Olms Verlag, 1973).

Le Grand, Antoine, *An Entire Body of Philosophy, According to the Principles of the Famous Renate Des Cartes*, tr. Richard Blome, 2 vols. (London, 1694).

Leibniz, Gottfried Wilhelm, *Die philosophischen Schriften von Gottfried Wilhem Leibniz*, ed. C. I. Gerhardt, 7 vols. (1875–90; facsimile edn.: Hildesheim: Georg Olms Verlag, 1978).

—— *New Essays on Human Understanding*, tr. and ed. J. Bennett and P. Remnant (Cambridge: Cambridge University Press, 1981).

—— *Philosophical Essays*, tr. and ed. Roger Ariew and Daniel Garber (Indianapolis: Hackett, 1989).

—— *Philosophical Papers and Letters*, tr. and ed. Leroy E. Loemker (Chicago: Chicago University Press, 1956).

—— *Theodicy*, tr. E. M. Huggard and ed. Austin Farrar (La Salle, Ill.: Open Court, 1985).

Locke, John, *An Essay Concerning Human Understanding*, ed. P. H. Nidditch (Oxford: Oxford University Press, 1975).

Lucretius, *De Rerum Natura*, tr. and ed. Cyril Bailey, 3 vols. (Oxford: Oxford University Press, 1947).

Malebranche, Nicolas, *The Search After Truth*, tr. Thomas M. Lennon and Paul J. Olscamp (Columbus: Ohio State University Press, 1980).

Martin, Benjamin, *The Philosophical Grammar, Being a View of the Present State of Experimented Physiology, or Natural Philosophy* (London: J. Noon, 1735).

Melvill, Thomas, 'Observations on Light and Colours', in *Essays and Observations, Physical and Literary. Read before a Society in Edinburgh, and Published by them* (Edinburgh: G. Hamilton & J. Balfour, 1756), ii. 12–90.

Meyer, Lodewijk, preface to Benedict Spinoza, *The Principles of Cartesian Philosophy*, tr. Samuel Shirley (Indianapolis: Hackett: 1988), 1–6.

More, Henry, *Philosophical Writings of Henry More*, ed. F. I. Mackinnon (New York and Oxford: Oxford University Press, 1925).

—— *A Collection of Several Philosophical Writings* (London: Joseph Downing, 1712).

—— *Enchiridium Metaphysicum* tr. Alexander Jacob. 2 vols. (Hildesheim: Georg Olms Verlag, 1995).

—— *The Immortality of the Soul*, ed. A. Jacob (Dordrecht: Kluwer, 1987).

Newton, Isaac, *Unpublished Scientific Papers of Isaac Newton*, tr. and ed. A. R. Hall and M. B. Hall (Cambridge: Cambridge University Press, 1962).

Newton, Isaac, 'Certain Philosophical Questions', tr. and ed. J. E. McGuire and Martin Tamny, in J. E. McGuire and Martin Tamny, *Certain Philosophical Questions: Newton's Trinity Notebook* (Cambridge: Cambridge University Press, 1983), 330–489.

—— 'Newton on Place, Time and Space: An Unpublished Source', tr. J. E. McGuire, *British Journal for the History of Science*, 11 (1978), 114–29.

—— *Sir Isaac Newton's Letters and Papers on Natural Philosophy and Related Documents*, ed. I. Bernard Cohen (Cambridge, Mass.: Harvard University Press, 1978).

—— *Principia Mathematica*, tr. Andrew Motte (Amherst, NY: Prometheus Books, 1995).

Nicholas of Autrecourt, *The Universal Treatise of Nicholas of Autrecourt* (Milwaukee, Wis.: Marquette University Press, 1971).

Norris, John, *An Account of Reason and Faith: In Relation to the Mysteries of Christianity* (London, 1697).

Plato, *Republic*, tr. G. M. A. Grube (Indianapolis: Hackett, 1992).

Price, Richard, *A Free Discussion of the Doctrines of Materialism and Philosophical Necessity* (1778; facsimile edn.: New York: Garland, 1978).

Reid, Thomas, *The Works of Thomas Reid*, ed. Sir William Hamilton, 2 vols. (1863; facsimile edn.: Bristol: Thoemmes Press, 1994).

Rohault, Jacques, *System of Natural Philosophy*, tr. John Clarke and Samuel Clarke (1723; facsimile edn.: New York: Johnson Reprint Corporation, 1969).

Sextus Empiricus, *Outlines of Skepticism*, tr. Julia Annas and Jonathan Barnes (Cambridge: Cambridge University Press, 1994).

Smollett, Tobias, *History and Adventures of an Atom* (Athens, Ga.: University of Georgia Press, 1989).

Spinoza, Benedict de, *The Collected Works of Spinoza*, ed. and tr. Edwin Curley (Princeton: Princeton University Press, 1985–).

—— *The Principles of Cartesian Philosophy*, tr. Samuel Shirley (Indianapolis: Hackett: 1988).

—— *Spinoza: The Letters*, tr. Samuel Shirley (Indianapolis: Hackett, 1995).

Sterne, Laurence, *Life and Opinions of Tristram Shandy, Gentleman* (Baltimore: Penguin Books, 1967).

Stillingfleet, Edward, *Origines Sacrae, or a Rational Account of the Grounds of Christian Faith, as to the Truth and Divine Authority of the Scriptures, And the matters therein contained* (London, 1662).

Vico, Giambatista, *On the Most Ancient Wisdom of the Italians*, tr. L. M. Palmer (Ithaca, NY: Cornell University Press, 1988).

Voltaire, *The Elements of Sir Isaac Newton's Philosophy*, tr. John Hanna (London: Cass, 1969).

—— *Œuvres Complètes*, ed. Louis Moland (Paris: Garnier Frères, 1877–80).

—— *Philosophical Dictionary*, tr. and ed. Theodore Besterman (Bungay, Suffolk: Penguin, 1971).
Watts, Isaac, *The Works of the Reverend and Learned Isaac Watts*, 6 vols. (London: John Barfield, 1810).
Wolff, Christian, *Gesammelte Werke*, ed. Joannes Ecole (Hildesheim: Georg Olms Verlag, 1962).
Wollaston, William, *The Religion of Nature Delineated* (1724; facsimile edn.: Delmar, NY: Scholars' Facsimiles and Reprints, 1974).

Other Sources

Adams, Robert Merrihew, 'Phenomenalism and Corporeal Substance in Leibniz', in Peter French, Theodore Uhling, and Howard Westein (eds.), *Midwest Studies in Philosophy VIII* (Minneapolis: University of Minnesota Press, 1983), 217–57.
—— *Leibniz: Determinist, Theist, Idealist* (New York: Oxford University Press, 1994).
Al-Azm, Sadik J., *The Origins of Kant's Arguments in the Antinomies* (Oxford: Oxford University Press, 1972).
Anapolitanos, D. A., 'The Continuous and the Discrete: Leibniz versus Berkeley', *Philosophical Inquiry*, 13 (1991), 1–24.
Ayers, Michael, and Garber, Daniel (eds.), *The Cambridge History of Seventeenth Century Philosophy* (Cambridge: Cambridge University Press, 1998).
Baltzly, Dirk, 'Who are the Mysterious Dogmatists of *Adversus Mathematicus* ix 352?', *Ancient Philosophy*, 18 (1998), 145–70.
Bartlett, Robert C., introduction to Pierre Bayle, *Various Thoughts on the Occasion of a Comet*, tr. and ed. Robert C. Bartlett (Albany, NY: State University of New York Press, 1999), pp. xxiii–xlvii.
Baxter, Donald L. M., 'Hume on Infinite Divisibility', in Stanley Tweyman (ed.), *Hume: Critical Assessments*, 6 vols. (London and New York: Routledge, 1995), iii. 16–24.
—— 'Hume on the Simplicity of Moments' (unpublished draft).
Bechler, Zev, *Newton's Physics and the Conceptual Structure of the Scientific Revolution* (Dordrecht: Kluwer, 1991).
Benaceraff, Paul, 'Tasks, Supertasks and the Modern Eleatics', *Journal of Philosophy*, 59 (1962), 765–84.
Bennett, Jonathan, *Kant's Dialectic* (Cambridge: Cambridge University Press, 1974).
—— *Learning from Six Philosophers: Descartes, Spinoza, Leibniz, Locke, Berkeley, Hume*, 2 vols. (Oxford: Oxford University Press, 2001).

Berkson, W., *Fields of Force* (London: Routledge & Kegan Paul, 1974).

Benardete, José A., *Infinity: An Essay in Metaphysics* (Oxford: Oxford University Press, 1964).

Blackburn, Simon, 'Filling in Space', *Analysis*, 50 (1990), 62–5. Reprinted in Simon Blackburn, *Essays in Quasi-Realism* (Oxford: Oxford University Press, 1993), 255–8.

Bolzano, Bernard, *Paradoxes of the Infinite*, tr. D. A. Steele (London: Routledge & Kegan Paul, 1950).

Bongie, Laurence L., introduction to Etienne Bonnot de Condillac, *Les Monades*, in *Studies on Voltaire and the Eighteenth Century*, 187 (Oxford: Cheney & Sons, 1980), 7–107.

Borges, Jorge Luis, *Labyrinths* (New York: New Directions, 1964).

—— *Selected Non-Fictions* (New York: Viking Penguin, 1999).

Bostock, David, 'Aristotle, Zeno and the Potential Infinite', *Proceedings from the Aristotelian Society* (1972–3), 37–51.

Bracken, Harry M., 'On Some Points in Bayle, Berkeley and Hume', *History of Philosophy Quarterly*, 4 (1987), 435–46.

Brandt, Frithiof, *Thomas Hobbes' Mechanical Conception of Nature* (Copenhagen: Levin & Munksgaard, 1928).

Broad, C. D., 'Hume's Doctrine of Space', Dawes Hicks Lecture on Philosophy, *Proceedings of the British Academy*, 47 (1961).

—— *Kant: An Introduction* (Cambridge: Cambridge University Press, 1978).

Burns, John V., *Dynamism in the Cosmology of Christian Wolff* (New York: Exposition Press, 1965).

Calinger, Ronald S., 'The Newtonian–Wolffian Controversy', *Journal of the History of Ideas*, 30 (1969), 319–30.

Cartwright, Richard, 'Scattered Objects', in Keith Lehrer (ed.), *Analysis and Metaphysics: Essays in Honor of R. M. Chisholm* (Dordrecht: D. Reidel Publishing Co., 1975), 153–71.

Casati, Roberto, and Varzi, Achille C., *Parts and Places: The Structure of Spatial Representation* (Cambridge, Mass.: MIT Press, 1999).

Charlton, William, 'Aristotle's Potential Infinities', in Lindsay Johnson (ed.), *Aristotle's Physics: A Collection of Essays* (Oxford: Oxford University Press, 1991), 129–49.

Clavelin, Maurice, *The Natural Philosophy of Galileo* (Cambridge, Mass.: MIT Press, 1974).

Craig, William Lane, *The Kalam Cosmological Argument* (London: Macmillan, 1979).

Craig, William Lane, and Smith, Quentin, *Theism, Atheism and Big Bang Cosmology* (Oxford and New York: Oxford University Press, 1993).

Cummins, Phillip, 'Bayle, Leibniz, Hume and Reid on Extension, Composites and Simples', *History of Philosophy Quarterly*, 7 (1990), 299–314.

Dauben, Joseph, *Georg Cantor: His Mathematics and Philosophy of the Infinite* (Cambridge, Mass.: Harvard University Press, 1979).

De Gandt, Francois, *Force and Geometry in Newton's Principia*, tr. Curtis Wilson (Princeton: Princeton University Press, 1995).

Des Chene, Dennis, *Physiologia: Natural Philosophy in Late Aristotelian and Cartesian Thought* (Ithaca, NY: Cornell University Press, 1996).

Dijksterhuis, E. J., *The Mechanization of the World Picture* (Oxford: Oxford University Press, 1961).

Duhem, Pierre, *Medieval Cosmology: Theories of Infinity, Place, Time, Void and Plurality of Worlds*, tr. and ed. Roger Ariew (Chicago: University of Chicago Press, 1985).

Earman, John, and Norton, John D., '"Infinite Pains: The Trouble with Supertasks', in Andrew Morton and Stephen P. Stitch (eds.), *Benacerraf and his Critics* (Oxford: Blackwell, 1996), 231–61.

Einstein, Albert, *Ideas and Opinions*, tr. Sonja Bargmann, ed. Carl Seelig (New York: Crown Publishers, 1954).

Ellington, James W., introduction to Immanuel Kant *Metaphysical Foundations of Natural Science*, in Immanuel Kant, *Philosophy of Material Nature*, tr. James W. Ellington (Indianapolis: Hackett, 1985), pp. v–xxix.

Faraday, Michael, *Experimental Researches in Electricity*, 3 vols. (London: Richard and John Edward Taylor, 1844).

Fieser, James (ed.), *Early Responses to Hume's Metaphysical and Epistemological Writings*, 2 vols. (Bristol: Thoemmes Press, 2000).

Flew, Anthony, 'Infinite Divisibility in Hume's *Treatise*', in J. T. King and D. W. Livingstone (eds.), *Hume: A Re-evaluation* (New York: Fordham University Press, 1967), 257–69.

—— '"Hume on Space and Geometry": One Reservation', *Hume Studies*, 8 (1982), 62–5.

Fogelin, Robert, *Hume's Skepticism in the Treatise of Human Nature* (London and Boston: Routledge & Kegan Paul, 1985).

—— 'Hume and Berkeley on the Proofs of Infinite Divisibility', *Philosophical Review*, 97 (1988), 47–69.

Forrest, Peter, 'Is Space–Time Discrete or Continuous? An Empirical Question', *Synthese*, 103 (1995), 327–54.

Forrester, James Wm., ' "If, in thought, all composition be removed ..." ', *Kantstudien*, 71 (1980), 406–17.

Foster, John, *The Case for Idealism* (London: Routledge & Kegan Paul, 1982).

Franklin, James, 'Achievements and Fallacies in Hume's Account of Infinite Divisibility', *Hume Studies*, 20 (1994), 85–101.

Frasca-Spada, Marina, 'Some Features of Hume's Conception of Space', *Studies in the History and Philosophy of Science*, 21 (1990), 371–411.

Frasca-Spada, Marina, 'Reality and the Coloured Points in Hume's *Treatise*: Part 2: Reality', *British Journal for the History of Philosophy*, 6 (1998), 25–45.
Friedman, Michael, 'Matter and Material Substance in Kant's Philosophy of Material Nature: The Problem of Infinite Divisibility', in Hoke Robinson (ed.), *Proceedings of the Eighth International Kant Congress*, 5 vols., 1/2 (Milwaukee: Marquette University Press, 1995), 595–610.
Garber, Daniel, *Descartes' Metaphysical Physics* (Chicago: University of Chicago Press, 1992).
Gennaro, Rocco J., and Huenemann, Charles (eds.), *New Essays on the Rationalists* (Oxford: Oxford University Press, 1999).
Grant, Edward, *Much Ado About Nothing: Theories of Space and Vacuum from the Middle Ages to the Scientific Revolution* (Cambridge: Cambridge University Press, 1981).
Greene, Brian, *The Elegant Universe* (New York: Random House, 2000).
Grünbaum, Adolf, 'A Consistent Conception of the Extended Linear Continuum as an Aggregate of Unextended Elements', *Philosophy of Science*, 19 (1952), 288–306.
—— *Philosophical Problems of Space and Time* (New York: Knopf, 1963).
—— *Modern Science and Zeno's Paradoxes* (Middletown, Conn.: Wesleyan University Press, 1967).
Hamlyn, D. W., *Metaphysics* (Cambridge and New York: Cambridge University Press, 1984).
Hare, Peter H. (ed.), *Doing Philosophy Historically* (Buffalo, NY: Prometheus Books, 1988).
Harman, P. M., *Energy, Force and Matter: The Conceptual Development of Nineteenth-Century Physics* (Cambridge: Cambridge University Press, 1982).
—— *Metaphysics and Natural Philosophy: The Problem of Substance in Classical Physics* (Brighton and Totowa, NJ: Barnes & Noble, 1982).
Harré, Rom, and Madden, E. H., *Causal Powers* (Oxford: Basil Blackwell, 1975).
Heinemann, P. M., and McGuire, J. E., 'Newtonian Forces and Lockean Powers: Concepts of Matter in Eighteenth Century Thought', *Historical Studies in the Physical Sciences*, 3 (Philadelphia: University of Philadelphia Press, 1971), 233–306.
Hempel, C. G., 'Geometry and Empirical Science', *American Mathematical Monthly*, 52 (1945). 7–17.
Herbert, Gary S., 'Hobbes's Phenomenology of Space', *Journal of the History of Ideas*, 48 (1987), 709–17.
Hesse, Mary B., *Forces and Fields* (London: Thomas Nelson & Sons, 1961).
Hoffman, Joshua, and Rosenkrantz, Gary S. *Substance Among Other Categories* (Cambridge: Cambridge University Press, 1994).
Holden, Thomas, 'Infinite Divisibility and Actual Parts in Hume's *Treatise*', *Hume Studies*, 28 (2002), 3–25.
—— 'Bayle and the Case for Actual Parts', *Journal of the History of Philosophy*, 42 (2004), 145–64.

Holton, Richard, 'Dispositions All the Way Round', *Analysis*, 59 (1999), 9–14.
Huby, Pamela, 'Kant or Cantor? That the Universe, if Real, must be Finite in Both Space and Time', *Philosophy*, 46 (1971) 121–32.
Jacquette, Dale, 'Kant's Second Antinomy and Hume's Theory of Extensionless Indivisibles', *Kantstudien*, 84 (1993), 38–50.
Jammer, Max, *Concepts of Force* (Cambridge, Mass.: Harvard University Press, 1957).
Jesseph, Douglas, *Berkeley's Philosophy of Mathematics* (Chicago: Chicago University Press, 1993).
—— *Squaring the Circle: The War Between Hobbes and Wallis* (Chicago: Chicago University Press, 1999).
Johnson, Lindsay (ed.), *Aristotle's Physics: A Collection of Essays* (Oxford: Oxford University Press, 1991).
Joy, Lynn Sumida, *Gassendi the Atomist* (Cambridge: Cambridge University Press, 1987).
Koyré, Alexander, *From the Closed World to the Infinite Universe* (Baltimore: Johns Hopkins University Press, 1968).
Kemp Smith, Norman, *A Commentary to Kant's Critique of Pure Reason* (London: Macmillan, 1918).
—— *The Philosophy of David Hume* (London: Macmillan, 1941).
Kenny, Anthony, Norman, Kretzman, and J. Pinsborg, (eds.), *The Cambridge History of Medieval Philosophy* (Cambridge: Cambridge University Press, 1982).
Kretzman, Norman (ed.) *Infinity and Continuity in Ancient and Medieval Thought* (Ithaca, NY: Cornell University Press, 1972).
Laird, John, *Hume's Philosophy of Human Nature* (Hamdon, Conn.: Archon Books, 1967).
Langton, Rae, *Kantian Humility* (Oxford Oxford University Press, 1998).
Lear, Jonathan, 'Aristotelian Infinity', *Proceedings from the Aristotelian Society* (1979–80), 188–210.
Leclerc, Ivor, 'The Problem of the Physical Existent', *International Philosophical Quarterly*, 9 (1969), 40–62.
—— *The Nature of Physical Existence* (London: George Allen & Unwin, 1972).
—— *The Philosophy of Nature* (Washington, DC: Catholic University of America Press, 1986).
Lehrer, Keith (ed.), *Analysis and Metaphysics: Essays in Honor of R. M. Chisholm* (Dordrecht: D. Reidel Publishing Co. 1975).
Levey, Samuel, 'Leibniz on Mathematics and the Actually Infinite Division of Matter', *Philosophical Review*, 107 (1999), 49–96.
—— 'Leibniz's Constructivism and Infinitely Folded Matter', in Rocco J. Gennaro and Charles Huenemann (eds.), *New Essays on the Rationalists* (Oxford: Oxford University Press, 1999), 134–62.

Lewis, David, *Parts of Classes* (Oxford: Blackwell, 1991).
Loemker, Leroy, introduction to G. W. Leibniz, *Philosophical Papers and Letters* (Chicago: Chicago University Press, 1956), 1–62.
McGuire, J. E., 'Newton on Place, Time and Space: An Unpublished Source', *British Journal for the History of Science*, 11 (1978), 114–29.
—— and Tamny, Martin, *Certain Philosophical Questions: Newton's Trinity Notebook* (Cambridge: Cambridge University Press, 1983).
Mach, Ernst, *Space and Geometry*, tr. Thomas J. McCormack (Chicago: Open Court, 1906).
Mackie, J. L., *The Miracle of Theism* (Oxford: Oxford University Press, 1982).
Marksosian, Ned, 'Simples', *Australasian Journal of Philosophy*, 76 (1998), 213–28.
Michel, P. H., *The Cosmology of Giordano Bruno*, tr. R. E. W. Maddison (Ithaca, NY: Cornell University Press, 1973).
Miller, Jr., Fred D., 'Aristotle against the Atomists', in Norman Kretzman (ed.), *Infinity and Continuity in Ancient and Medieval Thought* (Ithaca, NY: Cornell University Press, 1972), 87–111.
Mijuskovic, Ben, *The Achilles of Rationalist Arguments: The Simplicity, Unity, and Identity of Thought and Soul from the Cambridge Platonists to Kant* (The Hague: Martinus Nijhoff, 1974).
—— 'Hume on Space (and Time)', in Stanley Tweyman (ed.), *Hume: Critical Assessments*, 6 vols. (London and New York: Routledge, 1995), iii. 61–70.
Moore, A. W., *The Infinite* (London and New York: Routledge, 1990).
Morton, Andrew, and Stitch, Stephen P., (eds.), *Benaceraff and his Critics* (Oxford: Blackwell, 1996).
Murdoch, John E., 'Infinity and Continuity', in Anthony Kenny, Norman Kretzman, and J. Pinsborg (eds.), *The Cambridge History of Medieval Philosophy* (Cambridge: Cambridge University Press, 1982), 568–76.
Newman, Rosemary, 'Hume on Space and Geometry', *Hume Studies*, 7 (1981), 1–31.
North, J. D., 'Finite and Otherwise: Aristotle and Some Seventeenth Century Views', in William Shea (ed.), *Nature Mathematized* (Dordrecht: D. Reidel, 1983), 113–48.
Polonoff, Irving, *Force, Cosmos, Monads and Other Themes in Kant's Early Thought* (Bonn: Grundmann, 1973).
Popkin, Richard H., 'So, Hume Did Read Berkeley', *Journal of Philosophy*, 61 (1964), 773–8.
—— *The History of Scepticism: From Savonarola to Bayle* (Oxford: Oxford University Press, 2003).
Pressman, H. Mark, 'Hume on Geometry and Infinite Divisibility in the *Treatise*', *Hume Studies*, 23 (1997), 227–44.

Price, H. H., 'Universals and Resemblances', in Peter Van Inwagen and Dean W. Zimmerman (eds.), *Metaphysics: The Big Questions* (Oxford: Blackwell, 1998), 23–40.

Pyle, Andrew, *Atomism and its Critics* (Bristol: Thoemmes Press, 1995)

Quinton, Anthony, *The Nature of Things* (London: Routledge & Kegan Paul, 1973).

Radner, Michael, 'Unlocking the Second Antinomy', *Journal of the History of Philosophy*, 36 (1998), 413–41.

Ray, Christopher, *Time, Space and Philosophy* (London and New York: Routledge, 1991).

Raynor, David, '"*Minima Sensibilia*" in Berkeley and Hume', in Stanley Tweyman (ed.), *Hume: Critical Assessments* (London and New York: Routledge 1995), 370–3.

Rebondi, Pietro, *Galileo: Heretic*, tr. Raymond Rosenthal (Princeton: Princeton University Press, 1987).

Rees, D. A., 'Kant, Bayle and Indifferentism', *Philosophical Review*, 63 (1954), 592–5.

Robinson, Hoke (ed.), *Proceedings of the Eighth International Kant Congress*, 5 vols. (Milwankee: Marquette University Press 1995), 595–610.

Robinson, Howard, *Matter and Sense: A Critique of Contemporary Materialism* (Cambridge: Cambridge University Press, 1982).

Rosenberg, Jay F., *One World and our Knowledge of it* (Dordrecht: Reidel, 1980).

Russell, Bertrand, *Mysticism and Logic and Other Essays* (London: Allen & Unwin, 1917).

—— *Our Knowledge of the External World* (London: Allen & Unwin, 1926).

—— *The Analysis of Matter* (London: Kegan Paul, 1927).

Rutherford, Donald, *Leibniz and the Rational Order of Nature* (Cambridge University Press, 1995).

Scaltsas, Theodore, *Substance and Universals in Aristotle's Metaphysics* (Ithaca, NY: Cornell University Press, 1994).

Schmaltz, Tad, 'Spinoza on the Vacuum', *Archiv für Geschichte der Philosophie*, 81 (1999), 174–205.

Schopenhauer, Arthur, *The World as Will and Representation*, tr. E. F. J. Payne, 2 vols. (New York: Dover, 1969).

Shaffer, Simon, and Shapin, Stephen *Leviathan and the Air-Pump: Hobbes, Boyle and the Experimental Life* (Princeton: Princeton University Press, 1988).

Shea, Willliam (ed.), *Nature Mathematized* (Dordrecht: D. Reidel, 1983).

Sider, Theodore, 'Van Inwagen and the Possibility of Gunk', *Analysis*, 53 (1993), 285–9.

—— *Four-Dimensionalism* (Oxford: Oxford University Press, 2001).

Skyrms, Brian, 'Zeno's Paradox of Measure', in R. S. Cohen and L. Lauden (eds.), *Physics, Philosophy and Psychoanalysis* (Dordrecht and Boston: D. Reidel, 1983).

Sleigh, R. C., *Leibniz and Arnauld* (New Haven: Yale University Press, 1990).

Smith, A. D., 'Of Primary and Secondary Qualities', *Philosophical Review*, 99 (1990), 221–54.

Snow, A. J., *Matter and Gravity in Newton's Physical Philosophy* (Oxford: Oxford University Press, 1926).

Sorabji, Richard. 'Atoms and Time Atoms', in Norman Kretzman (ed.), *Infinity and Continuity in Ancient and Medieval Thought* (Ithaca, NY: Cornell University Press, 1982), 37–86.

—— *Time, Creation and the Continuum* (Ithaca, NY: Cornell University Press, 1983).

Stack, George J., 'Nietzsche and Boscovich's Natural Philosophy', *Pacific Philosophical Quarterly*, 62 (1981), 69–87.

Stones, G. B., 'The Atomic View of Matter in the XVth, XVIth and XVIIth Centuries', *Isis*, 10 (1926).

Thiel, Udo, 'Individuation', in Michael Ayers and Daniel Garber (eds.), *The Cambridge History of Seventeenth Century Philosophy*, 2 vols. (Cambridge: Cambridge University Press), i. 212–64.

Thijssen, J. M. M. H., 'David Hume and John Keill and the Structure of Continua', *Journal of the History of Ideas*, 53 (1992), 271–86.

Tweyman, Stanley (ed.), *Hume: Critical Assessments*, 6 vols. (London and New York: Routledge, 1995).

Vailati, Ezio, 'Clarke's Extended Soul', *Journal of the History of Philosophy*, 31 (1993), 387–404.

Van Bendegem, Jean Paul, 'In Defence of Discrete Space and Time', *Logique et Analyse*, 150–1 (1995), 127–50.

Van Cleve, James, 'Reflections on Kant's Second Antinomy', *Synthese*, 47 (1981), 481–94.

—— 'Inner States and Outer Relations: Kant and the Case for Monadism', in Peter H. Hare (ed.), *Doing Philosophy Historically* (Buffalo, NY: Prometheus Books, 1988), 231–47.

—— *Problems from Kant* (Oxford: Oxford University Press, 1999).

Van Inwagen, Peter, 'The Doctrine of Arbitrary Undetached Parts', *Pacific Philosophical Quarterly*, 62 (1981), 123–37.

—— *Material Beings* (Ithaca, NY: Cornell University Press, 1990).

—— and Zimmerman, Dean W. (ed.), *Metaphysics: The Big Questions* (Oxford: Blackwell, 1998)

Watkins, Eric, 'Kant's Justification of the Laws of Mechanics', *Studies in History and Philosophy of Science*, 29 (1998), 539–60.

Whitrow, G. J., *The Natural Philosophy of Time* (Oxford: Oxford University Press, 1961).

Whittaker, Edmund, *From Euclid to Eddington: A Study of Conceptions of the External World* (New York: Dover, 1958).

Williams, L. P., *The Origins of Field Theory* (New York: Random House, 1966).

Wilson, Catherine, *The Invisible World: Early Modern Natural Philosophy and the Invention of the Microscope* (Princeton: Princeton University Press, 1995).

Wittgenstein, Ludwig, *Philosophical Investigations*, tr. G. E. M. Anscombe (Oxford: Blackwell, 1953).

—— *Tractatus Logico-Philosophicus*, tr. D. F. Pears and B. F. McGuinness (London: Routledge & Kegan Paul, 1961).

Wojcik, Jan W., *Robert Boyle and the Limits of Reason* (Cambridge: Cambridge University Press, 1997).

Yolton, John W., *Thinking Matter: Materialism in Eighteenth Century Britain* (Minneapolis: University of Minnesota Press, 1983).

Zimmerman, Dean W., 'Theories of Masses and Problems of Constitution', *Philosophical Review* 104 (1995), 53–110.

—— 'Could Extended Objects be Made out of Simple Parts? An Argument for "Atomless Gunk" ', *Philosophy and Phenomenological Research*, 56 (1996), 1–29.

—— 'Indivisible Parts and Extended Objects: Some Philosophical Episodes from Topology's Prehistory', *The Monist*, 79 (1996), 148–80.

INDEX

abstractionist accounts of geometry 219–221
action at a distance 237, 252–7
aether 254, 271
aggregates, *see* composites
Anapolitanos, D. A. 233 n.45
anti-atomism, *see* gunk, atomless
anti-realism 96–8, 203
Aquinas, Thomas 19, 79
argument from actual parts to a determinate number of parts 27–8, 33–4, 58, 107, 132, 133–4, 135–41, 167
argument from composition 28, 35–6, 58, 90, 133, 146–7, 169–192
  objections to 192–204
argument from geometry 206, 208–13, 220, 226, 231, 275
  objections to 213–7
argument from the definiteness of parts to ultimate parts 34–5, 90–1, 126 n.104, 133–4, 141–7, 167, 169
Aristotle 30, 79, 142
  epistemology of geometry 220
  on infinite divisibility 41, 52, 103, 160, 161
  on infinity 41, 155 n.33, 160
  on potential parts 52, 95, 103, 125, 161, 236
Aristotelianism 19, 52, 92, 209, 274, 277–8, *see also* scholasticism

Arnauld, Antoine 7, 65, 66, 67, 70, 71, 209, 210, 211, 214, 225
Arriaga, Roderic 65
atomism
  metaphysical atomism 14, 33, 34–6, 37–9, 44–50, 53–5, 90–1, 111, 126, 141–7, 165, 170, 172–3, 175, 245, 274, 275, 276
    arguments for, *see* argument from composition *and* argument from the definiteness of parts to ultimate parts
  physical atomism 11, 12, 93, 238
  see also force-shell atoms
  see also gunk, atomless

Bacon, Roger 211
Baltzly, Dirk 98 n.44
Barrow, Isaac 241, 273
  on actual parts 57–8, 84–5, 108
  on geometry and geometrical structures 213, 214, 219 n., 220, 222
  on infinite divisibility 30, 31, 46, 57–8, 64–5, 67, 209, 213, 232
  on infinity 43, 65
  on motion 232
  on points 47 n.
  on Pythagoreanism 214
Baxter, Donald L. M. 23 n.27, 27 n.39, 28, n.40, 54 n.86
Bayle, Pierre 7, 8 n.5, 52 n.79, 105–6, 128, 241, 274

Bayle, Pierre (*cont.*)
  on actual parts 34, 35, 57–8, 87,
    88 n.25, 105–6, 107, 108, 109,
    111–4, 115, 116, 117, 138, 145,
    162 n.42
  on composites 88 n.25
  on idealism 9, 69, 70
  on the identification of space and
    matter 21, 50
  on infinite divisibility 30, 31, 38, 47,
    49, 50, 57–8, 65, 66, 67, 69, 71, 76,
    77, 145, 151, 157, 158, 162, 209,
    210, 211, 235
  on infinity 44, 138, 159
  on metaphysical atomism 35, 38, 145
  on points 48 n.72
  on potential parts 162 n.42
  on Pythaogreanism 215
  on scepticism 71, 72–3, 127
Bechler, Zev 60 n.99, 167 n.47
Benacerraf, Paul 154 n.30
Bennett, Jonathan 184 n., 244 n.19
Bentley, Richard 254
Berkeley, George 7
  on actual parts 34, 87, 104, 107, 145
  on idealism vii, 9, 69–70
  on infinite divisibility 30, 46, 54, 55,
    69, 145, 209, 225, 232
  on infinity 43
  on metaphysical atomism 35, 54,
    127, 145, 230, 275
  on minima sensibilia 14, 15, 15–6, 54,
    70, 127, 202, 230
  on 'positive' and 'relative' ideas 270
  on Pythagoreanism 215, 216
Berlin Academy controversy 8, 172,
  173 n.10, 194, 245
Bernadete, José 43 n.62, 155 n.32
Blackburn, Simon 261 n.55, 262 n.,
  267 n.
Bohr, Niels 249

Bolzano, Bernard 59 n.95
Bolyai, János 214
Bongie, Laurence L. 8, 170 n.7, 173 n.10
Borges, Jorge Luis 69 n.129
Boscovich, Roger Joseph 7
  on actual parts 87
  on dynamism viii, 9, 237, 238, 252, 273
  on force-shell atoms 14, 55, 81, 127,
    129, 191, 202, 207, 236–8, 239,
    244, 245–9, 251–2, 254–8, 261–5,
    271, 275
  on infinite divisibility 55, 66, 67
  on points 47 n.
Bostock, David 25 n.32, 103 n.54
Boyle, Robert, 6 n., 92, 259 n.49, 273
  on infinite divisibility 49, 67–8, 72,
    73, 149, 152
  on infinity 43
  on physical atoms 93, 238
  on scepticism 127
Bracken, Harry M. 54 nn.85–6
Brandt, Frithiof 98 n.43
Brentano, Franz 50 n.76, 77
Broad, C. D. 23 n.27
Brouwer, L. E. J. 41 n.60
Bruno, Giordano 36, 46, 49, 171 n.
Buridan, John 157, 215

Calinger, Ronald S. 173 n.10
Cambridge Platonists 87
Cantor, Georg 42 n.61, 44 n.67, 60, 61
Cartaud de la Vilate, Francois 211,
  213 n.10
Cartesianism 106, 209, 210, 256
  and actual parts 86, 106 n.60
  identification of space and
    matter 21, 50, 239–40, 259 n.49
  on problems of infinity 140–1
  see also Descartes
Cartwright, Richard 81 n., 84 n.3
Casati, Roberto 92 n., 123 n.99

Charles I 104
Charleton, Walter 6 n., 14, 16, 119, 122–3, 128
   on actual parts 34, 35, 54, 84, 104–5, 106, 114–5, 116–7, 137, 145
   on the argument from composition 170, 171, 188
   on infinite divisibility 30–1, 38, 46, 47, 48, 54, 55, 75, 137, 145, 232, 235 n.49
   on metaphysical atomism 15, 38, 54, 145, 170, 271, 275
   on motion 232
   on points 48 n.72
   on potential parts 95, 120, 125 n.101
   rejection of Pythagoreanism 214–5, 216
Charlton, William 92 n., 122–4
Chatton, Walter 157
Cheyne, George 239 n.1
Clarke, Samuel 7, 210
   on action at a distance 254
   on actual parts 17, 85, 86 n.11, 87 n.12, 88 n.25, 120
   on composites 88 n.25, 89 n.28
   on infinite divisibility 65, 211
   on infinity 43
   on mechanistic causation 254
   on space 10 n.13, 12 n. 14, 13 n.15
   on spirits 14, 15, 81, 239, 241, 245
Clavelin, Maurice 25 n.32, 103 n.54, 165 n.45
cohesion 246–7, 256
Coimbra, Jesuits of 113–14
Collier, Arthur 66, 67, 69, 73
Collins, Anthony 94 n.32, 118, 120–1, 122, 241
'common notions,' *see* innate knowledge
composites 28, 88–9, 90, 91, 92–3, 123–4, 174, 176–8, 186, 193–6, 205
   see also simples

compounds, *see* composites
conceptual argument (for infinite f-divisibility) 207, 208, 223–31, 275
Condillac, Etienne Bonnot de 36, 170
Constance, Council of 8
contact 77–8, 253
continua 39, 61, 63–4
Conway, Anne 159 n.40
Copernicus, Nicolaus 236
Cordemoy, Gerauld de 14
corpuscularianism 238, 258, 259, 271, 273
   relation to actual parts/potential parts dispute 277–8
cosmological argument 189–90
Craig, William Lane 42 n.62, 154 n.30
Cudworth, Ralph
   on actual parts 57, 87, 89 n.28
   on infinite divisibility 57
Cummins, Phillip 110 n.74

Darwinianism 222
Dauben, Joseph 45 n.
De Bergerac, Cyrano 8, 234
De Gandt, Francois 165 n.45
Dedekind, J. W. R. 60, 61
Des Chene, Dennis 86 n.9
Descartes, René 7, 44 n.64, 92, 113 nn.81–2, 156, 176
   on actual parts 86
   on the epistemology of geometry 221–2
   identification of space with matter 239–40
   on infinite divisibility 65, 225
   on infinity 141 n.15
   on mechanistic causation 253
   on the new science vii
   on the plenum 13, 271

Digby, Kenelm 6 n., 118–9, 128, 145
  on actual parts 35, 52, 125–6, 141–3, 150, 161
  on infinite divisibility 30, 47, 52, 141–3, 150, 225, 276
  on metaphysical atomism 35, 126, 141–3
  on points 47 n., 48 n.72
  on potential parts 18, 34 n.52, 52, 92 n., 94, 96, 102, 116, 118–9, 122, 125–6, 135 n., 276
discerpibility 17 n., see also divisibility, metaphysical
divisibility
  formal 207, 237
    defined 14–15
    infinite formal divisibility 52–3, 124–5, 173, 207–8; arguments for 232–5, see also argument from geometry and conceptual argument
    introduced 11
  kinds of 9–16
  intellectual 94–5, 96–8
    defined 15–6
    introduced 11
  metaphysical 80–1, 94–5, 106, 107, 142–3, 207, 237
    defined 12–14
    infinite metaphysical divisibility 2, 20–1, 22–4, 51, 53, 56, 67, 107–8, 124–5, 126–7, 137, 147–8, 148–67, 207–8; paradoxes of, see paradoxes of material structure
    introduced 1–2, 11
  physical, 94–5
    defined 11
    introduced 1–2, 11
division, infinite
  completion of 135, 148–52, 198
  all-at-once completion 135, 148, 151, 156–62, 182
  successive completion 135, 148, 151, 152–5, 181
  see also divisibility, infinite metaphysical
Drummond, William
  epistemology of geometry 220
  on infinite divisibility 49, 69
  on infinity 43
  on idealism 69, 70
Du Hamel, Jean B. 216
Duhem, Pierre 212 n.6, 215 n.15, 219 n.
Duns Scotus, John 211–2
dynamism viii, 9, 237, 238, 239, 252, 273, see also hollow world problem

Earman, John 154 n.31
Einstein, Albert 217, 228, 249
elastic impacts 55, 232–3, 256
Ellington, James W. 249 n.29
emanative causality 243
empiricism 219–23
Epicureanism 15, 30, 54, 55, 95, 105, 107, 140, 170, 209, 215–6, 226, 235, 277
Epicurus 46, 209, 215
Euclid 30, 212, 214, 216, 217, 218, 220
Eudoxus 30
Euler, Leonhard 7, 8, 173 n.10, 175, 194
  on actual parts 58, 87
  on infinite divisibility 38, 40, 49, 58, 172–3, 209
  on infinity 40, 44
  on metaphysical atomism 38, 40, 172–3
  on Pythagoreanism 216, 217, 220

f-divisibility, see divisibility, formal
Faraday, Michael 237, 244, 249, 264–5, 269
fideism 71–2
field theory, see dynamism

Friedman, Michael 255 n.44
Flew, Anthony 23 n.27, 24 n.29, 25–6, 30, 31, 56, 121 n.97
Fogelin, Robert 23 n.27, 25, 26, 30, 31, 56
force
    as essence of matter, *see* dynamism *and* hollow world problem
    attractive, 246–8, *see also* gravitation
    repulsive 246–8, *see also* impenetrability
    see also action at a distance
force-shell atoms 14, 55, 81, 127, 129, 173, 191, 202, 207, 236–8, 248–267, 271
    and action at a distance 252–7
    ontological status of cores and shells 250–2, 265, 275
Forrest, Peter 229–9
Forrester, James W. 198 n.
Foster, John 262 n.56, 266, 267
Foucher, Simon
    on infinite divisibility 38
    on metaphysical atomism 38
Franklin, James 27, 28
Frasca-Spada, Marina 23 n.27, 24 nn.28–9, 25, 54 n.86
Frederick the Great 8
Frege, Gottlob 60, 61

Galilei, Galileo vii, 7, 119, 277, 279
    on actual infinities 39, 40 n., 43, 59–62, 107, 127, 129, 139–40, 165, 186–7, 236, 271, 275
    on actual parts 17, 34, 35, 79, 87, 95, 104, 107, 108, 129, 139–40, 144–5, 166–7
    on infinite divisibility 59–62, 65, 139–40, 162–7, 168, 186, 189–90, 209
    on metaphysical atomism 39, 59–62, 64, 127, 144–5, 186–7, 189–90, 202, 271, 275
    and the new science vii, 84, 95
    on potential parts 18, 95, 163–5
    rejection of Aristotelian account of infinity vii, 9, 279
Gassendi, Pierre 105, 238, 259 n.49
    on actual parts 87
    epistemology of geometry 220–1
    on infinite divisibility 209
    rejection of Pythagoreanism 214, 220–1
Gell-Man, Murray 249
geometrical structures, instantiation of 219, 220
    see also space, physical vs. geometrical
geometrization of nature vii, 3, 54, 237
geometry 54, 208, 209–13, 214–7, 218
    epistemology of 218–23
Gerald of Odo 157
Glanvill, Joseph 79
    on infinite divisibility 30, 38, 46, 49, 51, 71–2
    on metaphysical atomism 38, 51
    on scepticism 71–2
    on spirits 240
Gorlaeus, David 171 n.
Grant, Edward 13 n.15, 44 n.67, 254 n.41
gravitation 246, 247, 253–4, 256
Greene, Brian 228, 249
Gregory of Rimini 212
Grünbaum, Adolf 39, 46 n.68, 61
gunk, atomless 51, 146, 181, 184–5, 194, 196, 271, 274, 276

Hamlyn, D. W. 200–201
Harré, Rom 244 n.18, 265 n.60
Heisenberg, Werner 249
Hempel, C. G. 217
Henry of Harclay 157
Herbert, Gary S. 98 n.44

Hesse, Mary 244, n.18
Hippasos of Metapontion 8
Hobbes, Thomas 10 n.12, 79, 175, 239, 263, 271
   anti-realism of 96–8, 203
   on individuation 96, 98–9, 130–1, 201, 203
   on infinite divisibility 40, 52, 76, 103, 125, 149, 160, 161, 209, 226, 276
   on infinity 40, 103, 160
   on potential parts 18, 19, 52, 96–9, 103, 118, 125, 130, 161, 236, 276, 278
   on simplicity 199–203
Hoffman, Joshua 226 n.38
hollow world problem 238, 263–72
Holton, Richard 268 n.63
Hooke, Robert 234
Huby, Pamela 43 n.62, 154 n.30
Hume, David vii, viii, 7, 9, 141
   on actual parts 32, 35, 87, 104, 138–9, 145, 171
   on the argument from composition 170, 172, 183, 189, 204
   on causation 256
   on the cosmological argument 190 n.25
   on impenetrability 266 n.61
   on infinite divisibility 5, 22–32, 38, 47, 48, 54, 66–7, 77, 138–9, 145, 209, 278–9
   on mereological essentialism 130 n.107
   on metaphysical atomism 35, 36, 38, 54, 55, 59, 145, 170, 203 n.41, 271, 275
   on minima sensibilia 32 n.50, 54, 202
   on simplicity 203 n.41
Huygens, Christiaan
   on impacts 233

i-divisibility, *see* divisibility, intellectual
idealism viii, 54, 69–70, 73
imagination 241–2
impenetrability 220, 238, 259–60
   absolute 246, 258, 261
   degrees of 246–8, 257–8, 261
   see also solidity
incommensurable magnitudes 8, 211–2
individuation 93–4, 96, 98, 113–4, 129–31, 203, 205, 274
infinite division, *see* division, infinite
infinitesimals 60
infinity
   actual infinities 34, 59–62, 90, 136–140, 145–6, 158–61, 187, 275–6
      alleged impiety of 43–4
      of metaphysical atoms 44–50, 275
      paradoxes of 39–44
   infinite regresses 100–101, 184–5, 189–90
   potential infinities 24, 52, 56, 90, 102–3, 134, 136–7, 146, 274
innate knowledge 221–3, 229
intensive magnitudes 257–8
intuitionism 41 n.60

Jackson, John 94 n.32
Jacquette, Dale 23 n.27
Jesseph, Douglas 54 n.85, 212 n.8, 219 n., 221 n.28
Joy, Lynn Sumida 87 n.14, 214 n., 221 n.27
Justi, J. H. G. 175, 194–5

Kant, Immanuel 3–4, 7, 64, 67, 68, 70, 70, 211, 241, 274
   on actual parts 21 n.26, 55, 87, 104, 110, 136, 139, 140, 176–8

Index / 301

on the argument from
composition 28, 36, 170–92
on dynamism viii, 9, 237, 238, 252, 273
on the epistemology of
geometry 223
on force-shell atoms 14, 55, 129, 173, 191, 202, 207, 236–8, 239, 244–5, 246–9, 250–1, 252, 254–5, 257–8, 261–3, 265–7, 271, 275
on infinite divisibility 34, 38, 52, 100–101, 139, 160, 161, 173, 209, 210, 276
on infinity 41, 100–101, 139, 160
on intensive magnitudes 258 n.47
on metaphysical atomism 36, 38, 59, 170–1, 176, 177, 192
on potential parts 52, 99–101, 103, 161, 236, 245, 255, 276, 278, 279
on Pythagoreanism 214, 218
on the rejection of transcendental
realism 7, 9, 53, 69–70, 99, 245
on the 'Second Antinomy' vii, 28, 38, 53, 99, 103, 110, 179, 188, 191, 197, 199 n.
on simplicity 173, 202
Keill, John 7, 10 n.12
on actual parts 144
on geometry 213, 220
on infinite divisibility 30, 31, 38, 46, 49, 52, 65, 67, 144, 151, 210, 211, 213, 232, 235
on infinity 43, 159–60
on metaphysical atomism 38, 61, 144
on motion 232
on potential parts 52, 162 n.42
on Pythagoreanism 216, 217, 220
Kemp Smith, Norman 31 n.45, 175, 192, 193–5
Koyré, Alexander 239 n.2
Kronecker, Leopold 41 n.60

Laird, John 23 n.27
Langton, Rae 258 n.47, 261 n.55, 262, 267
Law, Edmund 43, 112 n.78
Le Grand, Antoine 7, 14 n.
on actual parts 86 n.10, 137–8
on infinite divisibility 49, 65, 137–8, 140–1, 210, 211, 225
on infinity 43
Lear, Jonathan 25 n.32, 103 n.54
Leclerc, Ivor 91 n.30, 171 n.
Lee, Henry 10 n.13, 65
Leeuwenhoek, Anthony van 234
Leibniz, Gottfried Wilhelm 7, 67 n.121, 98–9, 251
on actual parts 35, 63–4, 88, 89–90, 104, 171
on the argument from
composition 28, 35, 90, 170, 171, 173, 183, 188, 204
on composites 89, 170
on impacts 233
on infinite divisibility 209
on infinity 62 n.107
on 'the labyrinth of the
continuum' vii, 7
on mechanistic causation 253
on metaphysical atomism 59, 52–4, 170, 202
on monads, see Leibniz on
metaphysical atomism
on points 47 n.
on potential parts 63–4
on simplicity 204 n.
on space 10 n.13, 209
Leucippus 109
Levey, Samuel 62 n.107
Lewis, David 146 n.22
literature and the problems of material
structure, see satire and the
problems of material structure
Lobavchevscky, Nicolay Ivanovich 214

Locke, John 7, 71
  on corpuscles 238, 258, 259 n.49
  on infinite divisibility 209
  on infinity 41
  on 'positive' and 'relative' ideas 270
  on solidity and impenetrability 259–60
  on space 13 n.15, 209
Lucretius 107

m-divisibility, *see* divisibility, metaphysical
Mach, Ernst 229, n.
Mackie, J. L. 189–90
Madden, E. H. 244 n.18, 265 n.60
Magnen, Jean Chrysotome 171 n.
Malebranche, Nicolas 7, 65, 209, 235
Malezieu, Nicolas de 38, 65, 170 n.6, 211
Markosian, Ned 92 n., 93 n.
Martin, Benjamin 210
materialism 88, 239
matter, space-filling property of, *see* solidity *and* impenetrability
  see also hollow world problem
Maxwell, James Clerk 237
McGuire, J. E. 54 n.84, 84, 226 n.38
mechanistic model of causation 252–6
Michelangelo 81
Melvill, Thomas 55 n.85
Meré, Antoine Gombauld, Chevalier de 206
mereological essentialism 130
Michel, P. H. 49 n.75
Mijuskovic, Ben Lazare 89 n.26
Miller, Fred D., Jr. 25 n.32, 103 n.54
More, Henry 10 n.12, 12 n.14, 236, 263–4
  on actual parts 34, 35, 87, 107, 108–9, 138–9
  on 'discerpibility' 14 n., 238
  on infinite divisibility 30, 38, 46, 62, 70–1, 138–9, 150–1, 152

  on metaphysical atomism 14, 15, 38, 62
  on 'physical monads' 14, 15, 62, 81
  on points 47 n.
  rejection of identification of space with matter 239–40
  on spirits 14, 15, 70–1, 81, 108, 238, 239–44, 245, 250
motion 232–3
Murdoch, John E. 157 n.
Mutakallimum school 215

nativism, *see* innate knowledge
neo-Epicureanism, *see* Epicureanism
new science vii, 84
Newton, Isaac 7, 13 n.15, 16, 40 n., 79, 213, 219 n., 258, 259 n.49
  on action at a distance 254
  on actual parts 34, 35, 86, 95, 105, 107, 115–16
  epistemology of geometry 220
  on gravitation 246, 247, 253–4, 256
  on infinite divisibility 30, 46, 54, 149, 152, 153, 209
  on infinity 40 n57, 43
  on mechanistic causation 254
  on metaphysical atoms 14, 15, 35, 36, 54, 170
  and the new science vii, 84, 95
  on physical atoms, *see* physical atomism
  on Pythagoreanism 220
  relation to Charleton 105
  on space 240
  see also space, absolute
Newtonianism 17, 21, 172, 209, 210, 211, 240, 245
  and actual parts 87, 95
  and gravitation 253–4, 256
  and mechanistic causation 253
  see also Newton

Nicholas of Autrecourt 48 n.72, 107 n.63, 115 n.86
Nicole, Pierre 65, 66, 70, 71, 209, 210, 211, 225
Nietzsche, Friedrich 236, 244, 263, 265
Norris, John 65
North, J. D. 40 n.
Norton, John D. 154 n.31

Ockam, William of 212

p-divisibility, *see* divisibility, physical
paradoxes of infinite divisibility, *see* paradoxes of material structure
paradoxes of material structure 66–7, 68–73, 124–7, 129–30, 158, 237, 246, 275
   mathematical reading of viii, 19–20, 27, 199
   metaphysical reading of viii, 2–3, 19, 20–2, 23, 27, 33, 37, 199, 275
     objections to 147–8
   survey of 36–50, 75–8
   responses to 51–73, 126–7, 274–6
particle physics, models of 249
parts
   actual parts 20, 21, 27–9, 32, 33–6, 274
     actual parts doctrine 20, 51, 55, 56, 67, 92, 147–8, 165–7, 168, 176–8, 237, 248, 276–8; arguments against 118–127, 128–30; arguments for 103–118, 128–30, 277; defined 16–17, 80–82; introduced 3, 79–80; and mereological essentialism 130; popularity in new science 84–8; relationship to doctrine of arbitrary undetached parts 82–3; and the inference to a determinate number of parts, *see* argument from actual parts to the determinate number of parts and the inference to ultimate parts, *see* argument from composition, *and* argument from the definiteness of parts to ultimate parts
     defined 80–1
   aliquot parts 28, 212
   arbitrary undetached parts ix, 82–3, 110
   potential parts 274, 166–7
     defined 18, 91
     potential parts doctrine 36, 37, 52, 108, 125, 147, 156–7, 276–8; arguments against 103–18, 128–30, 278; arguments for 118–127, 128–30; defined 18–19, 91–5; introduced 79–80
   proportional parts 28, 56
   of space, *see* space, parts of
   ultimate parts, *see* atomism
Pascal, Blaise 206
Plato 114, 221
plenum 13
points 47 n., 61, 77, 219
Polonoff, Irving 173 n.10, 198 n.
Popkin, Richard H. 51 n.45, 73 n.135
preformation 234–5
Pressman, Mark H. 26 n.37
Price, H. H. 130 n.108
Price, Richard 241
Priestley, Joseph 241, 244
problems of material structure, *see* paradoxes of material structure
Pyle, Andrew 116 n.90, 165 n.45, 215 n.15
Pythagoreanism 214, 217, 218–23, 224
   arguments for 219–23

Pythagoreanism (*cont.*)
  defined 214
  see also geometrization of nature

Quinton, Anthony 265

Radner, Michael 185 n.
rationalism 219, 221–3
  see also innate knowledge
Rebondi, Pietro 60 n.99
Reid, Thomas 7, 271
  on actual parts 17, 58, 85–6, 276
  on atomless gunk 58, 276
  on infinite divisibility 38, 58
  on metaphysical atomism 38
  on points 47 n.
Richelieu, Armand-Jean du Plessis 8
Robinson, Howard 261 n.55, 267, 268
Rohault, Jaques 7
  on actual parts 57, 86 n.10
  on infinite divisibility 49, 57, 76, 210, 211, 235
  on infinity 43
Rosenberg, Jay 258 n.47
Rosenkrantz, Gary S. 226 n.38
Russell, Bertrand 8 n.11, 27 n.38, 42 n.61, 155, 266, 267
Rutherford, Ernest 249

satire and the problems of material structure 5–6, 8, 234, *see also* Voltaire
Scaltsas, Theodore 96 n.
scattered objects ix, 82–4
scepticism viii, 9, 71–4
Schmaltz, Tad 112 n.80
scholasticism 19, 52, 93, 95–6, 103, 113, 142, 209, 225, 274, 277–8
Schopenhauer, Arthur 175, 192–5, 197
Sextus Empiricus 46

Sider, Theodore 111 n.75, 116 n.89, 146 n.22
simples 28, 92–4, 177–8, 193, 205
  different standards of simplicity 93–4, 175, 199–203
  see also atomism *and* composites
Skyrms, Brian 46 n.68
Sleigh, R. C. 90 n.29
Smith, A. D. 270 n.65
Smollett, Tobias 6, 8
Snow, A. J. 233 n.45
solidity 220, 258
  degrees of 257–8
  instrinsic vs. dispositional property 259–63
  see also hollow world problem
  see also impenetrability
Sorabji, Richard 154 nn.30–31, 155 n.35, 209 n.1
space
  absolute 10 n.13, 11, 12–13, 81, 271
  discrete, mathematical models of 229–30
  parts or regions of 10 n.13
  physical vs. geometrical 214, 217, 218–9, 222, 223, 224, 227, 228–9
  and transcendental idealism 223
Spinoza, Benedict de 7, 109, 111–2, 176
  on infinite divisibility 125, 156
  on infinity 141 n.16
  on potential parts 118, 125
  on the plenum 13
Stack, George J. 244 n.18
standard model 249
Sterne, Laurence 8
Stillingfleet, Edward 7 n., 21 n.26, 65
stoicism 98 n.44
Stones, G. B. 171 n.8
string theory 249
stuff ontologies ix

Suarez, Francisco 77
substances 89, 112–3, 250–2, *see also* substratum
substantial forms 52, 93, 95–6, 274
   relation to actual parts/potential parts debate 277–8
substantial presence 253, 254 n.41
substratum 270–2
supertasks, *see* division, infinite, successive completion of
surfaces 76–8, 219, 247, 257
   fuzzy 258, 273
Swammerdam, Jan 234
Swift, Jonathan 8, 234

Tamny, Martin 54 n.84, 84, 226 n.38
Thijssen, J. M. M. H. 26 n.37, 31 n.45

'ungeometrical philosophers' 54, 208, 216, 275

Vailati, Ezio 329 n.1
Van Bendegem, Jean Paul 154 n.30, 230
Van Cleve, James 185, 198 n., 199 n., 261 n.54
Van Goorle, David, *see* Gorlaeus, David
Van Inwagen, Peter 82 n., 130 n.107
Varzi, Achille 92 n., 123 n.99
Vico, Giambatista 218

virtual presence 253, 254 n.41
Voltaire 8, 105, 132, 169, 234

Watkins, Eric 255 n.43
Watts, Isaac 65, 241
White, Thomas 94 n.33, 118
Whitrow, G. J. 154 n.30
Whittaker, Edmund 225 n.31, 229 n.
Wilson, Catherine 234 n.
Wittgenstein, Ludwig 175, 199–203
Wojcik, Jan W. 73 n.136
Wolff, Christian
   on actual parts 35, 87, 104
   on the argument from composition 35, 170, 171, 173, 204
   on metaphysical atomism 35, 170, 172
Wolffians 245
Wollaston, William 14 n., 87 n.12
Wyclif, John 8–9

Zeno of Elea 45–6, 103, 105, 215
Zeno the Epicurean 215
Zeno's metrical paradox 47–8, 61, 103, 275
Zimmerman, Dean W. 48 n.73, 49 n.76, 59 n.95, 77 n.142, n.144, 110 n.74, 146 n.22, 219 n., 226 n.38

The manufacturer's authorised representative in the EU for product safety is
Oxford University Press España S.A. of el Parque Empresarial San Fernando de
Henares, Avenida de Castilla, 2 – 28830 Madrid (www.oup.es/en or product.
safety@oup.com). OUP España S.A. also acts as importer into Spain of products
made by the manufacturer.

www.ingramcontent.com/pod-product-compliance
Ingram Content Group UK Ltd.
Pitfield, Milton Keynes, MK11 3LW, UK
UKHW021317180426
11947UKWH00015B/1281